Global Approaches to Environmental Management on Military Training Ranges

Global Approaches to Environmental Management on Military Training Ranges

Edited by
Tracey J Temple and Melissa K Ladyman
Cranfield University, Shrivenham, Swindon, SN6 8LA

IOP Publishing, Bristol, UK

ISBN   978-0-7503-1605-7 (ebook)
ISBN   978-0-7503-1603-3 (print)
ISBN   978-0-7503-1938-6 (myPrint)
ISBN   978-0-7503-1604-0 (mobi)

DOI   10.1088/978-0-7503-1605-7

Version: 20191201

IOP ebooks

British Library Cataloguing-in-Publication Data: A catalogue record for this book is available from the British Library.

Published by IOP Publishing, wholly owned by The Institute of Physics, London

IOP Publishing, Temple Circus, Temple Way, Bristol, BS1 6HG, UK

US Office: IOP Publishing, Inc., 190 North Independence Mall West, Suite 601, Philadelphia, PA 19106, USA

Contents

Acknowledgements

We would like to thank IOP Publishing for giving us this opportunity to produce this collection of case studies that enables us to reach a wide readership. This book would not have been possible without the technical and practical expertise and willingness of the authors, we are very grateful for their hard work and determination. The editors would specifically like to thank Professor William Proud who initially suggested that we embark on this venture. Finally, we would like to thank Cranfield University for supporting the editors in creating this network of contributors.

Editor biographies

Tracey J Temple

Dr Tracey J Temple is a lecturer in environmental science at Cranfield University and has held this position since 2006. Prior to Cranfield, Tracey was an environmental consultant primarily for UK Defence, working in the UK and Cyprus. Before returning to academia, Tracey served in the RAF for 10 years. Tracey gained her PhD in the fate and transport of explosives from Cranfield University, and her MSc in Environmental Science and BA in Geography from Queen's University, Belfast. Tracey is currently the Course Director for the *Explosives Ordnance and Engineering* (EOE) MSc at Cranfield University. Tracey frequently chairs and attends NATO panel meetings, specifically related to the environmental impact of explosives, which include sampling for contamination, toxicity from military training ranges and environmental munition regulations.

Melissa K Ladyman

Dr Melissa K Ladyman graduated from Edinburgh University with an MChem degree in 'Chemistry with a Year in Europe' in 2010. Her undergraduate degree included a year placement at the Ecole Nationale Supérieure de Chimie de Lille where she undertook research into potential medications for tuberculosis and depression. She gained a PhD in 2014 in organic and medicinal chemistry focussing on the development of fluorescent assays for biological analysis, also from The University of Edinburgh. In 2014, Melissa joined the environmental science group at Cranfield University and is currently a lecturer where she is responsible for teaching and researching the fate and transport of explosives in the environment and environmental management.

Together, Tracey and Melissa lead the teaching and research of Environmental Science at Cranfield University at Shrivenham. They teach a diverse mix of students and actively engage with the explosives and defence industries. For example, they have coordinated a NATO soil sampling workshop showcasing the multi-increment sampling method, which is ideal for identifying explosive contamination on training ranges. They have also published a range of scientific research on the behaviour of energetic materials in soil, air and water. Tracey and Melissa are currently investigating the applicability of environmental risk assessment to explosive activities through life. Their aspiration is to develop a flexible and effective approach to support environmental management for the use of legacy and future energetic materials.

Contributors

M C Barbosa
Program of Civil Engineering, Federal
University of Rio de Janeiro, Rio de
Janeiro, Brazil

S Beal
U.S. Army
Cold Regions Research and Engineering Laboratory,
Hanover, NH, USA

M Bigl
U.S. Army Cold Regions Research and Engineering Laboratory,
Hanover, NH, USA

I Bortone
Cranfield School of Water, Energy and
Environment,
Cranfield University, MK43 0AL, UK

L Brennan
Directorate of Operations and Training Area Management,
Department of Defence, Australia

S Brochu
Valcartier Research Center
2459 de la Bravoure Road,
Quebec (Qc) Canada G3J 1X5

F Coulon
Cranfield School of Water, Energy and
Environment,
Cranfield University, MK43 0AL, UK

H Craig
U.S. Environmental Protection Agency
Region 10, Portland, OR, USA

T A Douglas
U.S. Army Cold Regions Research and Engineering Laboratory,
Fairbanks, AK, USA

W Fawcett-Hirst
Cranfield University, Centre for Defence
Chemistry, Defence Academy of the United
Kingdom, Shrivenham, SN6 8LA, UK

C Ferreira
Rua Luis Reis dos Santos 290,
Departamento de Engenharia Mecânica, Faculdade de Ciência e Tecnologia da
UC, 3030-194,
Coimbra, Portugal

F Freire
Rua Luis Reis dos Santos 290,
Departamento de Engenharia Mecânica, Faculdade de Ciência e Tecnologia da
UC, 3030-194,
Coimbra, Portugal

E F Galante
Chemical Engineering Department, Instituto
Militar de Engenharia, Rio de Janeiro, Brazil

P P Gill
Cranfield University, Centre for Defence
Chemistry, Shrivenham, SN6
8LA, UK

O Hausheer
Soil Comtetence center
Federal Department of Defence,
Cicil Protection and Sport, DDPS
armasuisse real estate
Environmental Management & Standards
Guisanplatz 1,
CH-3003 Bern, Switzerland

Competence Centre for Soil and
Contamination, Environmental Management,
Standards and Compliance,
Blumenbergstrasse 39, 3003 Bern,
Switzerland

M van Hulst
TNO Defence, Safety and Security,
PO Box 45, 2288 GJ Rijswijk, The
Netherlands

R Kaiser
Soil Comtetence center
Federal Department of Defence,
Cicil Protection and Sport, DDPS
armasuisse real estate
Environmental Management & Standards
Guisanplatz 1,
CH-3003 Bern, Switzerland

W P C de Klerk
TNO Defence, Safety and Security,
PO Box 45, 2288 GJ Rijswijk, The
Netherlands

M K Ladyman
Cranfield University, Centre for Defence
Chemistry, Shrivenham, SN6 8LA, UK

J Langenberg
TNO Defence, Safety and Security,
PO Box 45, 2288 GJ Rijswijk, The
Netherlands

G W Larsen
U.S. Army Cold Regions Research and Engineering Laboratory,
Fairbanks, AK,
USA

P de Lasson
Danish Defence Estate Agency,
Miljoesektionen
Arsenalvej 55,
9800 Hjørring, Denmark

N Mai
Cranfield University, Centre for Defence
Chemistry, Shrivenham, SN6 8LA, UK

M E S Marques
Program of Defense Engineering, Instituto
Militar de Engenharia, Rio de Janeiro, Brazil

R Martel
Institut National de la Recherche
Scientifique, Eau, Terre et
Environnement, 490 de la Couronne,
Québec, QC, Canada, G1K 9A9

D McAteer
Cranfield University, Centre for Defence
Chemistry, Shrivenham, SN6
8LA, UK

J Pons
Cranfield University, Centre for Defence
Chemistry, Shrivenham, SN6
8LA, UK

C Ramsey
Envirostat, Inc., Vail, AZ, USA

M M Reis
Program of Defense Engineering,
Instituto Militar de Engenharia, Rio de
Janeiro, Brazil

J Ribeiro
Rua Luis Reis dos Santos 290,
Departamento de Engenharia Mecânica, Faculdade de Ciência e Tecnologia da
UC, 3030-194, Coimbra, Portugal

T J Temple
Cranfield University, Centre for Defence
Chemistry, Shrivenham, SN6 8LA, UK

S Thiboutot
Defence R&D Canada
Valcartier Research Center
2459 de la Bravoure Road
Quebec (Qc) Canada G3J 1X5

M E Walsh
U.S. Army Cold Regions Research and Engineering Laboratory,
Hanover, NH, USA

M R Walsh
U.S. Army Cold Regions Research and Engineering Laboratory,
Hanover, NH, USA

Message from the editors

The idea behind compiling this book was to bring together professionals from around the world to share their expertise and experience. We hope it will provide an insight into the global practices for environmental management of military live-fire training ranges by combining scientific research with practical solutions to ensure continued training capability. The editors have thoroughly enjoyed working with the contributors to produce this book, and hope the network will continue to grow.

Abbreviations

μg	microgram
1,3,5-TNB	1,3,5-trinitrobenzene
2,4-DNT	2,4-dinitrotoluene
2,4-DNT	2,4-dinitrotoluene
2,6-DNT	2,6-dinitrotoluene
2A-DNT	2-amino-4,6-dinitrotoluene
2-ANAN	2-amino-4-nitroanisole
4A-DNT	4-amino-2,6-dinitrotoluene
4-ANAN	4-amino-2-nitroanisole
4-NT	4-nitrotoluene
AAD	Army ammunition depot
AC	activated carbon
ACIH	American Conference of Industrial Hygienists
ADRT	advective-dispersive-reactive transport
ALI	air–liquid interface
AOP	advanced oxidation processes
AP	ammonium perchlorate
APCI	atmospheric pressure chemical ionisation
API	atmospheric pressure ionisation
API	new atmospheric pressure ionisation
ARDEC	Armament Research, Development and Engineering Center
ASE	accelerated solvent extraction
ATSDR	Agency for Toxic Substances and Disease Registry
AWC	Australian Wildlife Conservancy
AWS	automatic weather stations
BAT	best available techniques
BC	bone char
BI	bullet impact
BMP	bushfire management plan
BOM	Australian Bureau of Meteorology
CAA	Clean Air Act
CAF	Canadian Armed Forces
CC	coconut coir
CCME	Canadian Council of Ministers of the Environment
CE	capillary electrophoresis
CERCLA	Comprehensive Environmental Response Compensation and Liability Act
CFB	Canadian Force Base
CH	compositional heterogeneity
CI	chemical ionisation
CIO_4	perchlorate
CL20	2,4,6,8,10,12-hexanitro-2,4,6,8,10,12-hexaazaisowurtzitane
CN-HPTLC	HPTLC modified with a cyano propyl group
CN-HPTLC	cyano propyl group
COCO	contractor owned contractor operated
Comp B	Composition B
CPT	core penetration test
CRREL	Cold Regions Research and Engineering Laboratory

CSF	cancer slope factor
DALY	disability-adjusted life years
DART	direct analysis in real time
DAT	downward advective time
DC	combined direct current
DEHA	di(2-ethyl hexyl) adipate
DEM	discrete elements method
DESI	desorption electrospray ionisation
DF	drought factor
DH	distributional heterogeneity
DMSO	dimethyl sulfoxide
DNAN	2,4-dinitroanisole
DNP	dinitrophenol
DNX	hexahydro-1,3-nitroso-5-dinitro-1,3,5-triazine
DOTAM	Directorate of operations and training area management
DQO	data quality objectives
DRDC	Defence Research and Development Canada
DRE	destruction and removal efficiency
DTA	Donnelly training area
DTRA	Defense Threat Reduction Agency
DU	decision unit
DWEL	drinking water equivalent
EC_{50}	half effective concentration
ECA	European Chemicals Agency
ECD	electron capture detector
EDA	ethylene diamine
EGDN	ethylene glycol dinitrate
EI	electron ionisation
ELPI	electrical low pressure impactor
EM	energetic materials
EMS	environmental management systems
EOD	explosives ordnance disposal
EPA	Environmental Protection Agency
Epi Suite	Estimation Programs Interface Suite
EQS	Environmental Quality Standards
ER	explosive residue
ERF	Eagle River Flats
ESI	electrospray ionisation
ESI/MS	electrospray ionisation mass spectrometry
ETPE	energetic thermoplastic elastomer
EVO	emulsified vegetable oil
FDF	Finnish Defence Forces
FEM	finite elements method
FFDI	forest fire danger index
FI	fragment impact
FSE	fundamental sampling error
FTIR	Fourier transform infra-red
FWA	Fort Wainwright
GAC	granular activated carbon
GC	gas chromatography

GC-ECD	gas chromatography with an electron capture detector
GCMS	gas chromatography mass spectrometry
GFDI	grassland fire danger index
GIM	green insensitive munitions
GLP	good laboratory practice
GOCO	government owned contractor operated
GOGO	government owned government operated
GPS	global positioning system
GSE	grouping and segregation error
GW	groundwater
HA	Health Advisories
HCL	hydrochloric acid
HDPR	high density polyethylene
HI	hazard index
HMX	1,3,5,7-tetranitro-1,3,5,7-tetrazoctane octahydro-1,3,5,7-tetranitro-1,3,5,7-tetrazocine high melting explosive
HNS	2,2',4,4',5,5'-hexanitrostilbene
HPLC	high pressure liquid chromatography
HPTLC	high-performance thin layer chromatography
HTPB	hydroxyl-terminated polybutadiene
IC	ion exchange chromatography
IHE	insensitive high explosive
IM	insensitive munitions
IMS	ion mobility spectrometry
INRS	Institut national de la Recherche Scientifique
IO	ion exchange
IR	infra-red
ISB	*in situ* bioremediation
IT	ion trap
JBER	Joint Base Elmendorf-Richardson
JHTD	Joule heating thermal desorption
KTA	key technical area
L	litre
LAP	load, assemble, pack
LC	liquid chromatography
LCA	life cycle assessment
LD_{50}	median lethal dose
LLM-105	2,6-diamino-3,5-dinitropyrazine-1-oxide
LLNL	Lawrence Livermore National Laboratory
LOAEL	lowest observed adverse effect level
LoD	limit of detection
LSGT	large scale gap tests
MAA	mutual aid agreement
MAE	microwave assisted extractions
MALDI	matrix assisted laser desorption/ionisation
MC	munitions constituents
MD	molecular dynamics
ME	materialisation errors
MEG	military exposure guidelines
min	minute

MIS	multi-increment sampling
mL	millilitre
MMR	Military Massachusetts Reservation
MNX	hexahydro-1-nitroso-3,5-dinitro-1,3,5-triazine
MOU	memorandum of understanding
MRL	minimal risk level
MS	mass spectrometry
MSIAC	Munitions Safety Information Analysis Center
MTA	Marrangaroo Training Area
NAD	Naval Ammunition Depots
NATO	North Atlantic Treaty Organization
NC	nitrocellulose
NCI	negative ionisation
NG	nitroglycerine
NIOSH	National Institute for Occupational Safety and Health
nm	nanometre
NOAEL	no observed adverse effect level
NOP	Naval Ordnance Plant
NP-HPLC	normal phase HPLC
NQ	nitroguanidine
NTO	3-nitro-1,2,4-triazol-5-one
OB	open burning
OD	open detonation
OECD	Organisation for Economic Co-operation and Development
OSBR	overall site bushfire risk
OSHA	Occupational Safety and Health Administration
P_4	white phosphorus
PA	polar polyacrylate
PAF	potentially affected fraction
PBEC	primary bronchial epithelial cells
PCI	positive ionisation
PDA	photo diode array
PDF	potentially disappeared fraction
PDMS	polydimethylsiloxane
PEG	polyethyleneglycol
PETN	pentaerythritol tetranitrate
pg	picogram
PGC	porous graphitic carbon
Picric acid	2,4,6-trinitrophenol
POHC	principal organic hazardous constituents
ppb	part per billion
ppm	part per million
ppt	part per trillion
PRB	permeable reactive barrier
PSI	pounds per square inch
PTFE	polytetrafluoroethylene
PVC	polyvinyl chloride
Q	quadrupole
QC	quality control
QqQ	triple quadrupole

QSAR	quantitative structure–activity relationship
RCO	range control officer
RCRA	Resource Conservation and Recovery Act
RDX	1,3,5-trinitro-1,3,5-triazinane hexahydro-1,3,5-trinitro-1,3,5-triazine royal demolition explosive
REACH	Registration, Evaluation, Authorisation and Restriction of Chemicals
RF	radiofrequency
R_f	retardation factor
RF	radiofrequency
RfD	reference dose
RG	remediation goals
RIGHTTRAC	Revolutionary Insensitive, Green and Healthier Training Technology with Reduced Adverse Contamination
RMGB	Risk Management Governance Board
RMS	root mean square
RP-HPLC	reversed-phase HPLC
RSL	risk based screening levels
RTA	range training area
RTAOC	range and training area operations course
RTAs	range and training areas
S/S	solidification/stabilisation
SAE	sonicated assisted extractions
Sb	antimony
SCJ	shaped charge jet
SCO	slow cook-off
SD	sympathetic detonation
SE	Soxhlet extraction
SEM	scanning electron microscope
SERDP	Strategic Environmental Research and Development Program
SFE	supercritical fluid extraction
SPE	solid phase extraction
SPME	solid phase micro-extraction
SPR	source-pathway-receptor
SQC	sample quality criteria
STEL	short term exposure limits
SUMATECS	sustainable management for trace elements contaminated soils
SVOC	semi-volatile organic compounds
TATB	1,3,5-triamino-2,4,6-trinitrobenzene
TCLP	toxicity characteristic leaching procedure
TDP	Technology Demonstration Program
Tetryl	N-methyl-N-(2,4,6-trinitrophenyl)nitramide
TLC	thin layer chromatography
TLV	threshold limit values
TNT	2,4,6-trinitrotoluene
TNT	1,2,4-trinitrotoluene
TNX	hexahydro-1,3,5-nitroso-1,3,5-triazine
ToF	time of flight
TOS	theory of sampling
UPLC	ultra-high performance liquid chromatography
US EAP	U.S. Environmental Protection Agency

US EPA	United States Environmental Protection Agency
UTLC	ultra-thin-layer chromatography
UV	ultra violet
UXO	unexploded ordnance
VCCT	variable confinement cook-off test
VOC	volatile organic compound
VUV	vacuum ultra violet
WCL	Webster clay loam
WHS	workplace health and safety
XRF	X-ray fluorescence
YTA	Yukon Training Area
ZVI	zero valent iron

Introduction: a global approach to environmental management on military training ranges

Melissa K Ladyman and Tracey J Temple

The main purpose of this book is to collate examples of environmental management from military live-fire training ranges to demonstrate that environmental best practice is compatible with operational activities. The book is divided into four sections; the first provides background information that will help the reader to understand the scientific principles behind environmental management. The second comprises methodologies for the environmental risk assessment of explosives and munitions. The third collates case studies and innovative management techniques that have been applied to reduce remediation costs, enhance public perception and promote environmental best practice. The final section considers the design of 'greener or insensitive munitions' to reduce environmental impact.

Clear examples of how to identify soil and groundwater contamination from explosive activities have been brought together in this book. It outlines identification, monitoring and mitigation methodologies that prevent adverse environmental impact on military training ranges. It is the first time that examples of practical experiences from global perspectives have been brought together in one comprehensive volume. Historically this information has been inaccessible where it has been held in closed forums, published in different languages or restricted.

This book has been written for a non-technical audience, making it particularly beneficial to all those with a responsibility for environmental management of military training ranges and who are required to ensure sustainable long-term training capability.

The aim of this introduction is to summarise environmental management in the military context. It also outlines the chapter content by linking core concepts to the relevant individual chapters.

Introduction to environmental management

Environmental management is the process of managing, mitigating and preventing environmental impact from an organisations' activities (ISO14001). Since the middle of the 19th century there has been increasing awareness regarding the impact of human activities on the environment and a concurrent increase in environmental legislative requirements in many countries. As well as being more stringent, environmental legislation has evolved to become more proactive; for example, older legislation was directed at specific issues, such as disease transmission in contaminated water. More recent legislation aims to minimise the generation of hazardous chemicals and prevent their access to the environment (e.g. the Regulation Evaluation Authorisation and restriction of Chemicals Regulation (REACH)).

Increasingly, legislation embeds key concepts such as 'Polluter Pays', i.e. ensuring that the organisation or person responsible for causing contamination must pay for remediation or clean-up. Or, the best available technology (BAT) principle that makes it a legislative requirement for large installations to keep up with industry standards of environmental protection. Chapter 11 specifically discusses the use of the BAT approach to select appropriate mitigation for small arms shooting ranges.

Environmental management systems have been developed to aid organisations demonstrate their compliance, and implement appropriate mitigation to minimise their environmental impact. Most nations have their own versions of environmental management systems, but large organisations commonly use internationally recognised standards such as ISO14001 (Environmental Management) and 14040 (Life Cycle Assessment) that are audited by a third-party (chapter 5 and chapter 13). In addition, the system will have a continuous review process that aims to ensure that the system is a 'living' requirement and incorporates plan, do, check act elements [1, 2].

Environmental management for defence

Defence related activities spanning all domains are not exempt from complying with environmental law, and must therefore be managed to ensure compliance with national and international legislation. However, the benefits of environmental management for defence can extend beyond compliance; for example, these practices assist with the prevention of penalties and fines for bad practice, improve perception issues and promote best practice. In addition, remediation costs for contaminated land can be minimised and ensure that environmental risks remain low. Frequent review and updating of management systems also ensures that new legislation is identified before an activity becomes non-compliant. Under current defence policy it is essential for military activities to comply with all environmental legislation. Where compliance is not possible exemptions may be obtained for specific activities and equipment that are non-compliant as they are essential for defence.

Environmental management for military training ranges

Compliance with environmental legislation is also applicable to live-firing on military training ranges. This is especially relevant as long-term live-training with munitions can lead to cumulative environmental impacts as the heavy metals and explosives deposited are known toxins and carcinogens [3, 4]. Historically, the impact of military training with live-fire munitions has not been extensively considered and training has been carried out wherever suitable locations have been found. This means landscapes around training ranges areas can vary enormously in and between nations. For example, in Australia many training ranges are situated in areas prone to bushfires (chapter 14), while in North America training ranges may be in deserts or on Marshland (chapter 8). In the UK, where space is at a premium, training ranges may be adjacent to residential areas, and in some areas have public access. These diverse settings can make environmental management difficult and this is due to different soil types, weather and topography. However, contextualised tools such as the source-pathway-receptor (SPR) pollutant linkage concept, which describes links between contaminants and receptors in the eco-system (e.g. flora and fauna) can be used to characterise training ranges and identify where mitigation is required as described in chapter 1 and chapter 12.

Explosive residue from live-firing with munitions has been considered minimal; however, explosives such as RDX, trinitrotoluene (TNT) and nitroguanidine (NG) are frequently detected in soils at heavily used firing and impact areas [5]. It is now known that although first order detonations consume almost all the explosive material, second-order-detonations, blow-in-place and open burning may result in the deposition of significant quantities of explosives. In addition, many training ranges have hundreds or thousands of buried unexploded ordnance (UXO), which are slowly corroding and leaching their chemical contents into the environment. The behaviour of explosives in the environment is discussed in chapter 1, and chapter 6 summarises techniques for assessing the health hazards posed by munitions constituents.

Once contamination has been identified in training areas it is essential to determine the extent by soil and water sampling. However, sampling may be undertaken to varying extents of efficiency depending on resource availability and motivation. The work described in chapter 2 and chapter 3 demonstrates the importance of using a reproducible and representative sampling method in order to ensure effective characterisation of the extent of contamination on training ranges. Similar efforts have been made to identify the extent of heavy metal contamination, as discussed in chapters 9 and 10. Suitable characterisation of contamination is required to ensure that risk is not under- or overestimated, and so that suitable remediation is selected (if appropriate). Chapter 7 summarises some of the remediation techniques used by the US EPA for explosively contaminated sites. The importance of suitable and efficient analytical techniques is outlined in chapter 4.

All of the above methods have been developed based on environmental impacts from legacy explosives and munitions currently in use. Although, emerging scientific approaches are being used to design munitions that have a reduced impact on the environment (chapter 15).

A summary of contributions

Section 1	Identification of explosive contamination on live-fire training ranges
Chapter 1 **Principles of environmental range management (UK)**	It is necessary to understand the impact of live-fire military training on the environment to ensure compliance with increasingly stringent

legislation, maintain a positive public image and minimise the risk of incurring large clean-up costs. However, the impact of a particular munition product can vary significantly between sites depending on the physicochemical properties of the contaminant, the local geology and the local climate. This chapter outlines how simple conceptual models can be used to link activities to potential receptors, and summarises the computational and experimental methods used to investigate the behaviour of contaminants in the environment.

It is becoming increasingly important to assess military training areas for contamination in order to ensure compliance with environmental legislation, to baseline contamination and to ensure the land is safe for use, sale or redevelopment. Characterisation of the concentration of contaminants in the soil across a whole site can be daunting, and is prone to error. Poor sampling strategies can result in up to 1000% error in the results due to inappropriate choice of sampling method, area and tools. This chapter summarises the development of the multi-increment (MI) sampling approach for accurately and reproducibly characterising explosively contaminated military training ranges. The results of a comprehensive research programme into effective sampling are summarised describing how to plan and conduct site characterisation, and how to interpret the data. Deposition of energetic and heavy metals from live-fire training may result in contamination of groundwater by the transport of contaminants from the soil surface. Therefore, an understanding of the hydrogeology of training sites is essential to ensure contamination is avoided, appropriately managed or, if necessary, remediated. Groundwater is rarely directly accessible and therefore observation wells are installed to enable sampling. This chapter summarises the procedure for installing wells for sampling groundwater on military

Chapter 3
Hydrologeological characterization of military training ranges and production of maps for land management (Canada)
R Martel, S Brochu

training ranges from choosing a location to laboratory analysis.

The second half of the chapter focusses on the use of risk maps to visually demonstrate areas overlaying aquifers vulnerable to energetic material contamination within a training range. Risk is assessed based on the vulnerability of aquifers, and the hazard of the munitions. Risk maps can be used to improve environmental management of training ranges and to aid land-use decisions, e.g. situating high impact activities away from vulnerable areas.

Chapter 4
Analysis of explosives in the environment (UK)
N Mai, J Pons, D McAteer, P P Gill

There are a number of analytical techniques available for providing detailed qualitative and quantitative information on soil and water contamination. No single technique is suitable for all situations or, indeed, is capable of detecting every type of explosive. This chapter discusses the main techniques used for the analysis of explosives in soils and water with a particular focus in the areas of sample preparation, spectroscopic, spectrometric and chromatographic procedures. A broad introduction is given to these major techniques, their applications and advantages, at a level that does not require the reader to have a detailed technical knowledge.

Section 2	**Environmental risk assessment for munitions**

Chapter 5
Environmental management of military ranges with the support of a life cycle assessment approach (Portugal)
C Ferreira, J M Baranda Ribeiro

Due to the complexities of environmental fate and transport it is not possible to experimentally determine all potential toxicity impacts for all locations. Life cycle assessment (LCA) can be employed in conjunction with munition emissions data to quantify environmental consequences from military training activities. The aim of LCA is to identify contributing factors to environmental toxicity of emissions in order to reduce their environmental impact. This can be achieved using a combination of emissions data, experimental data and predictive modelling software. This chapter summarises the advantages of LCA when used in conjunction with other management tools to assess the environmental impacts of military training.

A summary of significant findings from over fifteen years of range characterisation and monitoring focusing on the species and concentration of energetic material deposition. The data are compiled from two Alaskan live-fire training areas and provide case studies of the magnitude of contamination from different range activities such as firing, impact, small arms training and demolition. The studies found that high concentrations of propellant were observed at the propellant burn site, demolition range and some firing points. High explosives were identified at impact areas, particularly around impact craters from low order detonations and demolition ranges. Groundwater and surface impacts were not evident.

Due to the ubiquity of lead, antimony and copper in small arms ammunition, contamination of shooting ranges has become a widespread issue. Lead is of particular concern due to quantity, public awareness and the well-known toxicity. Moreover, due to technical advances the detection and quantification of lead in soils and groundwater has become easier. With increasing environmental concentrations, these metals may pose a significant risk to local ecosystems and ultimately human health. Therefore, it is essential to be able to identify and quantify areas with high levels of contamination, and in consequence reduce the environmental risk associated to pollution from shooting practices. This chapter summarises the method for identifying and quantifying contamination on Swiss military shooting ranges in order to determine whether, and to what extent, remediation is required.

The Borris Shooting range in Denmark has been used for a range of military activities from small arms to anti-tank training for at least fifty years. Due to the high usage, and sensitive location – on sandy soil near to a stream

	Defence approach to bushfire management, which must ensure safety, comply with governmental and defence policy and consider public perception. Case studies of wildfire incidents are given to demonstrate how improvements in land management can reduce environmental and safety risks.
Section 4	**Environmental considerations in munition design**
Chapter 14 **Greener or insensitive munitions: Selecting the best option (Canada)** S Brochu	The requirement for munitions to be high performing, safe to handle and have low environmental impact often requires a trade-off in properties. For example, insensitive high explosives meet the required safety profile, but have been found to deposit more residue on ranges through low order detonations compared to their predecessors. On the other hand, it may be possible to improve the environmental performance of current formulations during the manufacturing process, but the safety profile would remain the same. With many, often contradictory, factors to consider it can be difficult to make an informed decision on which formulation to use. This chapter describes the development and use of a decision matrix tool to select an explosive candidate to replace in-service formulations that most successfully meets the safety, environmental and performance criteria.

References

[1] ISO - International Organization for Standardization 1996 *ISO 14001 Environmental Management Systems—Specification with Guidance for Use* (Geneva, Switzerland: International Organization for Standardization)

[2] ISO - International Organization for Standardization 2006 *ISO 14040 International Standard Environmental management—Life cycle assessment—Principles and framework* (Geneva, Switzerland: International Organisation for Standardization)

[3] Johnson M S and Salice C J 2009 *Ecotoxicology of Explosives* ed G I Sunahara, G Lotufo, R G Kuperman and J Hawari (CRC Press)

[4] Ryu H, Han J K, Jung J W, Bae B and Nam K 2007 Human health risk assessment of explosives and heavy metals at a military gunnery range *Environ. Geochem. Health* **29** 259–69

[5] Pennington J C, Jenkins T F, Ampleman G, Thiboutot S and Brannon J M Technical report 2003 *Distribution and Fate of Energetics on DoD Test and Training Ranges* Interim Report 3 http://oai.dtic.mil/oai/oai?verb=getRecord&metadataPrefix=html&identifier=ADA417819

Chapter 1

Scientific principles of environmental management

I Bortone, F Coulon, W Fawcett-Hirst, M Ladyman and T Temple

Military training ranges are essential for live-fire training; however, with increasingly stringent environmental legislation it is also necessary to ensure that the environmental impact of live-fire activities is mitigated and managed. The environmental impact of live-fire training includes noise nuisance, heavy metal contamination and land degradation. For a long time it was thought that explosive residue from live-fire training was limited as all materials would be consumed during detonation, but research has shown that a significant amount of explosive residue may be deposited during partial detonations and through unexploded ordnance. In addition, contamination of soil and groundwater resulting from explosive residue deposition at training ranges is now well documented. Therefore, knowledge of the explosives likely to be deposited at training ranges and their behaviour in the environment is essential in order to appropriately mitigate or manage their impact. This chapter outlines how simple conceptual models can be used to summarise the training range environment including soil types, geology, hydro-geology and topography in order to link training activities to potential receptors such as ecosystems, humans and animals. An overview of the impact of commonly used explosives in the environment, and how computational and experimental methods can be used to determine and predict the behaviour of contaminants in the environment is also given.

1.1 Introduction

Military training ranges are essential to maintaining defence capability and environmental issues need to be appropriately managed to ensure continuing legal compliance and training capacity [1]. Severe contamination incidents, such as groundwater contamination, arising from poorly managed sites may result in a

1-1

limitation on activities such as use of particular munitions [2] or closure of the site. In addition, many training ranges may undergo a change of use, or release to the public, which under current law would require the site to be returned to its original state [3]. In some cases, the cost of remediation may be so great that it is more economical to continue to maintain the land even though it is no longer in use. Environmental management is a widely used framework that enables identification and mitigation of environmental impacts. However, effective environmental management requires a thorough understanding of the site (above and below the surface), knowledge of the sources of potential contamination and areas, flora and fauna that are potentially affected (adversely or positively) by the contaminants. Often the source of potential impacts is readily identifiable, e.g. chemicals from munitions or spillages. However, once a chemical enters the environment it becomes more difficult to determine where it might travel (through soil and water) and how it may interact with the environment (degradation and absorption to soil). The environmental impact may be far removed from the source of contamination.

This chapter therefore aims to outline the critical environmental parameters and physicochemical properties of explosives that influence the behaviour of contaminants in military training environments. Focussing on how to progress from a simple conceptual diagram to a comprehensive knowledge of the potential environmental impacts from a selection of high explosives on military training ranges, the initial conceptual approach enables preliminary evidence-based estimations to be made, which will highlight any data gaps and guide the next steps, e.g. field sampling, laboratory experiments or predictive modelling.

1.2 Contextualising military training environments

For security and safety reasons military training ranges are usually located in remote open spaces. This means they are often situated in sensitive or protected areas, which require careful management to ensure that military training can co-exist with the natural environment. In some countries due to space limitations military ranges may allow recreational access to the public, be used for farmland, or be situated near to residential areas.

Regular activities undertaken on training ranges may include transporting munitions around the site, live-fire training, tank manoeuvres, demolitions and open-burning of unused and out-of-date ordnance. In addition, abnormal activities may also occur such as blow-in-place of unexploded ordnance (UXO). As described in this chapter, historic explosive activities can have significant impacts on the environment such as contamination of soil and groundwater leading to ecological toxicity.

In order to manage and mitigate the environmental impact of explosive related activities on military training ranges it is necessary to understand how contaminants travel through the environment. Awareness of the fundamental principles that govern the behaviour of explosives in the environment allows their effects to be described and estimated for particular environments. For example, how explosives, combustion products and metals can be transferred from the point of use or disposal, to flora, fauna and ecosystems that may be adversely affected.

The transport pathways of contaminants can be conceptualised using the source-pathway-receptor (SPR) pollution linkages model [4]. The SPR model describes how a contaminant can enter the environment from a source such as, for example, deposited explosive residue, and moves through the environment by air, soil or water (pathway) to a receptor. A receptor is defined as an entity that may be adversely affected by interaction with a contaminant. An 'entity' can include flora and fauna, humans, the ecology of an area such as lake ecosystems and man-made structures and dwellings.

The point at which a contaminant reaches and affects a receptor is when the contaminant can be said to cause an environmental impact. If the SPR linkages between the source and receptor are broken at any point, i.e. if the contaminant does not reach a receptor, then no environmental impact will be observed. SPR conceptual diagrams can therefore be used to illustrate the pathways between sources and potential receptors in and around training ranges to predict possible environmental impacts.

Figures 1.1–1.3 outline basic SPR linkages for typical military training range activities that may lead to environmental impact. Simple diagrams can illustrate the main features of the training ranges such as basic topography, locations and depths of surface water, prevailing wind direction and nearby residential and recreational areas. Figure 1.1 illustrates the SPR linkage between live-fire training (source) and

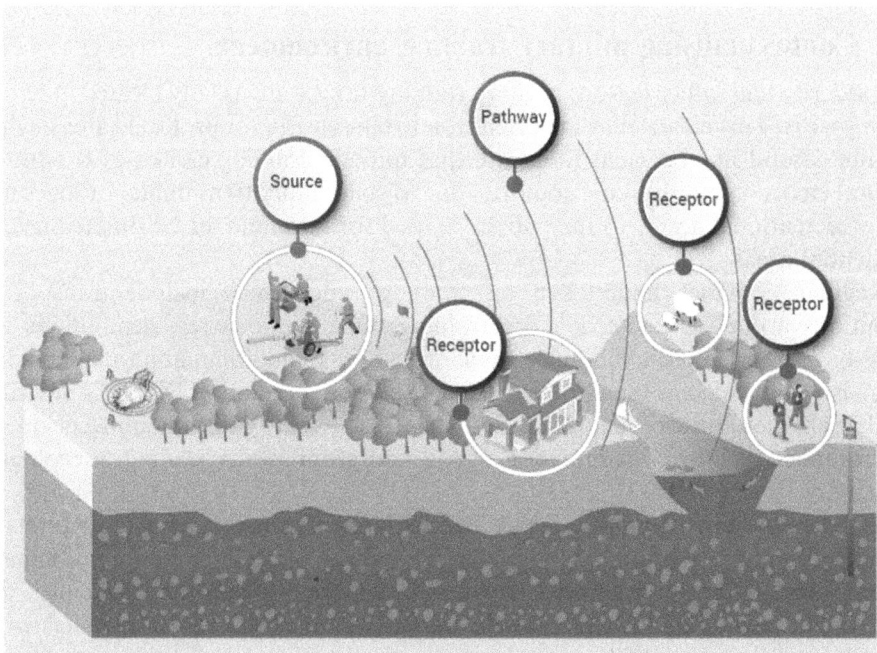

Figure 1.1. A conceptual diagram of the SPR linkages between the firing of guns and receptors.

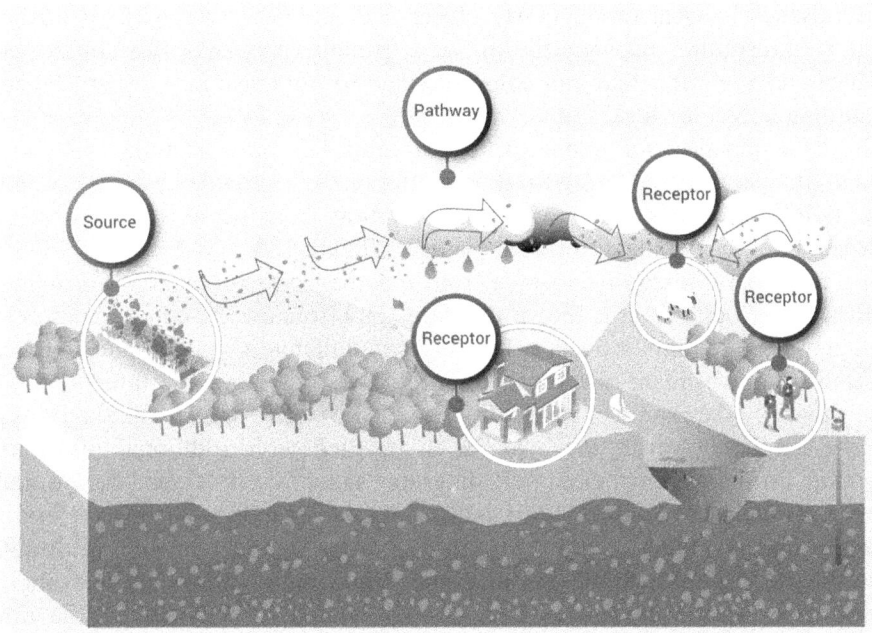

Figure 1.2. SPR linkage between open-burning of explosives/munitions and receptors.

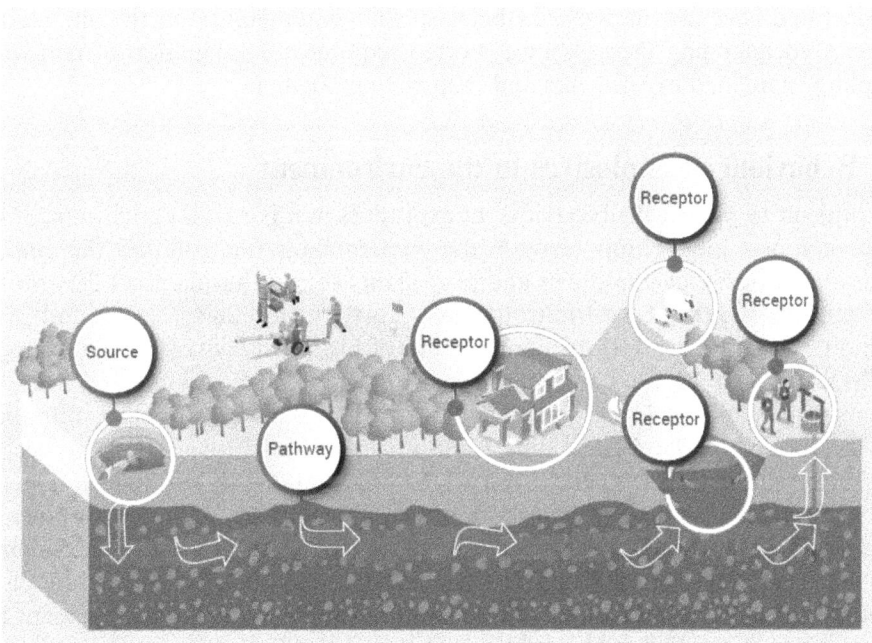

Figure 1.3. SPR linkages between buried UXO and receptors.

noise nuisance to local residents, structures and ecology (receptors). In this case, sound is transmitted through the air and ground (pathways) leading to noise nuisance and vibration that can cause damage to buildings and disrupt breeding and nesting habits for local fauna.

The SPR linkages for open-burning of munitions may include transmission through air (pathway) to cause local air pollution and nuisance from black smoke generation (source), as well as ground pollution from deposition of explosive, carbonised and heavy metal residues (sources) (figure 1.2).

The transport of explosive residue from a damaged UXO (source) through soil (pathway) to groundwater is shown in figure 1.3. Groundwater contamination may affect humans and animals (receptors) using the aquifer as a potable water source, or contaminate downstream rivers or lakes and affect local flora and fauna.

As shown by the examples above (figures 1.1–1.3), SPR conceptual diagrams can be very simple; however, they can be augmented with additional information depending on the circumstance. SPR diagrams should be developed before undertaking any practical work such as air, land and water sampling on the site to ensure a thorough understanding of the environment, or to highlight missing information about the site, e.g. the activities taking place. When undertaking an SPR evaluation the source is usually evident as it is directly related to activities on the range. Receptors can be identified through observation of the site, particularly where there have been historic contamination incidents. However, understanding the pathway is more complex as it depends on how the contaminant enters the environment and the physical and chemical interactions between the contaminant and the air, land or water. Understanding these pathways often requires a combination of real-world sampling, simulated experiments and predictive modelling.

1.3 Behaviour of explosives in the environment

It is difficult to predict the behaviour of explosives in a particular environment due to the complex interactions between the contaminants, the soil and the climatic conditions. For legacy explosives and propellants such as hexahydro-1,3,5-trinitro-1,3,5-triazine (RDX), 1,2,4-trinitrotoluene (TNT), nitrocellulose (NC) and nitro-guanidine (NQ) (figure 1.4) the combination of historical contamination examples and research can be used to inform environmental impact. However, with the increasing demand for Insensitive Munitions (IM) it is likely that legacy munitions will be replaced by those containing Insensitive High Explosive (IHE) fills such as 2,4-dinitrotoluene (DNAN) and 3-nitro-1,2,4-triazol-5-one (NTO) (figure 1.4). It is important that the fate and transport of these explosives in the environment is understood before their widespread use to prevent any adverse environmental impacts. Therefore, the following sections summarise current understanding of the fate and transport of legacy and new generation explosives in the environment and their potential environmental impacts.

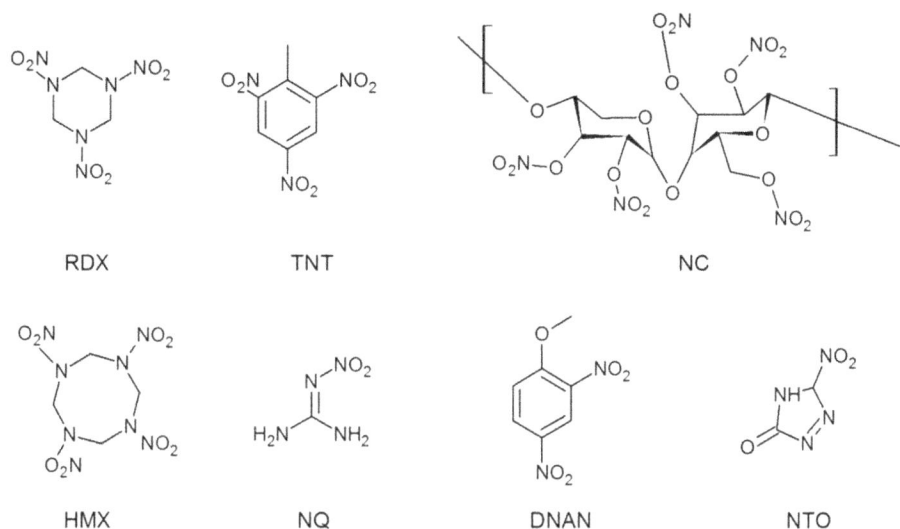

Figure 1.4. Structures of explosives discussed in this chapter: RDX, HMX, TNT, NQ, NC, DNAN and NTO.

1.3.1 Nitramine explosives

RDX and octahydro-1,3,5,7-tetranitro-1,3,5,7-tetrazocine (HMX) are nitramine explosives with similar chemical structures and properties commonly used in combination with TNT in formulations such as Comp B (40:60/TNT:RDX) and Octol (25:75/TNT:HMX) [5, 6]. In addition, military grade RDX may contain up to 10% HMX impurity as a side-product of the manufacturing process [3]. RDX is also used as a constituent in several IM fills such as PAX-21 (RDX, DNAN and ammonium perchlorate (AP)) and IMX-104 (RDX, DNAN and NTO) [7, 8]. RDX and HMX are common contaminants at military training ranges and have been identified in soils at demolition, burning and impact areas with higher concentrations found in areas where intense firing occurs towards a compact target area [3, 9].

Research into residue deposition from munitions containing RDX, e.g. mortars, artillery rounds and grenades containing Comp B, has shown that second-order detonations from blow-in-place or partial detonations are responsible for the majority of residues [10]. First order, live-fire detonations have been shown to consume on average 99.997% of the energetic materials [11–14], which may be an under-estimate in some cases due to the method of detonation, i.e. simulated live-fire [15, 16]. Similarly, for munitions with the IHE fill Pax-21 high order detonations resulted in deposition of less than 0.001% RDX [8].

On the other hand, low order blow-in-place detonations have been shown to deposit up to 50% of the Comp B fill from mortar (60 mm, 81 mm, 120 mm) and artillery (105 mm and 155 mm) rounds, with detectable concentrations of HMX [17]. Explosive residue from low order detonations was found to be deposited in large friable chunks of the original formulation (up to 5.5 g for Pax-21), the highest concentration of explosive on training ranges is usually found near these types of deposits [8, 12]. Although the number of low order detonations is thought to be less than 1%, this can be equivalent to several hundred rounds depending on the number fired per year, which over time can culminate in kilograms of explosive residue being deposited on training ranges [18].

UXO is also a likely source of contamination as the safest course of action is often to leave them *in situ* on training ranges, particularly if they are buried. The percentage of duds can be up to 3% and given that some training ranges have been in use for almost one hundred years, the amount of buried ordnance could be significant [19]. For example, estimates for some North American training areas suggest that there may be tens of UXO per acre (4047 m^2) [18]. However, contamination from UXO is dependent on the rate of leaching of the explosive from the munition, which can take between 10 and 90 years depending on the acidity of the soil [18]. Therefore, UXO is unlikely to be a large source of current contamination, but may pose a contamination issue in the long term [14].

Depositions of RDX and HMX on the soil surface, particularly large chunks of explosive, may pose a risk to users of the site if the material is handled with bare hands or ingested. RDX is a suspected human carcinogen [20], and if ingested targets the central nervous system with the potential to cause seizures in humans and animals [21, 22]. RDX has a median lethal dose (LD$_{50}$) of 119 mg kg^{-1} in rats [23], while HMX is much less toxic due poor bioavailability (not adsorbed by mammalian gastrointestinal tract) [24, 25]. The minimal risk level (MRL) for non-cancer health effects by oral RDX and HMX exposure is 0.006 mg kg^{-1} day^{-1}, exposure above this level should be considered a danger to health [26]. The United States Environmental Protection Agency (USEPA) has established residential and industrial soil screening levels of 5.6 mg kg^{-1} and 24 mg kg^{-1}, which can be used as a guideline for safe exposure limits [20].

RDX and HMX have low solubility in water (42 mg l^{-1} and 4.5 mg l^{-1} at 20 °C–25 °C, respectively) [27, 28], and therefore tend to accumulate in the top layers of soil where they are likely to be exposed to sunlight and may undergo photodegradation [3]. RDX is susceptible to photodegradation as a solid and in solution, though photodegradation is far slower in the solid (half-life of 2–3 months, compared to 2–3 days in solution) [29]. One of the main degradation products is nitrate (NO^{3-}), accounting for 20% of the photo-degraded RDX and contributing to pandemic nitrate groundwater contamination in North America [30–32]. HMX has also been shown to photo-degrade under laboratory conditions at an order of magnitude slower than RDX. However, in this case HMX was dissolved in an aqueous solution and exposed to consistent UV irradiation forcing degradation to occur faster than in the real environment [28, 33].

Although RDX and HMX have low solubility, their detection in groundwater at several military training ranges demonstrates that significant quantities of the explosive are solubilised [3, 9, 34–36]. The limiting factor for the rate of transport from an RDX/HMX source is the rate of dissolution from the formulation matrix, which increases with increasing rainfall [21, 37, 38]. Once solubilised, they tend to be transported at the infiltration rate due to their low sorption to soil [38, 39].

RDX degrades slowly in soil environments to form methanol, hydrazine and dimethylhydrazine via anaerobic transformation [40–42]. RDX has two proposed degradation routes as investigated in municipal anaerobic sludge. The first is initial reduction of the nitro groups to nitroso groups to form hexahydro-1-nitroso-3,5-dinitro-1,3,5-triazine (MNX), hexahydro-1,3-nitroso-5-dinitro-1,3,5-triazine (DNX) and hexahydro-1,3,5-nitroso-1,3,5-triazine (TNX) followed by ring cleavage. The second mechanism is initial ring cleavage to give methylenedinitramine and bis (hydroxylmethyl)nitramine followed by further degradation into carbon dioxide, methane and water. The nitroso derivatives of RDX have been detected in soil and water samples, suggesting that the reductive pathway may be preferred under environmental conditions, or that these compounds may be stable in solution for environmentally significant periods of time, i.e. long enough to impact a receptor [43–45]. HMX is more stable than RDX in the environment, and no HMX degradation products have been detected in environmental samples. However, in laboratory studies, the nitro groups of HMX can be reduced to nitroso groups under anaerobic conditions [42]. Of the two compounds, RDX tends to present more of a problem on training ranges due to its use in large quantities in munitions.

1.3.2 Nitroaromatic explosives

Nitroaromatic explosives such as TNT and DNAN have been used in high explosive fills and propellants for many years. TNT in particular is often used in combination with RDX and is therefore commonly detected in surface soils around impact areas [11, 27, 45, 46]. Similarly to RDX, deposition on military training ranges is most likely from second-order detonations, open-burning and UXO [17]. The nitro-aromatic 2,4-dinitrotoluene (2,4-DNT) is found in smokeless powders and as a plasticizer in propellants and, along with 2,6-dinitrotoluene (2,6-DNT), is a by-product of TNT manufacture and a common TNT biodegradation product. Both 2,4-DNT and 2,6-DNT are frequently detected on training ranges [3, 36].

The impact of TNT on the environment can be influenced by its joint deposition with RDX in formulations. TNT is more soluble in water than RDX and HMX (130 mg l^{-1} at 20 °C), but in Comp B formulations TNT dissolution is limited by RDX, which is slower to dissolve and protects TNT deeper inside the particle [27]. Unlike RDX, very little TNT has been detected in groundwater at military training ranges though it is commonly detected in surface soils [3, 47]. However, TNT degradation products such as 2-amino-4,6-dinitrotoluene (2A-DNT) and 4-amino-2,6-dinitrotoluene (4A-DNT) are frequently detected suggesting that TNT transport is limited by biodegradation [36]. In addition, TNT and its degradation products are more likely to sorb to soil compared to RDX, and the amino groups on 2A-DNT and 4A-DNT may covalently bind to functional groups in the soil, immobilising the contaminants [41, 42].

In recent decades, attention has turned to the development of IHE fills resulting in renewed use of DNAN as replacement for TNT in melt-cast formulations (up to 40%) [48, 49]. In the past, DNAN has not been found in large quantities on training ranges, mainly due to its limited use in ordnance. However, as IM are increasingly adopted by militaries worldwide, DNAN may present a future environmental issue. Similar to TNT and RDX, very little DNAN has been found to be deposited in high order detonations with only a few milligrams per tested round (60 and 81 mm mortars), amounting to less than 0.005% of the original mass [50]. Low order or blow-in-place detonations are much more likely to result in significant deposition, with up to 20% of the initial mass recovered after simulated detonations. This can be as much as 34 g for the 60 mm mortar, and 49 g for the 81 mm mortar [50].

Unlike Comp B residues, IHE containing DNAN (such as IMX-101 and IMX-104) dissolve throughout their volumes as DNAN and the other major constituents are soluble in water (DNAN: 276 mg l^{-1}) and therefore dissolve more quickly [51]. Dissolution rates are dependent on average rainfall, as well as the size of the residue particles, making it difficult to accurately predict the rate of dissolution in a given environment [38].

Research into DNAN has shown that it undergoes rapid aerobic biodegradation with complete mineralisation to nitrites possible within 100 h in culture [52]. In soil environments, DNAN has been shown to be degraded by both anaerobic and aerobic processes [53]. However, under aerobic conditions very little chemical transformation was observed, with DNAN more likely to be absorbed to the soil — particularly soils with high organic content [54–57]. Under anaerobic conditions DNAN is reductively bio-transformed to 2-amino-4-nitroanisole (2-ANAN) and 2,4-diaminoanisole (DAAN). In soil column studies 2-ANAN has been shown to be the most common degradation product, with trace dinitrophenol (DNP) also detected [57]. DNP has been identified as a common photodegradation product of DNAN, and is more likely to be produced in solid IMX-104 residues on the surface, or in surface waters [58].

DNAN has an oral LD_{50} of 199 mg kg^{-1} and a LD of 300 mg kg^{-1} in rats, suggesting that DNAN would also be toxic to humans [59, 60]. The common degradation products 2-ANAN, DNP and DAAN are associated with slightly lower toxicity compared to DNAN in standard toxicity models [53]. DNAN and its degradation products exhibit significant toxicity in aquatic environments with a half effective concentration (EC50) of 60.3 mg l^{-1} as determined by Microtox [61]. Whilst limited to model species, the repeated evidence of toxicity suggests that DNAN presents a significant risk to freshwater aquatic species such as micro-organisms, amphibians and fish [62, 63].

1.3.3 Propellants

Nitroglycerin (NG), NQ, 2,4-DNT and 2,6-DNT have been detected at firing points on many US training ranges [9]. Studies have also found high concentrations of perchlorates [36]. The presence of NC on firing ranges has not been extensively investigated as it tends to be deposited as discrete fibres that are difficult to quantify [9]. An added complication is the tendency for other explosives to sorb onto NC, and so areas where high quantities of NC propellants are in use have been more difficult to analyse [64].

NC is insoluble in water and tends to be deposited on the soil surface in fibrous strands, rather than particles. NC is susceptible to biodegradation to a non-energetic polymer through the loss of nitrogen [64]. NC is not toxic towards humans, animals or plants due to the difficulty of absorption through the intestines and insolubility in water [65]. Deposition of NC on ranges does not have a significant environmental impact on receptors, other than an unnatural presence in the environment, and therefore has not been subject to intense scrutiny compared to high explosives such as TNT and RDX.

NG is soluble in water, and does not tend to sorb significantly to soil so it is quite mobile in soil environments [64]. Under favourable soil conditions NG undergoes several stages of biodegradation and can ultimately form glycerol, and in some cases carbon dioxide. However, in reality not all NG will be completely biodegraded, and in sandy soils where attenuation is poor it is unlikely that NG will be significantly biodegraded before migration to groundwater [66]. Photodegradation of NG occurs more quickly when it is solubilised in water, with a half-life of weeks. Some degradation products are more toxic than NG itself, therefore if NG contamination is suspected it is recommended that samples are also tested for mono and di-nitroglycerine [29].

NQ is a common component of propellants, but has also recently been used in the IHE IMX-101 in combination with DNAN and NTO [67]. Research has suggested that a higher percentage of residue may be deposited from IHEs compared to conventional explosives such as Comp B, even for high order detonations [51, 68]. Therefore, higher quantities of NQ may be deposited on ranges in the future. NQ is fairly water soluble (2.6 g l^{-1}) and does not sorb significantly to soil. NQ has been found to be biodegraded in some soils to ammonia and nitrous oxide, particularly soils with high organic carbon content but it may migrate too quickly through soil for any significant degradation to occur before reaching groundwater [69]. Photodegradation of NQ gives guanidine, urea and cyanoguanidine.

Perchlorate is a naturally occurring and man-made anion that contains one chlorine atom bonded to four oxygen atoms (ClO_4^-). It is commonly produced when energetic compositions in the form of ammonium perchlorate (AP), sodium perchlorate and potassium perchlorate dissolve in water. Ammonium perchlorate is a key component of PAX-21, an IHE fill also containing DNAN and RDX, which now has restrictions on use due to the significant quantities of AP deposited during detonation (averaging 15% during high order detonations) [8, 70].

Perchlorates are generally soluble and stable in water and do not sorb significantly to soil and therefore have a great potential to contaminate surface and groundwater. Incidences of significant groundwater contamination from military use of perchlorates in North America have been well documented [70, 71]. A risk assessment of UK use of perchlorate highlighted that monitoring for AP in potable water sources is not widespread in the UK, but that high levels of AP have been detected in the groundwater at Shoeburyness disposal range where 14.5 tonnes of propellant have been disposed of since 1993 [72]. While drinking and groundwater limits have not been set for the UK, where contamination has occurred the UK uses the limit of detection (1 μg l^{-1}) as an indication of acceptable levels. The presence of perchlorate in drinking water is a particular concern as it is known to cause abnormal thyroid function in humans [73, 74].

1.3.4 3-Nitro-1,2,4-triazol-5-one

With the increasing importance of IM, previously unused energetic materials such as NTO are becoming more common. NTO has been used in IHE fills such as PAX-21, IMX-101 and IMX-104 in combination with more conventional munition constituents such as RDX and NQ [7, 67]. There are some concerns with the use of NTO due to the increased residue deposition from both high and low order detonations of ordnance using IHE fill [50, 75]. In particular, sampling after blow-in-place detonations of IMX-104 60 mm and 81 mm mortars demonstrated that up to 50% NTO may be deposited on the range [50].

NTO is highly soluble in water (16 g l^{-1}) and dissolves in direct correlation with the volume of incident water, suggesting that it could quickly enter soil environments with transport highly dependent on the volume of rainfall [76, 77]. In addition, NTO does not sorb significantly to soil and therefore has the potential to be highly mobile [20]. Batch adsorption experiments with NTO have demonstrated very low adsorption to a variety of soils within the pH range 4.4–8.2, organic matter contents of 0.34%–5.25% and specific surface areas of 1.7–38.3 m^2 g^{-1} [78].

Although NTO has the potential to be mobile, it is also susceptible to biodegradation with complete degradation observed within five weeks [57]. Degradation of NTO in aerobic soil cultures has been shown to occur through loss of nitrate followed by transformation to 1,2-dihydro-3H-1,2,4-triazol-3-one (ATO) (figure) [79]. However, NTO degradation by-products have yet to be detected in real soil environments, although ATO is expected to be the main intermediate before complete mineralisation [80]. Soil column studies with NTO have demonstrated that a significant percentage of NTO is rapidly biodegraded, particularly in soils with high organic content where in some cases it has not been possible to recover any of the energetic material [57, 81, 82]. Whereas in soil with low organic content, high proportions of NTO can be quickly transported with up to 100% recovery within a week [57, 81].

NTO is one of the least toxic explosives, with an oral LD$_{50}$ of >5000 mg kg^{-1} in rats [83]. Rats exposed to 184 μg l^{-1} for 4 h exhibited no adverse effects, which is a good indicator of low chronic toxicity in humans [84]. However, both NTO and ATO have anomalous effects on the development of zebrafish embryos, which have neurodevelopment processes homologous to those in humans, suggesting that NTO may cause reproductive toxicity [85]. NTO exposure to skin causes mild, short-term irritation [83], and can penetrate skin at 332 μg cm^{-2} h^{-1} [86]. Although considered less toxic than TNT, the higher solubility of NTO in water compared to other explosive compounds may cause water discoloration, even at low concentrations.

1.4 Predicting environmental behaviour of explosives

It is possible to understand the behaviour of contaminants in the environment by studying contaminated environments; however, this can only be achieved after contamination has occurred. With increasing emphasis of pollution prevention in environmental law, the ability to predict and prevent environmental impact is becoming increasingly important [87]. The behaviour of contaminants in the environment, such as explosives, is affected by a multitude of interacting physical, chemical and biological processes. Therefore, controlled laboratory experiments have been developed to investigate their environmental behaviour, and are frequently used in conjunction with predictive modelling techniques to determine the potential impact of specific activities. The aim of modelling is typically to: (a) detect a site-specific problem; (b) conceptualise important features, processes and events so that proximity to receptors can best be determined (by defining the conceptual model); (c) implement a quantitative description; (d) collect and embrace field/experimental data that can be used to calibrate the model and assess its analytical assumptions; and (e) predict long term contaminant migration and transformation.

1.4.1 Laboratory experiments

Laboratory fate and transport experiments are often designed to simulate environmental conditions to ensure the generation of robust, representative and reproducible data under controlled conditions. Often, the dissolution, transport, degradation and toxicity of explosives are investigated independently to isolate the effect of environmental variables such as rainfall, UV exposure, soil types, temperature, physical degradation and biodegradation. Dissolution studies are used to determine the mechanism and rate of dissolution from the formulation matrix under environmental conditions by simulating rainfall [51]. Soil columns are designed to mimic the flow paths of contaminants in soil, and are often used to simulate the transport of contaminants in different soil types and climates [88]. Finally, photodegradation and biodegradation can be investigated in aqueous solutions and soil suspensions to isolate the effect of UV, biodegradation and adsorption, making it possible to identify associated breakdown products.

Rate of dissolution
The primary mechanism by which explosives migrate into the environment is by dissolution during precipitation [21]. However, only a limited number of methods are available to determine the dissolution rate of explosives in the environment often using different experimental approaches, making the direct comparison of results difficult. For example, methods used in the pharmaceutical industry, where dissolution rates are determined by introducing a solid to a large volume of agitated water and then measuring the time taken to achieve dissolution by sub-sampling at frequent intervals have been used to investigate the effect of variable temperatures (10 °C, 20 °C and 30 °C) on the dissolution rate of RDX and TNT [21]. However, explosives in the environment are unlikely to be submerged continually in agitated water, and are more likely to be dissolved through the action of sporadic rainfall. Further studies to determine changes to the three-dimensional structure of explosive formulations during dissolution by precipitation have been undertaken by

submerging samples in water without agitation [89]. The samples were submerged in deionised water (DI), which is not representative of natural rainfall because the pH of the water is controlled (pH 7) and submerging the sample does not simulate the mechanical action of rainfall. This approach allowed the samples to be imaged by digital microscopy, which measured the dimensions of the samples as they dissolved.

To simulate the mechanical action of rainfall, indoor dripping tests have been carried out using glass frits holding pieces of explosive underneath the simulated water flow [76]. Water can be delivered using a gravity-fed system [79], or by rate-controlled systems such as peristaltic or syringe pumps [38, 90, 91]. The dripping is designed to mimic rainfall, although the rate of delivery is averaged against real weather conditions, e.g. to a consistent time and droplet size. However, artificial droplets are generally larger than natural rain falling at the same rate, highlighting the difficulties of simulating rainfall precisely [37]. Indoor rainfall simulations may be continuous or may include pauses for drying time, which increases the mechanical stress on the sample and is more representative of natural weather [90]. For example, drying the sample introduces cracks that allow the ingress of water, and may increase the rate of dissolution, but it is difficult to isolate and measure the effect of this variable in the laboratory. In addition, these indoor experiments only investigate the relationship between rainfall and dissolution, and do not consider other environmental variables such as temperature and sunlight. To account for all variables, similar experiments have been undertaken outside and have shown similar dissolution rates, but significant differences were observed in the mass of explosive recovered [90, 92]. This suggests that the dissolution rate may be largely dependent on water volume, but the explosives may be degraded through other mechanisms such as photodegradation when on the soil surface, which cannot be determined by laboratory-based dripping experiments.

Mass balance

Soil columns have been used as a standard method to evaluate the fate and transport of various chemicals and contaminants, ranging from pesticides to landfill waste [88, 93–99]. Soil columns are usually vertical tubes containing soil packed at a specific density, allowing the soil to be spiked with a contaminant. There is no standardised format, but the literature suggests best practice for building soil columns with different lengths and diameters, with the largest columns weighing up to 50 tonnes [88].

Columns can be extracted from the ground (monolithic), thus retaining the natural flow path characteristics such as micropores, root cavities and cracks. The method used to extract the monolithic column may influence performance, especially if there is any deformation or compression during extraction. Alternatively, homogeneous columns can be used, which are assembled manually from well-characterised soil samples. Preparation involves full soil categorisation (organic content, pH, particle size, etc), sieving the soil before filling, and tamping the soil in the column to ensure homogeneous packing without voids or inclusions. This method does not simulate the soil structure found naturally, but it standardises the soil structure making columns more reproducible and allowing comparisons to be drawn between different soil types [100].

Soil columns have been used to investigate the fate and transport of explosives in different soils and environments, although there are no standardised methods [56, 80, 101, 102]. For example, soil column experiments have been used to measure the transport of individual explosives through varying soil types with differing levels of organic matter [56], as well as investigation formulations [57, 80]. At the end of the experiment, soil columns are often dissected to enable full calculation of the mass balance to account for all spiked materials and their transport pathways. This would not be possible in the real environment as contamination concentrations are unknown, and the soil is uncontained allowing far greater sideways transport. Although soil columns are not fully representative of the real environment they provide a controlled, reproducible system that can be used to predict contaminant transport rates and pathways.

Degradation

As seen in section 1.3, explosives are likely to degrade in the environment, particularly when exposed to sunlight or soil that is rich in minerals, organic compounds and microbes [103]. Degradation in soil may occur via physical mechanisms (e.g. metal ions as catalysts) or due to biodegradation. The latter can be investigated by exposing specific strains of bacteria to explosive materials at low concentrations for hours or days [104]. This is not representative of a real soil environment, therefore soil microcosms have been used where spiked soils are suspended in a basal medium to encourage bacterial growth [55, 105]. This highlights the effect of different soil types on degradation, specifically looking at characteristics such as organic carbon content, iron availability and soil texture. However, under these conditions it is difficult to separate the effect of biodegradation from physical degradation and adsorption. One way to eliminate the effect of biodegradation is to sterilise the soils. Autoclaving at 120 °C or irradiating with gamma rays destroys bacteria, but keeps the organic carbon content the same, isolating its effect [81, 104]. More destructive methods (such as incineration at 400 °C) remove both microbes and organic carbon, so generic degradation pathways in different soil types can be determined. A more representative method is to monitor the effluent from soil columns for spiked explosives and their degradation products, but such methods need to be supported by additional degradation studies if the mechanism of degradation is not understood [56, 81].

1.4.2 Computational modelling

The most important parameter in explosives' environmental fate and transport studies is the contaminant concentration (C), which corresponds to the amount of that substance in a specific volume or mass of air, water, soil, or other material.

Many processes affect the fate, transport and transformation of a particular contaminant in the subsurface. The overall composition of leachates produced reflects the degree of aerobic and anaerobic decomposition as well as retention within the soil matrix. Understanding how this material can become mobilised and what happens to it is difficult and complex as the contaminant transport subsurface is affected by a large number of nonlinear and often interactive physical, chemical and biological processes (section 1.3) [106, 107]. Such processes can be modelled by mathematical equations, which describe the dependence of unknown quantities of interest (e.g. TNT concentration at some point and time) on known or estimable parameters (e.g. permeability, reaction coefficients, mass loading). The models are developed from conceptual models derived from laboratory and/or field observations of parameters to be interrelated. Concentration is a key quantity in fate and transport equations as a substance's concentration in an environmental medium also determines the magnitude of its biological effect. Because of the complexity of processes, simplifying assumptions can be introduced to make the model more tractable numerically, i.e. neglecting processes considered to be of minor influence such as volatility.

The most common conceptual models in the literature are based on the assumption that solid explosive residues (ER) in soils are dissolved by infiltrating precipitation at or near the ground surface and subsequently leached through the subsurface as a dissolved phase [107, 108].

The main transport processes occurring into the subsurface are advection, dispersion and diffusion in the liquid and gaseous phases that lead to solute transport and sorption [109]; therefore advective-dispersive-reactive transport (ADRT) is the underlying process governing the migration of explosives though a porous medium [4, 109], which can be written as (1.1)

$$n\frac{\partial c}{\partial t} = nD\frac{\partial^2 c}{\partial z^2} - nv_a\frac{\partial c}{\partial z} \pm R_f \qquad (1.1)$$

where c = concentration, $[M/L^3]$; t = time, $[T]$; n = effective porosity; D = hydrodynamic dispersion, $[L^2/T]$; z = coordinate system, $[L]$; v_a = average linear velocity, $[L/T]$; R_f = reactive term.

The ADRT equation describes the time-varying change in concentration of reactive dissolved contaminants in saturated, homogeneous, isotropic material under steady-state and uniform flow conditions. The term $nD\frac{\partial^2 c}{\partial z^2}$ represents dispersive transport and $nv\frac{\partial c}{\partial z}$ corresponds to advective transport. The concentration can change if there is a different concentration elsewhere in the flowing fluid and this different concentration is carried by advection to the fixed point of interest. The concentration can also change by Fickian transport if there is a spatially varying concentration gradient in the fluid due to the dispersion. Changes in the concentration also can occur if a source or sink process, such as a chemical or biological reaction, is introducing or removing the compound of interest (R_f).

Additional physical processes in the subsurface may include capillary and gravitational forces that cause transient water flow, dissolution, volatilisation, biotransformation, abiotic reaction, and convection and conduction that result in heat transport [110]. The mechanisms of the main processes mentioned are described in the following.

Advection and dispersion processes

Advection is the most important dissolved chemical migration process active in the subsurface and reflects the migration of dissolved chemicals along with groundwater flow, i.e. the mass of actual mass movement of a fluid through the porous media. In fact, it refers to the passive movement of a solute with flowing water (or other solvent) [111].

In a porous medium, the mass flow through a unit section perpendicular to the flow is equal to equation (1.2).

$$F_z = v_a n c \tag{1.2}$$

where F_z = mass flow, $[M/L^2T]$; v_a = average linear velocity, $[L/T]$; n = effective porosity; c = concentration, $[M/L^3]$.

Dispersion is the general term applied to the observed spreading of a solute plume and is generally attributed to hydrodynamic dispersion and molecular diffusion. Hydrodynamic dispersion is a physical process in which macroscopic spreading arises from the multiple variations in flow path velocity and tortuosity. Molecular diffusion is a physicochemical process resulting from the Brownian motion of molecules, resulting in a net migration down a chemical gradient. As a result of the kinetic-chemical activity, contaminants have the tendency to move from the areas of greater concentration to those of less concentration, a process that is known as diffusion. In diffusion, the dissolved substances are moved by a concentration gradient. For its study it is assumed that there is no movement of the fluid. The mass flow by diffusion is governed by the first Fick's law [112] (1.3).

$$F = -D\left(\frac{dc}{dz}\right) \tag{1.3}$$

where F_z = mass flow, $[M/L^2T]$, D = effective diffusion coefficient, $[L^2/T]$; c = concentration, $[M/L^3]$; z = distance over which changes in concentration are considered $[L]$.

The flow expressed in Fick's first law does not consider time: It expresses a permanent flow as long as the variables on which it depends are constant. If there is a point with a constant concentration of a substance (application of a contaminant) and it is desirable to know how the concentration in another point situated at a distance z varies with time, the second Fick's law is used equation (1.4).

$$\frac{\partial c}{\partial t} = D\left(\frac{\partial c}{\partial z^2}\right) \tag{1.4}$$

where t = time, $[T]$; D = effective diffusion coefficient, $[L^2/T]$; c = concentration, $[M/L^3]$; z = distance over which changes in concentration are considered $[L]$.

The effective diffusion coefficient is expressed as follows (1.5):

$$D = \tau_a D_0 \tag{1.5}$$

where D = effective diffusion coefficient, $[L^2/T]$; τ_a = apparent tortuosity; D_0 = aqueous diffusion coefficient for the solutes, $[L^2/T]$.

At macroscopic scale, it is the porous medium that regulates the flow rate and its direction. However, at a microscopic scale the porous medium is composed of discrete solid particles and voids. Water does not flow through the interconnected empty spaces, but around it. When encountering solid particles, water must alter its course, repeating this process millions of times. The result is a mixture of the flux known as mechanical dispersion. As a result of the mechanical dispersion, the mass of contaminant expands in a progressively larger volume, facilitating its mixing with water devoid of this substance. This causes a decrease in contaminant concentration, or dilution. In this way, the contaminant is transported essentially by advection, while its concentration varies due to the dispersion [113].

The ability of the porous medium to mechanically disperse a fluid flowing through it is reflected in a coefficient called dynamic dispersivity α (units: L), function of the porosity, tortuosity, grain shape, etc.

The mechanical dispersion is equal to the product of this coefficient by the average linear velocity (units: L^2/T) (1.6):

$$D' = \alpha v_a \tag{1.6}$$

where α = dynamic dispersivity, $[L]$; D' = mechanical dispersion, $[L^2/T]$; v_a = average linear velocity, $[L/T]$.

The propagation due to mechanisms of mechanical dispersion and molecular diffusion is known as hydrodynamic dispersion, which is a process in which the transport of contaminants is due to variations in velocity (magnitude and direction) in the porous media.

Both mechanisms are taken into account via the coefficient of hydrodynamic dispersion D, which is written as follows (1.7):

$$D = D' + D \tag{1.7}$$

where D = hydrodynamic dispersion, $[L^2/T]$; D' = mechanical dispersion, $[L^2/T]$; D = effective diffusion coefficient, $[L^2/T]$.

As the mechanical dispersion is more pronounced in the longitudinal direction than in the transverse direction, a longitudinal hydrodynamic dispersion coefficient D_{Li} and a transverse hydrodynamic dispersion coefficient D_{Ti} are introduced. The hydrodynamic longitudinal (1.8a) and transverse diffusion (1.8b) coefficients are defined as:

$$D_L = \alpha_{aL} v + D \tag{1.8a}$$

$$D_T = \alpha_{aT} v + D \tag{1.8b}$$

where D_L, D_T = hydrodynamic dispersion (longitudinal, transverse), $[L^2/T]$; $\alpha_{aL} v$, $\alpha_{aT} v$ = dynamic dispersivity (longitudinal, transverse), $[L]$; v = average linear velocity, $[L/T]$; D = effective diffusion coefficient, $[L^2/T]$.

1.4.3 Transformation and reaction processes

In addition to the physical phenomena mentioned above, there are processes of a chemical or biochemical nature that affect the destination of contaminants transported in groundwater.

Generally the reactive processes are described via a measure of the attenuated transport (i.e. retardation) of a reactive contaminant species compared to the advective behaviour of groundwater [114]. The retardation in the ADRT equation can be a consequence of the processes that prevent the transfer of contaminant substances by immobilisation. Among the examples of chemical reactions there are:

Sorption: mainly involves an accumulation of the contaminant molecules on the pore surfaces through physical adsorption to surface and absorption into the body of the sorbent material.

Precipitation: consists in the fact that any excess dissolved substance is transformed into solid because its concentration exceeds the solubility of a particular component.

Filtration: is a physical process of retardation produced by the obstruction of porous spaces resulting from the accumulation of solid particles, although it may also be due to precipitation and the accumulation of dissolved matter.

Ion exchange: consists of a process in which the ions are sorbed in solutions that accumulate on discrete surface areas with opposite charge of the soil particles.

Volatilisation: occurs when the intermolecular forces, holding together molecules in the liquid and solid state, when transferring from solid to liquid, liquid to vapour, or solid to vapour, are overcome by absorbing energy, usually thermal energy from the environment. Volatility depends upon the temperature and the chemical composition of the compound. At any particular temperature, a volatile liquid or solid has a characteristic vapour pressure that is the equilibrium partial pressure in the atmosphere or air space surrounding the compound.

It should be noted that the immobilised contaminants are not transformed and that these processes are reversible. Then, the delayed contaminants can return to their soluble state after a prolonged period of time, producing a long trail of contamination. The process of retardation in the transport of a contaminant is caused mainly by the adsorption processes.

Adsorption refers to the attachment of a chemical to the surface of a soil particle. In soil, this process is exhibited when a molecule of gas, free liquid, or contaminants dissolved in water is attached to the surface of an individual soil particle (often in the form of organic carbon). This surface attachment can be physical (very weak, caused by van der Waals forces), chemical (much stronger, which often requires significant effort to separate) and exchange, i.e. characterised by electrical attraction between the sorbate and the surface (exemplified by ion-exchange processes). Since sorption is primarily a surface phenomenon, its activity is a direct function of the surface area of the solid as well as the electrical forces active on that surface. When a pollutant is adsorbed onto soil, it can be released only when the equilibrium between it and the passing fluid (water or air) is disrupted [109, 115].

The process is complex, for which a delay factor is defined that slows down the transport velocity of the contaminant according to the following expression (1.9):

$$V_c = \frac{V_w}{R_d} \qquad (1.9)$$

where V_c = contaminant velocity, $[L/T]$; V_w = water velocity, $[L/T]$; R_d = retardation factor.

The retardation factor describes the apparent discrepancy between the actual migration rate of aquifer water and that of a dissolved organic chemical (somewhat slower). The difference in travel rates is the result of sorption of the chemical onto the aquifer matrix and release into water by the concentration gradient and time of contact. A general equation used for gross estimation of the retardation factor R_d is (1.10):

$$R_d = 1 + \frac{\rho}{n} K_d \qquad (1.10)$$

where R_d = retardation factor, ρ = bulk density of soil, $[M/L^3]$; K_d = partitioning coefficient, $[L^3/M]$; n = effective porosity.

The partitioning coefficient K_d can be calculated by (1.11):

$$K_d = K_{oc} f_{oc} \qquad (1.11)$$

where K_{oc} $[L^3/M]$ is the organic carbon equilibrium coefficient and f_{oc} is the fraction of organic carbon.

1.5 Conclusion

Explosive contamination is frequently found in soil and groundwater at military training ranges. In order to prevent contamination, minimise remediation costs and ensure continued operation it is essential to understand the potential SPR linkages at a given site. This chapter has shown how the use of conceptual models based on the training range environment can be augmented by laboratory experiments and computational models to understand and predict explosive fate and transport to support identification of pollutant linkages.

References

[1] Oglanis A 2017 Study of environmental management systems on defence *Glob. J. Environ. Sci. Manag.* **3** 103–20
[2] Walsh M E, Walsh M R, Collins C M and Racine C H 2014 White phosphorus contamination of an active army training range, water *Air, Soil Pollut.* **225** 2001
[3] Clausen J, Robb J, Curry D and Korte N 2004 A case study of contaminants on military ranges: Camp Edwards, Massachusetts, USA *Environ. Pollut.* **129** 13–21
[4] Fetter C W, Boving T and Kreamer D 2018 Derivation of the advection-dispersion equation for solute transport *Contaminant Hydrogeology* 3rd edn (Long Grove: Waveland Press), p 647

[5] Akhavan J 2011 *The Chemistry of Explosives* 3rd edn (Cambridge: The Royal Society of Chemistry)

[6] Dorbratz B M 1972 Technical Report: Properties of chemical explosives and explosive simulants, (UCRL-51319), Lawrence Livermore Laboratory, USA

[7] Singh S, Jelinek L, Samuels P, Di Stasio A and Zunino L IMX-104 Characterization for DoD qualification *2010 Insensitive Munitions Energ. Mater. Technol. Symp.* (*Munich, Germany, October 2010*)

[8] Walsh M R, Walsh M E, Taylor S, Ramsey C A, Ringelberg D B, Zufelt J E, Thiboutot S, Ampleman G and Diaz E 2013 Characterization of PAX-21 insensitive munition detonation residues *Propell. Explosives, Pyrotech.* **38** 399–409

[9] Jenkins T F, Hewitt A D, Grant C L, Thiboutot S, Ampleman G, Walsh M E, Ranney T A, Ramsey C A, Palazzo A J and Pennington J C 2006 Identity and distribution of residues of energetic compounds at army live-fire training ranges *Chemosphere* **63** 1280–90

[10] Hewitt A D, Jenkins T F, Ranney T A, Stark J A and Walsh M E 2003 Technical Report: Estimates for explosives residue from the detonation of army munitions ERDC/CRREL TR-03, 16 Hanover, USA

[11] Hewitt A D, Jenkins T F, Walsh M E, Walsh M R and Taylor S 2005 RDX and TNT residues from live-fire and blow-in-place detonations *Chemosphere* **61** 888–94

[12] Taylor S, Campbell E, Perovich L, Lever J and Pennington J 2006 Characteristics of composition B particles from blow-in-place detonations *Chemosphere* **65** 1405–13

[13] Walsh M R, Walsh M E, Ramsey C A, Rachow R J, Zufelt J E, Collins C M, Gelvin A B, Perron N M and Saari S P 2006 Technical Report: Energetic residues deposition from 60-mm and 81-mm mortars ERDC/CRREL TR-06-10, Hanover, USA

[14] Walsh M R 2007 Technical Report: Explosives Residues Resulting from the Detonation of Common Military Munitions: 2002-2006 ERDC/CRREL TR-07-02, Hanover, USA

[15] Jenkins T F, Walsh M E, Miyares P H, Hewitt A D, Collins N H and Ranney T A 2002 Use of snow-covered ranges to estimate explosives residues from high-order detonations of army munitions *Thermochim. Acta* **384** 173–85

[16] Walsh M R, Walsh M E and Ramsey C A 2012 Measuring energetic contaminant deposition rates on snow *Water Air Soil Pollut.* **223** 3689–99

[17] Pennington J C, Silverblatt B, Poe K, Hayes C A and Yost S 2008 Explosive residues from low-order detonations of heavy artillery and mortar rounds *Soil Sediment Contam.* **17** 533–46

[18] Taylor S, Lever J, Walsh M, Walsh M E, Bostick B and Packer B 2004 Technical Report: Underground UXO: Are They a Significant Source of Explosives in Soil Compared to Low- and High-Order Detonations? TR-04-23, Hanover, USA

[19] Taylor S, Bigl S and Packer B 2015 Condition of *in situ* unexploded ordnance *Sci. Total Environ.* **505** 762–69

[20] United States Environmental Protection Agency 2014 Technical Fact Sheet—Hexahydro-1,3,5-trinitro-1,3,5-triazine (RDX)

[21] Lynch J C, Brannon J M and Delfino J J 2002 Dissolution rates of three high explosive compounds: TNT, RDX, and HMX *Chemosphere* **47** 725–34

[22] Kaplan D L and Kaplan A M 1985 Technical Report: Degradation of nitroguanidine in soils (NATICK/TR-58/047), Natick, Massachusetts, USA

[23] Cholakis J M, Wong L C K, Van Goethem D L, Minor J and Short R 1980 Technical Report: Mammalian toxicological evaluation of RDX, Kansas City, USA

[24] Bannon D I and Williams L R 2015 Wildlife toxicity assessment for 1,3,5-trinitrohexahy-dro-1,3,5-triazine (RDX) ed M A Williams, G Reddy, J M Quinn and S M Johnson *Wildlife Toxic Assessments Chemical of Military Concern* (Amsterdam: Elsevier), pp 53–86

[25] Johnson M S and Reddy G 2015 *Wildlife Toxicity Assessments for Chemicals of Military Concern* 1st edn (Amsterdam: Elsevier)

[26] Chou C J S, Holler J, De Rosa C T and Chou S 1998 Minimal risk levels (MRL) for hazardous substances *Environ. Toxicol. Occup. Med.* **7** 1–24

[27] Walsh M E, Taylor S, Hewitt A D, Walsh M R, Ramsey C A and Collins C M 2010 Field observations of the persistence of Comp B explosives residues in a salt Marsh impact area *Chemosphere* **78** 467–73

[28] Monteil-Rivera F, Halasz A, Groom C, Zhao J-S, Thiboutot S, Ampleman G and Hawari J 2009 Fate and transport of explosives in the environment: a chemist's view *Ecotoxicology of Explosives* ed G Sunahara, G Lutofo, R Kuperman and J Hawari (Boca Raton, FL: CRC Press, Taylor and Francis Group LLC) ch 2, pp 5–33

[29] Bordeleau G, Martel R, Ampleman G and Thiboutot S 2013 Photolysis of RDX and nitroglycerin in the context of military training ranges *Chemosphere* **93** 14–9

[30] Mahbub P and Nesterenko P N 2016 Application of photo degradation for remediation of cyclic nitramine and nitroaromatic explosives *RSC Adv.* **6** 77603–21

[31] Hawari J, Halasz A, Groom C, Deschamps S, Paquet L, Beaulieu C and Corriveau A 2002 Photodegradation of RDX in aqueous solution: a mechanistic probe for biodegradation with Rhodococcus sp *Environ. Sci. Technol.* **36** 5117–23

[32] Bouchard D C, Williams M K and Surampalli R Y 1992 Nitrate contamination of groundwater: sources and potential health effects *J. Am. Water Works Assoc.* **84** 85–90

[33] Balakrishnan V K, Halasz A and Hawari* J 2003 Alkaline hydrolysis of the cyclic nitramine explosives RDX, HMX, and CL-20: new insights into degradation pathways obtained by the observation of novel intermediates *Environ. Sci. Technol.* **37** 1838–43

[34] Best E P H, Sprecher S L, Larson S L, Fredrickson H L and Bader D F 1999 Environmental behavior of explosives in groundwater from the Milan Army Ammunition Plant in aquatic and wetland plant treatments. Uptake and fate of TNT and RDX in plants *Chemosphere* **39** 2057–72

[35] Fuller M E, Hatzinger P B, Condee C W and Togna A P 2007 Combined treatment of perchlorate and RDX in ground water using a fluidized bed reactor *Groundw. Monit. Remediat.* **27** 59–64

[36] Pichtel J 2012 Distribution and fate of military explosives and propellants in soil: a review *Appl. Environ. Soil Sci.* **2012** 1–33

[37] Taylor S, Lever J H, Fadden J, Perron N and Packer B 2009 Simulated rainfall-driven dissolution of TNT, Tritonal, Comp B and Octol particles *Chemosphere* **75** 1074–81

[38] Lever J H, Taylor S, Perovich L, Bjella K and Packer B 2005 Dissolution of composition B detonation residuals *Environ. Sci. Technol.* **39** 8803–11

[39] Sharma P, Mayes M A and Tang G 2013 Role of soil organic carbon and colloids in sorption and transport of TNT, RDX and HMX in training range soils *Chemosphere* **92** 993–1000

[40] Hawari J, Halasz A, Sheremata T, Beaudet S, Groom C, Paquet L, Rhofir C, Ampleman G and Thiboutot S 2000 Characterization of metabolites during biodegradation of hexahydro-1, 3,5-trinitro-1,3,5-triazine (RDX) with municipal anaerobic sludge *Appl. Environ. Microbiol.* **66** 2652–57

[41] Pennington J C and Brannon J M 2002 Environmental fate of explosives *Thermochim. Acta* **384** 163–72

[42] Brannon J M and Pennington J C 2002 *Technical Report: Environmental Fate and Transport Process Descriptors for Explosives* (Vicksburg: US Army Engineer Research and Development Center)

[43] Sheremata T W *et al* 2001 The fate of the cyclic nitramine explosive RDX in natural soil *Environ. Sci. Technol.* **35** 1037–40

[44] Halasz A, Manno D, Perreault N N, Sabbadin F, Bruce N C and Hawari J 2012 Biodegradation of RDX nitroso products MNX and TNX by cytochrome P450 XplA *Environ. Sci. Technol.* **46** 7245–51

[45] Pennington J C 2006 Technical Report: Distribution and fate of energetics on DOD test and training ranges ERDC TR-06-12, Hanover, USA

[46] Jenkins T F, Ranney T A, Miyares P H, Collins N H and Hewitt A D 2000 Technical Report: Use of surface snow sampling to estimate the quantity of explosives residues resulting from land mine detonations, ERDC TR-00-12, Hanover, USA

[47] Gauthier C, Lefebvre R, Martel R, Ampleman G, Thiboutot S, Lewis J and Parent M 2003 Assessment of the impacts of live training on soil and groundwater at Canadian Forces Base Shilo, Manitoba *56th Can. Geol. Conf. 4th Jt. IAH-CNC/CGS Conf. (Winnipeg)*

[48] Klapötke T M and Witkowski T G 2016 Covalent and ionic insensitive high-explosives *Propellants Explos. Pyrotech.* **41** 470–83

[49] Ravi P, Badgujar D M, Gore G M, Tewari S P and Sikder A K 2011 Review on melt cast explosives *Propellants Explos. Pyrotech.* **36** 393–403

[50] Walsh M R M E, Walsh M R M E, Ramsey C A, Thiboutot S, Ampleman G, Diaz E and Zufelt J E 2014 Energetic residues from the detonation of IMX-104 insensitive munitions *Propellants Explos. Pyrotech.* **39** 243–50

[51] Taylor S, Dontsova K, Walsh M E M R and Walsh M E M R 2015 Outdoor dissolution of detonation residues of three insensitive munitions (IM) formulations *Chemosphere* **134** 250–56

[52] Karthikeyan S and Spain J C 2016 Biodegradation of 2,4-dinitroanisole (DNAN) by nocardioides sp. JS1661 in water, soil and bioreactors *J. Hazard. Mater.* **312** 37–44

[53] Liang J, Olivares C, Field J A and Sierra-Alvarez R 2013 Microbial toxicity of the insensitive munitions compound, 2,4-dinitroanisole (DNAN), and its aromatic amine metabolites *J. Hazard. Mater.* **262** 281–87

[54] Olivares C, Liang J, Abrell L, Sierra-Alvarez R and Field J A 2013 Pathways of reductive 2,4-dinitroanisole (DNAN) biotransformation in sludge *Biotechnol. Bioeng.* **110** 1595–604

[55] Olivares C I, Abrell L, Khatiwada R, Chorover J, Sierra-alvarez R and Field J A 2016 (Bio) transformation of 2,4-dinitroanisole (DNAN) in soils *J. Hazard. Mater.* **304** 214–21

[56] Arthur J D, Mark N W, Taylor S, Šimunek J, Brusseau M L and Dontsova K M 2017 Batch soil adsorption and column transport studies of 2,4-dinitroanisole (DNAN) in soils *J. Contam. Hydrol.* **199** 14–23

[57] Temple T, Ladyman M, Mai N, Galante E, Ricamora M, Shirazi R and Coulon F 2018 Investigation into the environmental fate of the combined Insensitive high explosive constituents 2,4-dinitroanisole (DNAN), 1-nitroguanidine (NQ) and nitrotriazolone (NTO) in soil *Sci. Total Environ.* **625** 1264–71

[58] Taylor S, Walsh M E, Becher J B, Ringelberg D B, Mannes P Z and Gribble G W 2017 Photo-degradation of 2,4-dinitroanisole (DNAN): an emerging munitions compound *Chemosphere* **167** 193–203

[59] Williams L R, Eck W and Johnson M S 2014 Toxicity of IMX formulations and components: what we know and path forward *JANNAF Work. Proc.* pp 92–104

[60] Lent E M, Crouse L C B, Hanna T and Wallace S 2012 *The Subchronic Oral Toxicity of DNAN and NTO in Rats* 87-XE-0DBP-10 US Army Institute of Public Health

[61] Dodard S G, Sarrazin M, Hawari J, Paquet L, Ampleman G, Thiboutot S and Sunahara G I 2013 Ecotoxicological assessment of a high energetic and insensitive munitions compound: 2,4-Dinitroanisole (DNAN) *J. Hazard. Mater.* **262** 143–50

[62] Kennedy A J, Poda A R, Melby N L, Moores L C, Jordan S M, Gust K A and Bednar A J 2017 Aquatic toxicity of photo-degraded insensitive munition 101 (IMX-101) constituents *Environ. Toxicol. Chem.* **36** 2050–57

[63] Lotufo G R, Rosen G, Wild W and Carton G 2013 Tehcnical Report: Summary review of the aquatic toxicology of munitions constituents ERDC/EL TR-13-8, Hanover, USA

[64] Mirecki J E, Porter B, Weiss J and Charles A 2006 Technical Report: Environmental transport and fate process descriptors for propellant compounds ERDC/EL TR-06-7, Hanover, USA

[65] Hartley W R, Roberts W C and Brower E 1992 *Drinking Water Health Advisory: Munitions* ed W C Roberts and W R Hartley (Boca Raton, FL: Lewis Publishers)

[66] Bordeleau G, Martel R, Drouin M, Ampleman G and Thiboutot S 2014 Biodegradation of nitroglycerin from propellant residues on military training ranges *J. Environ. Qual.* **43** 441

[67] Lee K E, Balas-Hummers W A, Di Stasio A R, Patel C H, Samuels P J, Roos B D and Fung V 2010 Technical Report: Qualification testing of the insensitive TNT replacement explosive IMX-101, Picatinny Arsenal, NJ, USA

[68] Walsh M R, Temple T, Bigl M F, Tshabalala S F, Mai N and Ladyman M 2017 Investigation of energetic particle distribution from high-order detonations of munitions, propellants *Prop. Explos. Pyrotech.* **42** 602–88

[69] Mulherin N D, Jenkins T F and Walsh M E 2005 Technical Report: Stability of nitroguanidine in moist, unsaturated soils ERDC/CRREL-TR-05-2, Hanover, USA

[70] Walsh M R, Walsh M E, Ramsey C A, Brochu S, Thiboutot S and Ampleman G 2013 Perchlorate contamination from the detonation of insensitive high-explosive rounds *J. Hazard. Mater.* **262** 228–33

[71] Smith P N, Theodorakis C W, Anderson T A and Kendall R J 2001 Preliminary assessment of perchlorate in ecological receptors at the longhorn army ammunition plant (LHAAP), Karnack, Texas *Ecotoxicology* **10** 305–13

[72] Blake S, Hall T, Harman M, Kanda R, McLaughlin C and Rumsby P 2009 Technical report: perchlorate – risks to UK drinking water sources Department for Environment, Food & Rural Affairs No. 7845, Swindon, UK

[73] Leung A M, Pearce E N and Braverman L E 2014 Environmental perchlorate exposure: potential adverse thyroid effects *Curr. Opin. Endocrinol. Diabetes. Obes.* **21** 372–76

[74] Steinmaus C M 2016 Perchlorate in water supplies: sources, exposures, and health effects *Curr. Environ. Heal. Reports* **3** 136–43

[75] Walsh M, Thiboutot S and Gullett B 2017 Technical report: characterization of residues from the detonation of insensitive munitions distribution statement ER-2219. Defence Research and Development, Ottawa, Canada

[76] Dontsova K, Brusseau M, Arthur J, Mark N, Taylor S, Pesce-rodriguez R, Walsh M, Lever J and Šim J 2014 Technical Report: Dissolution of NTO, DNAN and insensitive munitions formulations and their fates in soils ERDC/CRREL TR-14-23, Hanover, USA

[77] Braida W J, Wazne M, Ogundipe A, Tuna G S, Pavlov J and Koutsospyros A 2012 Transport of nitrotriazolone (NTO) in soil lysimeters *Prot. Restor. Environ. Conf. XI* (*Thessaloniki, Greece*)

[78] Mark N, Arthur J, Dontsova K, Brusseau M and Taylor S 2016 Adsorption and attenuation behavior of 3-nitro-1,2,4-triazol-5-one (NTO) in eleven soils *Chemosphere* **144** 1249–55

[79] Richard T and Weidhaas J 2014 Biodegradation of IMX-101 explosive formulation constituents: 2,4-dinitroanisole (DNAN), 3-nitro-1,2,4-triazol-5-one (NTO), and nitro-guanidine *J. Hazard. Mater.* **280** 561–69

[80] Arthur J D, Mark N W, Taylor S, Šimůnek J, Brusseau M L and Dontsova K M 2018 Dissolution and transport of insensitive munitions formulations IMX-101 and IMX-104 in saturated soil columns *Sci. Total Environ.* **624** 758–68

[81] Mark N, Arthur J, Dontsova K, Brusseau M, Taylor S and Šimůnek J 2017 Column transport studies of 3-nitro-1,2,4-triazol-5-one (NTO) in soils *Chemosphere* **171** 427–34

[82] Mark N W 2014 *Batch and ColumnTransport Studies of Environmental Fate of 3-nitro-1,2,4-triazol-5-one (NTO) in Soils* (The University of Arizona)

[83] London J O and Smith D M 1985 Technical Report: A toxicological study of NTO, Los Alamos, National Lab New Mexico, USA

[84] Crouse L C B, Lent E M and Leach G J 2015 Oral toxicity of 3-nitro-1,2,4-triazol-5-one in rats *Int. J. Toxicol.* **34** 55–66

[85] Madeira C L, Speet S A, Nieto C A, Abrell L, Chorover J, Sierra-Alvarez R and Field J A 2017 Sequential anaerobic-aerobic biodegradation of emerging insensitive munitions compound 3-nitro-1,2,4-triazol-5-one (NTO) *Chemosphere* **167** 478–84

[86] McCain W, Williams L and Grunda R 2013 Technical Report: Toxicology portfolio *in vitro* dermal absorption of insensitive munitions explosive 101 (IMX-101) and components US Army Public Health Command, Aberdeen Proving Ground, Maryland, USA

[87] Berry M A and Rondinelli D A 1998 Proactive corporate environmental management: A new industrial revolution *Acad. Manag. Perspect.* **12** 38–50

[88] Lewis J and Sjöstrom J 2010 Optimizing the experimental design of soil columns in saturated and unsaturated transport experiments *J. Contam. Hydrol.* **115** 1–13

[89] Taylor S, Ringelberg D B, Dontsova K, Daghlian C P, Walsh M E and Walsh M R 2013 Insights into the dissolution and the three-dimensional structure of insensitive munitions formulations *Chemosphere* **93** 1782–88

[90] Taylor S, Park E, Bullion K and Dontsova K 2015 Dissolution of three insensitive munitions formulations *Chemosphere* **119** 342–8

[91] Kumar M, Ladyman M K, Mai N, Temple T and Coulon F 2017 Release of 1,3,5-trinitroperhydro-1,3,5-triazine (RDX) from polymer-bonded explosives (PBXN-109) into water by artificial weathering *Chemosphere* **169** 604–08

[92] Taylor S, Lever J H, Fadden J, Perron N and Packer B 2009 Outdoor weathering and dissolution of TNT and Tritonal *Chemosphere* **77** 1338–45

[93] Paluch J, Mesquita R B R, Cerdà V, Kozak J, Wieczorek M and Rangel A O S S 2018 Sequential injection system with in-line solid phase extraction and soil mini-column for determination of zinc and copper in soil leachates *Talanta* **185** 316–23

[94] Bergström L 1990 Use of lysimeters to estimate leaching of pesticides in agricultural soils *Environ. Poll.* **67** 325–47

[95] Krüger O, Kalbe U, Berger W, Simon F-G and Meza S L 2012 Leaching experiments on the release of heavy metals and PAH from soil and waste materials *J. Hazard. Mater.* **207–208** 51–5

[96] Barry D A 2009 Effect of nonuniform boundary conditions on steady flow in saturated homogeneous cylindrical soil columns *Adv. Water Resour.* **32** 522–31

[97] Di Palma L and Mecozzi R 2010 Batch and column tests of metal mobilization in soil impacted by landfill leachate *Waste Manag.* **30** 1594–99

[98] Schoen R, Gaudet J and Elrick D 1999 Modelling of solute transport in a large undisturbed lysimeter, during steady-state water flux *J. Hydrol.* **215** 82–93

[99] López Meza S, Kalbe U, Berger W and Simon F-G 2010 Effect of contact time on the release of contaminants from granular waste materials during column leaching experiments *Waste Manag.* **30** 565–71

[100] Wefer-Roehl A and Kübeck C 2014 Guidelining protocol for soil-column experiments assessing fate and transport of trace organics Title: Guidelining protocol for soil-column experiments assessing fate and transport of trace organics CETaqua, Barcelona, Spain

[101] Morley M C, Yamamoto H, Speitel G E and Clausen J 2006 Dissolution kinetics of high explosives particles in a saturated sandy soil *J. Contam. Hydrol.* **85** 141–58

[102] Boopathy R, Widrig D L and Manning J F 1997 *In situ* bioremediation of explosives-contaminated soil: a soil column study *Bioresour. Technol.* **59** 169–76

[103] Spark K M and Swift R S 2002 Effect of soil composition and dissolved organic matter on pesticide sorption *Sci. Total Environ.* **298** 147–61

[104] Hawari J, Monteil-Rivera F, Perreault N N N, Halasz A, Paquet L, Radovic-Hrapovic Z, Deschamps S, Thiboutot S and Ampleman G 2015 Environmental fate of 2,4-dinitroanisole (DNAN) and its reduced products *Chemosphere* **119** 16–23

[105] Krzmarzick M J, Khatiwada R, Olivares C I, Abrell L, Sierra-Alvarez R, Chorover J and Field J A 2015 Biotransformation and degradation of the insensitive munitions compound, 3-nitro-1,2,4-triazol-5-one, by soil bacterial communities *Environ. Sci. Technol.* **49** 5681–88

[106] Kalderis D, Juhasz A L, Boopathy R and Comfort S 2011 Soils contaminated with explosives: Environmental fate and evaluation of state-of-the-art remediation processes (IUPAC Technical Report) *Pure Appl. Chem.* **83** 1407–84

[107] Lavoie B, Mayes M A and McKay L D 2012 Transport of explosive residue surrogates in saturated porous media *Water, Air, Soil Pollut.* **223** 1983–93

[108] Comfort S D, Shea P J, Hundal L S, Li Z, Woodbury B L, Martin J L and Powers W L 1995 TNT transport and fate in contaminated soil *J. Environ. Qual.* **24** 1174

[109] Bear J 1988 *Dynamics of Fluids in Porous Media* (New York: Dover)

[110] McGrath C J 1995 Technical Report: Review of formulations for processes affecting the subsurface transport of explosives WES/TR/IRRP-95-2Vicksburg, USA

[111] Sophocleous M 1979 Analysis of water and heat flow in unsaturated-saturated porous media *Water Resour. Res.* **15** 1195–206

[112] Bear J 1961 On the tensor form of dispersion in porous media *J. Geophys. Res.* **66** 1185–97

[113] Gelhar L W, Welty C and Rehfeldt K R 1992 A critical review of data on field-scale dispersion in aquifers *Water Resour. Res.* **28** 1955–74

[114] Berkowitz B, Dror I and Yaron B 2014 Characterization of the subsurface environment *Contaminant Geochemistry* (Berlin: Springer), pp 3–28

[115] Yamamoto H, Morley M C, Speitel G E and Clausen J 2004 Fate and transport of high explosives in a sandy soil: adsorption and desorption *Soil Sediment Contam An Int. J* **13** 361–79

Chapter 2

Characterization of soils on military training ranges

Michael R Walsh, Marianne E Walsh, Charles A Ramsey, Matthew F Bigl and Samuel A Beal

The characterization of military training ranges for energetics and metals is becoming more common as defense agencies work to preserve ranges and prevent the migration of potentially dangerous compounds into surface and ground waters and off installations. Past characterization efforts were plagued by unreliable and conflicting data, often resulting in flawed assessments and high clean-up liabilities [1, 2]. In the 1980s, a concerted effort was begun to characterize both training ranges and the munitions used on them with an emphasis on post-detonation residues containing energetics and other munitions-related constituents. These early efforts highlighted the shortcomings of the sampling and sample processing methods then in use. Through a comprehensive research program, the US Army Cold Regions Research and Engineering Laboratory and Envirostat, Inc., have developed methodologies for soil characterization that apply the theory of sampling and adapt methods developed in the mining and agricultural industries to the field of environmental sampling in general, and sampling for energetics on military training ranges in particular. This chapter describes the methods necessary to obtain reproducible range characterization data for soils. However, these sampling concepts have also been applied to surface water [3].

2.1 Introduction

The environmental management of any defense installation will require reliable information on the presence and concentration of metallic and chemical residues resulting from tactical training. Reliable information can only be obtained through the proper collection, documentation, storage, processing and analysis of samples. Reliable data is reproducible data, and only reproducible data is robust enough to

doi:10.1088/978-0-7503-1605-7ch2

stand up to both scientific and legal scrutiny. Through extensive research, field experience, and application of lessons learned to actual field characterizations, a methodology has been developed that will enable the collection of reproducible data. This methodology, known as the *Multi-Increment*® sampling method, will be described in this chapter.

2.2 Background

The multi-increment sampling (MIS) method was adapted by scientists from the US Army Cold Regions Research and Engineering Laboratory (CRREL) to overcome the problem of data irreproducibility during field investigations carried out for the US Army Alaska at the Eagle River Flats (ERF) indirect-fire impact area on Fort Richardson, AK, USA [4]. Ducks dabbling in the estuarine salt marsh within which ERF is located were dying by the thousands with no apparent external injuries. Examination of tissues and gizzard contents from dead ducks, swans and shorebirds recovered from the impact area pointed to the ingestion of white phosphorus (WP, P_4), a man-made form of elemental phosphorus used as an obscurant by artillery units during live-fire training [2]. Discrete (grab) sampling of the sediments in which the waterfowl were dabbling resulted in widely varying results, making quantification of the P_4 and delineation of contaminated areas extremely difficult. Further investigation of the post-detonation characteristics of the contaminant indicated that the P_4 was composed of small, discrete particles that were heterogeneously distributed near impact points throughout the permanently wet areas of the flats [5]. Discrete sampling, the common field sampling method for soils in use at the time, was an unreliable and inefficient method for finding and characterizing the areas of contamination within the impact area.

There was one method for finding P_4 that appeared grimly efficient: the multiple-location dabbling for food that was conducted by the ducks. By 'sampling' many locations within a contaminated area, the ducks were finding (and ingesting) P_4 particles and dying. The dabbling activity of the ducks was roughly imitated by field researchers by sampling along transects that were established in areas of high waterfowl mortality. Increments within the transects were taken at grid points based on the diameter of the crater formed by the detonation of an 81-mm WP round. Much more reliable data was possible using the transects, but some of the contaminated areas were quite large and a better characterization method was needed [2].

Similar data reproducibility issues were occurring with CRREL's investigations of munitions residues on ranges [6]. In the late 1990s, CRREL researchers involved in the ERF project and munitions residues research program attended a short course on sampling given by Chuck Ramsey of Envirostat, Inc. The course described a method of sampling similar to that used by the agricultural and mining industries to characterize soils in delineated areas for the presence of elements such as nitrogen, phosphorus, copper and iron. The method is based on the theory of sampling developed by Pierre Guy and included the collection of many increments within a defined area to build a sample that will represent the environment from which it was obtained [7]. The course covered sources of error, defining those errors and describing in general terms how to improve data quality.

Following the sampling course, CRREL started a long-term collaborative effort with Envirostat to develop a comprehensive method of soil sampling that encompasses not only the collection of samples but how to set up a sampling plan, tools for sampling that will reduce sampling error while increasing efficiency, comminution of samples, the splitting of samples, and subsampling. All aspects of the process have been analyzed scientifically through a comprehensive research program on energetics residues on military training ranges. The methods developed at the laboratory and through field research have been and are presently being successfully applied to real-world characterization projects in the US, Canada and in several countries in Europe [8]. Sampling methods that incorporate multiple small increments in each sample have also been adapted to the sampling of energetics' residues on snow surfaces, which has been invaluable in research characterizing munitions residues, in one case saving billions of dollars in potential environmental liabilities [9–11].

This chapter will walk the reader through methods needed for obtaining data that will enable the derivation of a reproducible mean analyte concentration within a designated area. We will start with the theory of sampling, upon which the *Multi-Increment*® environmental sampling method is based. We will then cover the types and sources of error inherent in soil sampling, methods of error reduction and quantification, how to obtain a representative sample, how to obtain a representative subsample and the current methods of analyzing for energetics. A subsequent chapter in this book presents a number of case studies for various types of training ranges that will give the reader a feel for how MIS is applied to real-world situations.

The methods and methodology described in this chapter must be implemented in full to obtain reproducible data [12]. The authors have conducted extensive research into methods that were and are commonly used, and almost all of these methods do not result in reproducible data. Omitting one of the steps would be like leaving a wheel off an automobile: You can do it and maybe still get to where you want to go, but the results will be catastrophic. Even modifying one of the steps could result in bad data. Think of having all four wheels but modifying the tire pressure at one of the wheels by deflating the tire.

2.3 Steps in the multi-increment sampling process

The MIS process can be described as a series of five steps, all of which must be executed in order to obtain reproducible and defensible data [13, 14]. These steps are listed below:

Step 1: Develop sample quality criteria
Step 2: Determine material properties
Step 3: Apply the principles of TOS to develop sampling protocol
Step 4: Evaluate all data, including QC data
Step 5: Make inference from analytical results to decision unit.

2.3.1 Sample quality criteria

The sample quality criteria (SQC) is formulation of the project objectives in such a manner that a sampling protocol can be developed [15]. The USEPA has also developed a similar approach called data quality objectives (DQO). The DQO process can be a bit complicated and makes many assumptions that may not be valid when investigating military ranges; therefore, the less complicated and more universal approach of the SQC process will be described. The three main parts of the SQC process are: (1) the *Question* that the study attempts to answer, (2) the *Decision Unit* (DU), and (3) the *Confidence* desired in the final decision.

    1) *Question*
- What is the analyte(s) of interest?
  - On a training range these may be pyrotechnics, energetics, or metals.
- What is the concentration of concern?
  - CoCs are risk-based and may be receptor-specific, i.e. human health or environmental risk.
- How are the data going to be used to make inference?
  - Data that is collected must be of sufficient quality and quantity to make the appropriate inferences (e.g. confidence intervals, means, medians, etc).

    2) *Decision Unit*

        The DU is the specific mass/volume of material from which increments are collected and to which inference will be made [16, 17]. There may be one or many decision units within a military training complex, such as firing points, target areas, engineer training ranges and fuel or ammunition storage sites.

    3) *Confidence*

        This is the assurance that the final decision is correct. Confidence can be based on statistical calculations or other techniques. Higher human health risk, public exposure and liability issues all impact the degree of confidence needed for a data set.

For those new to sampling, the most difficult part of the SQC process is determining the parameters (physical and temporal boundaries) of a decision unit. Decision units are typically based either on risk or potential source areas. Risk-based DUs are based on exposure to a receptor in the area in which the receptor typically exists. In cases where there are multiple receptors, multiple-sized DUs may exist within the same area or the DUs may be based on the most sensitive receptor. While the concept of DUs is sometimes ignored, it is critical that the DU be identified or incorrect decisions will result. Much of the work performed by CRREL in applying the concept of MIS to military ranges was based on specifying DUs to best achieve the goals specific to the study. Earlier studies (before 2006) are not to be used as a starting point for determining DUs for an actual military range characterization. The other type of DU is a source-based DU. These DUs are based on the source area/volume and not on risk.

2.3.2 Material properties

There are two types of material properties that must be determined. This first is the nature of the elements: are they finite or infinite element materials. In the case of characterizing military ranges, all materials can be assumed to be infinite element materials and therefore the theory of sampling (TOS) must be applied.

The second material property is the nature of the heterogeneity. There are two types of heterogeneity: compositional and distributional [18].

Compositional heterogeneity exists when the individual elements (particles) that make up the DU do not have exactly the same concentration of the analyte of interest. Compositional heterogeneity (CH) always exists to some degree.

Distributional heterogeneity (DH) exists when the individual elements that make up the decision unit are not randomly distributed throughout the entire DU. Examples of distributional heterogeneity are the settling of small, dense fines to the bottom of a container of soil and the uneven distribution of propellant residue across a weapons firing point. Like compositional heterogeneity, DH almost always exists and can be influenced by physical conditions. The magnitude and nature of CH and DH will vary for different materials.

2.3.3 Theory of sampling

The theory of sampling (TOS) is used to develop a sampling protocol. This protocol is dependent on the SQC and the material properties. The TOS is used to determine the appropriate mass, number of increments and types of tools that must be employed. The sampling mass, number of increments and types of acceptable tools are driven by the amount of acceptable error, which in turn is based on the desired confidence determined in the SQC. Sampling errors can broadly be classified into three types: (1) fundamental sampling error, (2) grouping and segregation error, and (3) materialization error. Fortunately, for military bases, research done by CRREL, Envirostat, and Defence Research and Development Canada, Val Cartier (DRDC), has identified the starting point for mass, increments and tools.

Fundamental sampling error

Error attributable to compositional heterogeneity, the non-uniform composition of each particle in the DU or within the sample leads to a fundamental sampling error (FSE). FSE is a precision error that is controlled through the collection of sufficient mass to represent all the particles of varying composition. There are various formulas to estimate FSE, some quite complicated and others quite simple if certain assumptions can be made. The basic relationship of FSE to particle size, sample mass and compositional heterogeneity is as follows:

$$FSE^2 \propto \frac{Cd^3}{m_s}$$

where:
FSE = Fundamental sampling error
C = Sampling constant (g cm^{-3})

d = diameter of largest particles (cm)
m_s = mass of sample (g).

This equation is used to determine the mass necessary to control the FSE. The sampling constant (C) is unique for each type of material and needs to be determined. However, research done through the US Strategic Environmental Research and Development Program (SERDP) has determined that the mass necessary to control the FSE at an adequate level to make the appropriate and defensible inferences is approximately 2 kg.

Grouping and segregation error
Error attributable to distributional heterogeneity, the non-uniform distribution of particles of concern in the DU and the collected samples is called a grouping and segregation error. A grouping and segregation error (GSE) is a precision error that is controlled through the collection of an adequate number of random increments to make up the sample. An increment is defined as a group of particles selected during the single operation of a sampling tool. The greater the heterogeneity of the distribution of the contaminant in the DU, the larger the GSE and the larger the number of increments that must be selected and combined for the sample. If the mass required to control the FSE could be collected one particle at a time at random, GSE would not exist. An increment may contain thousands of particles, but it is still located at only one random location within a much larger DU or sample. As more and more random increments are included in the sample, the GSE decreases.

There is no magic number of increments that will control the GSE for all sampling situations, but 60 increments is suggested as a minimum for soils containing residues of energetics at military installations. However, this is completely dependent on the degree of distributional heterogeneity of the material within the DU. Thus, there will exist materials and/or sample quality criteria where 100 or more increments are required [19, 20]. The number of increments is never based on what is easy to collect but is based on the number required to minimize the GSE where it does not impact the confidence in and the ability to legally and scientifically defend the decision. It may be appropriate to use 50 increments; however, this number must be justified.

Materialization error
Materialization errors (MEs) are caused by a lack of sample correctness that creates a sampling bias. While analytical biases are routinely estimated for analytical measurements, sampling bias is very difficult, if not impossible, to estimate. Sampling bias is inconsistent and therefore conventional bias correction techniques to estimate and correct bias cannot be employed. It is therefore critical that errors of bias are minimized as much as possible so the effect on sampling error is minimized to an insignificant level.

The correct increment shape for surface soils is a cylinder. Incorrect shaped increments will cause a bias in the concentration estimation. The concentration of the analyte in the increment will differ between the examples shown below for analytes that reside primarily on the surface. Only the mass collected in a core configuration will give the true concentration to the desired depth. Increments also need to be approximately the same mass to ensure uniform representation of the areas within the DU from which they are taken. Consistent increment mass is more easily controlled by using a coring device than a scoop or trowel. Scoops should only be used in the rare instance when a coring device will not work. Trowels are strongly discouraged because of their gross inaccuracy. When a scoop is used, great care must be taken to ensure that an approximately cylindrical shape is collected and that the increment masses are consistent. CRREL has developed a tool for the collection of unbiased surface soil (0–5 cm) samples (see the next section on error). This tool allows for the rapid and consistent collection of multiple increments [21].

2.3.4 Data evaluation and inference

Once the samples have been properly subsampled and analyzed, the analytical data needs to be reviewed to ensure that SQC was met. This will include not only the standard analytical quality control (QC) but also the QC associated with estimations of sampling and sample processing errors [22, 23]. A good test of the adequacy of the mass and the number of increments collected is to determine the confidence limits for replicate samples (3–5 reps are needed). If confidence limits are not met, the sampling error must be reduced by increasing mass and/or increments. For example, if the number of increments being collected in a project's DUs is 50 increments and the confidence limits or relative standard deviation goals are not being met, it is likely that the number of increments is not being properly considered and will need to be increased. For similar sized increments, an increase in the number of increments will result in an increase in mass per sample, further reducing the sampling error.

After the data has passed the data evaluation step, inferences can be made. There may be only one inference or there may be many [16]. Each inference is performed separately. The inference may be as simple as using the single result from the laboratory to estimate the concentration in the DU or it may involve calculations such as the average from multiple test portions.

2.4 Error and error reduction

The previous section discussed sampling theory and the primary sources of error described by that theory. There are two types of errors associated with all sampling: FSE, which is caused by compositional heterogeneity of the contaminant, and GSE, caused by the heterogeneous spatial distribution of the contaminant particles in the area to be characterized (DU) [7]. The DU cannot be sampled by simply going to the field, obtaining a few grams of soil, and analyzing the sample, unless there is no compositional or distributional error associated with the contaminant. FSE and GSE associated with a contaminant in a DU will render a discrete or grab sample meaningless. Inferences associated with any sample are based on distributional and compositional error and, for a grab sample, these inferences are quite limited. Rasemann goes into great detail on developing a range of values quantifying these errors for the sampling and analysis of industrial waste dumps [24]. His conclusion at the time of publication was that the greatest source of error in characterization is sampling error.

2.4.1 Magnitude of error

In the past, the majority of effort to reduce error in sampling has been misdirected. Rasemann has found that the greatest error in environmental characterization occurs not in the analytical lab, where 95% of the effort to reduce error occurs, but in the field while collecting the original sample. Little or no real effort had been expended in the field to reduce sampling error in the past. Much of the reason for this is because little or no quality assurance (QA), such as replicate sampling, had occurred in the field. It is also much easier to go to the field, fill a jar with 250 g of soil, and analyze 5 g taken from the top of the sample while back in the lab. With no QA, the results of the analysis are often presented as the definitive, usually to five or six significant figures. The perception of the immutability of the grab sample began to come into question when sites declared clean (or dirty) turned out, at great expense, to be just the opposite. Rudimentary QA, generally consisting of obtaining co-located 'replicate' samples, had contaminant concentrations that were sometimes 3–4 orders of magnitude different from each other. Rasemann estimated that sampling error is around 1000%, sample preparation error is around 300%, and analytical error, where the greatest effort to reduce error occurs, amounts to only about 3% (table 2.1).

2.4.2 Controlling for error

Table 2.2 illustrates the effect that controlling for various error sources has on the root mean square (RMS) global estimation error (GEE) of the final estimation of the mean concentration of a contaminant in a DU. Sampling error obviously has the greatest effect on the GEE until the maximum feasible sampling error reduction is achieved, at which point sample preparation error begins to dominate. Error reduction effects are illustrated in figure 2.1. Note that there will always be error associated with every step of the characterization process, and that sampling error is

Table 2.1. Three common sources of measurement error [24].

Process	Error (% of true value)
Sampling	1000%
Sample preparation	100%–300%
Analytical measurement	2%–20%

Table 2.2. Effect of controlling for maximum error on global estimation error (GEE).

| | Error Source | | | | Error from |
Case	Sampling	Sample prep	Analytical	GEE	sampling
1	1000%	300%	20%	1044%	92%
2	1000%	300%	3%	1044%	92%
3	1000%	5%	20%	1000%	>99%
4	1000%	5%	3%	1000%	>99%
5	25%	300%	20%	302%	1%
6	25%	5%	20%	32%	61%
7	25%	5%	3%	26%	95%

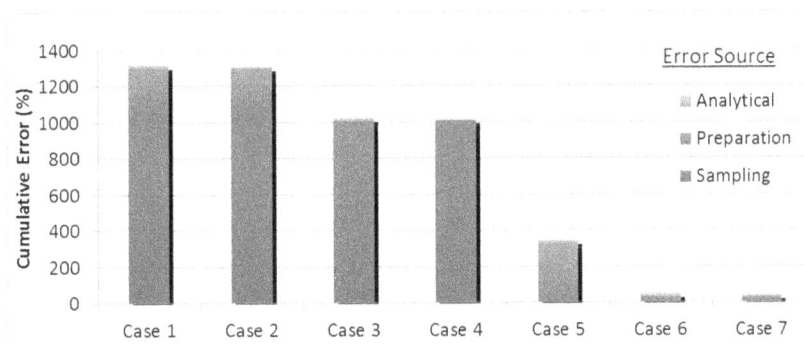

Figure 2.1. Graphical representation of results for sample error reduction efforts.

likely to be the largest source, even when all steps of the process are controlled to the greatest extent feasible (figure 2.2). The exception is if no effort is put into reducing sample preparation error, which is unlikely in an ideal world as it is much easier to reduce this error than sampling error. The problem with this assumption is if the sample preparation is done separately from the sample collection, as often happens as different contractors typically handle these two processes. The effort put into minimizing sampling error should not be perceived as excessive, but sample preparation is also critical to achieving reproducible data.

2.4.3 Other sources of error

There are other sources of error associated with sampling and characterization. A few of these will be discussed here.

Materialization error, a component of sampling error, relates to how a sample is collected. It is associated with the tools used for the collection of the sample. If the tool is not large enough to collect all particle sizes (delimitation error) or difficulty is

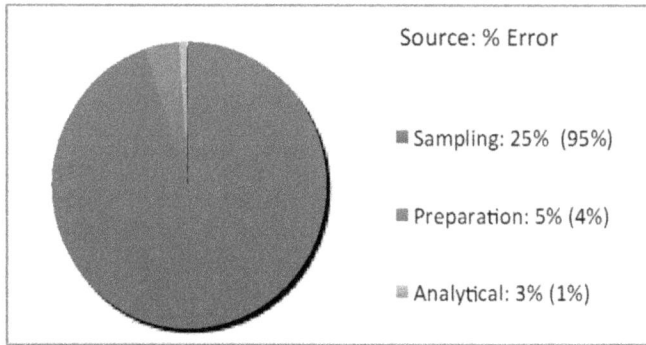

Figure 2.2. Total error distribution when error sources are controlled to currently achievable levels. Global estimation error (GEE) is 26% (Case 7, table 2.2).

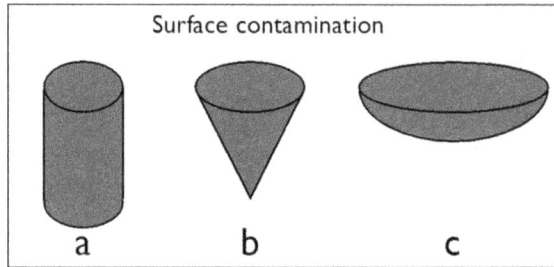

Figure 2.3. Increment shapes from different tools: (a) corer, (b) trowel and (c) spoon.

encountered in obtaining or extracting the material from the sampling device (extraction error), the consistency of the sample increments cannot be maintained and the quality of the sample will be degraded. Consistency of the shape of the increment throughout its depth is also very important, as is the uniformity of the increments that make up a sample (figure 2.3). Munitions constituents such as energetics and metals tend to initially accumulate on the surface of ranges. A coring tool that samples at a fixed diameter and can be set for a pre-determined depth will provide reasonably consistent increments. Sampling tools such as spoons and trowels will skew the sample data by overemphasizing the very top of the sample and not consistently reaching the target depth. Thus, they should only be used in circumstances where a coring tool will not work, such as in very loose sandy soils or soils with stones larger than the coring bit diameter. The increment geometry must allow the concentration of the contaminant to be represented consistently throughout the depth of the increment.

Sample handling can also affect the sample quality. Temperature, holding times, cross contamination and sample preservation are all factors that need to be controlled when handling samples. Sample processing errors associated with activities such as field splitting of samples, sample drying, comminution and subsampling need to be controlled as well as quantified through lab QA procedures. Sample and subsample extraction can be classified as a sample preparation procedure or an analytical lab procedure. In either case, QA also needs to be performed to ensure no loss of the analyte in the subsample. Finally, data analysis errors may also occur. Any of these errors may be fatal to the quality of the sample and the characterization effort, reinforcing the need to be vigilant during every step of the process. Figure 2.4 illustrates these errors.

2.4.4 Minimizing sampling error

The global estimation error can be minimized (but not eliminated) using practices derived from sampling theory. GSE can be addressed by taking many increments from the area being sampled (DU) and combining them into a single sample. Each increment should represent a similar amount of area within the DU to best characterize the complete area being sampled. The increments should also be uniform in mass. FSE can be addressed by sampling a sufficient mass to overcome the differences in composition in the particles collected in the samples. It is important that each increment mass, depth and surface area sampled be as close and uniform as possible to ensure that each increment can be considered equal.

Research at CRREL and Envirostat over the last 30 years has demonstrated what is required to minimize error associated with characterizing surface soils for energetics' residues. We have developed the *Multi-Increment®* sampling (MIS) method for the environmental characterization of surface soils for heterogene-ously-distributed munitions constituents. Sample increments are collected in a systematic fashion starting from a random location in one corner or location of the DU (figure 2.5). Reproducibility between samples has increased dramatically since MIS was implemented. Whereas prior sampling efforts resulted in proximate discrete samples varying by three orders of magnitude or more in a DU, using MIS has reduced variability to less than one order of magnitude, sometimes less than a factor of two between replicates. Whole population sample collection of high-order detonation residues on ice was compared to MIS of post-detonation residues on snow for the same operation and munition with no significant difference between the data sets [25].

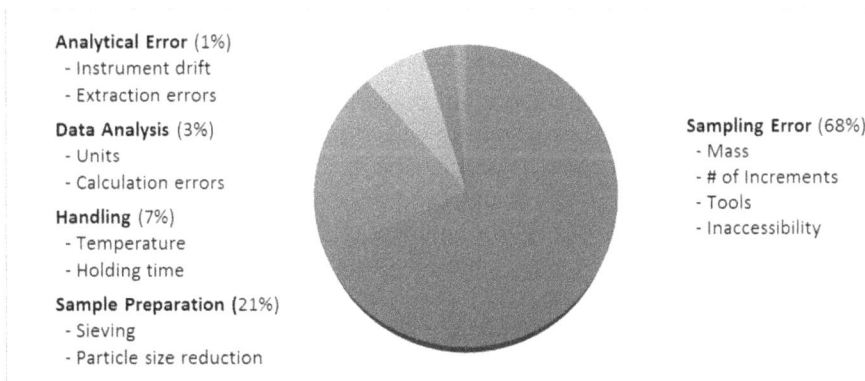

Figure 2.4. Error sources associated with a land-based DU characterization.

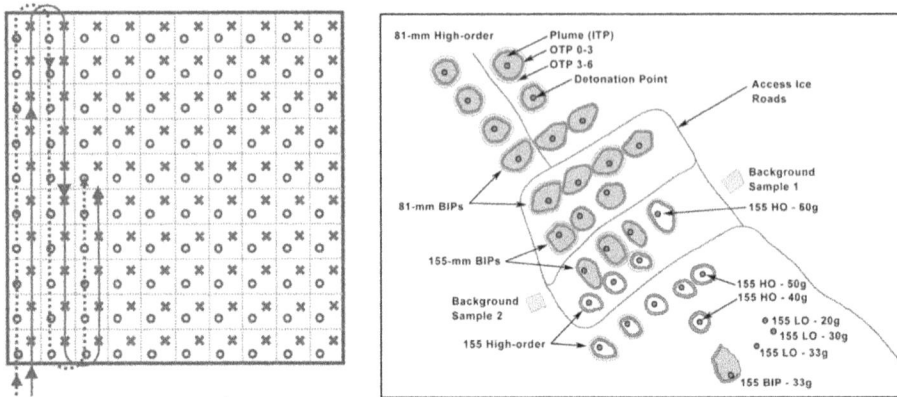

Figure 2.5. A replicate random-systematic sampling pattern in a square decision unit (left). Map of non-square detonation DUs from field research on munitions (right).

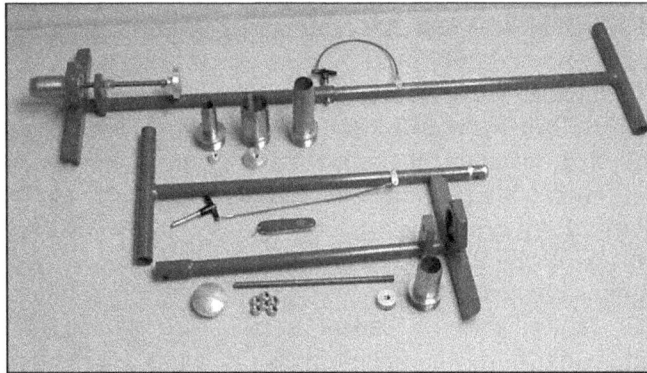

Figure 2.6. An example of a sampling tool for obtaining uniform soil increments.

Using specially designed tools, DU surface soils can be characterized quickly, efficiently, reproducibly and economically. These tools enable the collection of near-uniform increments in a variety of soils. In the case of the tool shown in figure 2.6, the diameter and depth of the increment can be varied in accordance with the desired final sample mass and number of increments [21]. The tool also enables the sampler to collect the increments from an upright position, greatly facilitating the process. Sampling bias is also reduced as the coring bit is pushed into the soil by the sampler's foot, aiding in the penetration of the coring bit through surface vegetation, a situation that will cause a sampler using a spoon or trowel to seek an easier location to sample. The coring tool will help minimize ME during sampling, reducing the total sampling error.

Sample handling is a very important aspect of sampling that is often overlooked, especially in the field. Most effort is focused on a chain of custody procedures, which are necessary but do not address sample integrity. Each sample collected in the field represents not only the DU from which it was obtained but a significant investment on the part of the entity tasked with sample collection. Sample labeling and documentation are thus a very important step in the field sampling process. Most MI samples should be collected in lab-grade clean polyethylene bags. The samples are labeled on the outside of the bag with an ink marker and recorded in a field notebook. Before sealing the bag with a plastic wire tie, we fill out a plastic tag with a sealable cover as a secondary label for each sample (figure 2.7). If the notation on the bag rubs off or becomes illegible, the tag can be used to identify the sample. If the tag is lost, the bag may retain enough information to identify the sample. The tag is also be used to 'track' the sample if it is rebagged for any reason. By transferring the tag to the new bag, all the original information is retained. During a field campaign at a closed military range, a contractor would not allow the employees to fill out and attach a tag to each sample. During shipping, the nomenclature rubbed off three sample bags, one of which was for a triplicate sample. The samples had to be reacquired at a cost of over USD 150 000 to the project.

Many soil samples will contain very low concentrations of munitions constituents. Cross contamination between samples from potentially high concentration DUs with samples known to be from areas of low potential contamination is very possible. A sample must be double-bagged in the field after documentation and stored separately during transportation and when in the lab. This precaution significantly improves the sample quality by reducing the variance between sample replicates.

Using the correct sample preparation techniques is also critical in maintaining sample quality [26, 27]. Subsampling of the DU sample must be done carefully. A large source of error occurs during comminution, the process used to reduce the variance in particle sizes in the sample, which allows more uniform distribution of the analytes within the total mass of the sample. Diminution typically is conducted with a grinder. Puck grinders, used in the mining industry, have been found to achieve the best results based on lab, field and project studies. It is important to follow the correct grinding procedures, as the analyte can be lost during grinding, as was found when using a ball mill. Subsampling should be done in the lab following grinding, never in the field using sample splitting. The process must ensure that the subsamples represent the sample, much as the sample must represent the DU from which it was taken. If not, the subsample will be no better than a grab sample. In

Figure 2.7. Field labeling of samples: bags with tags (left) and close-up of tag (right).

section 2.7 we will go into detail on field splitting, sample preparation in the processing lab, grinding to reduce sample particles to a more consistent size (comminution) and subsampling.

2.5 Sampling

Sampling is the process of obtaining a portion of a total population that represents that population as a whole. That portion is called the sample. Samples must be representative of the area as a whole from which they are collected or they will be meaningless, representing only themselves. How is a representative sample collected? This section will guide the reader through the process required to ensure that the samples collected will fulfill the objectives that the sample is required to meet. Subsampling, the process of obtaining a portion of a collected sample that represents the sample as a whole and thus the total population, will be discussed in section 2.7.

2.5.1 Objectives

Before going to the field to obtain samples, a sampling plan must be created. A key element of the sampling plan is the establishment of objectives that the samples are to meet. In the US, the Environmental Protection Agency has established a method based on what it calls the 'data quality objectives (DQO) process' [28]. The DQO process is iterative and will evolve as the project progresses. The SQC process, described in section 2.3.1, is a more direct approach that pares the development of objectives down to its essentials. Other countries have processes similar to these. The key elements for any sampling plan are that the sample must represent the mean concentration of the analytes in the DU, the samples must be reproducible (replicate samples from the field, not subsamples in the lab), and the data resulting from the analyses of the samples must be defensible. Without a robust sampling plan, the samples collected will not fulfill the objectives of the project.

At this point, it is worth saying something about 'hot spots'. The concept of hot spots has been around since samples were first collected. However, the definition of a hot spot has not [29, 30]. There are no mass, distribution, or availability criteria associated with a hot spot. Thus, hot spots are meaningless. If there is any analyte present in a DU, the highest concentration, the hottest hot spot, will be a million parts per million: a particle of pure substance (if it is present in that form). How many hot spots by this definition will occur within a DU? Many! Is there enough mass associated with these 'hot spots' to pose a risk to the receptor within the DU? The only way to determine this is to estimate a mean value of the analyte within the DU. Thus, the most important estimate that can be made within a DU is the mean concentration of the analyte. Hot spots are irrelevant.

2.5.2 Sampling

The process of obtaining a sample is quite straightforward. Once the correct geometry for the DU has been established, between 60 and 100 increments must be taken to the depth required to capture the analyte in the soil. The number of sample increments in a DU will depend on the relative standard deviation goals and contaminant distribution within the DU and may exceed 100 increments in some cases. The sample must represent the DU, so spacing of the increments is important. The sample must be protected from cross contamination in the field and throughout the chain of command all the way to the processing lab. Finally, QA procedures must be established and carried out in the field to determine the robustness of the sample.

The decision unit
Before sampling can begin, the DU must be established. A DU that is based on exposure must represent an area adequate to the exposure area of the receptor. Sampling will mimic in a general way the behavior of the receptor within the DU.

Collecting a sample
The sampling method must enable the collection of a sample that will enable the estimation of a mean concentration of the analytes within the DU. Single or multiple grab samples will not enable the derivation of a mean concentration estimate. The samples must consist of multiple increments collected in a 'random-systematic' fashion throughout the DU such that each increment represents approximately the same area within the DU and that all of the area of the DU is represented (see figure 2.5). By random-systematic, we mean that the starting point for each sample must be a random location within the first 'cell', as shown in figure 2.5, and that its location within the first cell must be carried through to similar locations in each subsequent cell. Each increment should penetrate an equal area of soil and be an equal depth as all others: In other words, the increment should be as consistent as possible. The coring tool shown in figure 2.6 is the best tool we have found for consistent increments. The number of increments necessary will vary, but field experience has shown that 100 increments is a good starting point to obtain reproducible data. The depth and diameter of the increments should be such that around 2 kg of material is collected for each sample. Research conducted by CRREL, Envirostat and DRDC has demonstrated that for energetics, the depth of the increment should be 2–3 cm in most soils. Samples should be collected in lab-grade polyethylene bags or their equivalent, which should be tied off (sealed) and labeled immediately after collection.

Quality assurance
Quality assurance *must* be conducted in the field to enable the determination of the robustness or reproducibility of the data. The best QA procedure is replicate sampling. Three to five replicate samples should be collected from randomly-determined DUs, the number of replicates and DUs sampled determined by the data quality objectives of the project. We recommend between three and five replicates in a DU that is designated for replicate sampling. Two replicates in a DU does not provide adequate data to determine reproducibility. On larger projects with many (>50) DUs, one set of replicates per 10 DUs may be adequate. For smaller projects or those requiring more robust data, a set of replicates every five or fewer DUs may be advisable.

2.5.3 Sources of error (see section 2.4)

The previous sections on the Multi-Increment® sampling process and error reduction have thoroughly described the innate sources of error associated with the sampling of soils and methods to reduce these errors to enable reproducibility of the samples. To reinforce these lessons, the sources of error are listed below.

- Inadequately-sized DU: The DU is too large or too small for the receptor; usually caused by a fixation on 'hot spots'.
- Sample error: Caused by inadequate sample mass, the number of increments is too low, increments are not geometrically consistent (incorrect sampling tool).
- Representativeness of increments: Each increment must represent an equal area and mass within the DU).
- Field splitting of samples: A recipe for disaster that has been shown to produce very inconsistent results. Field splitting *cannot* be used as a substitute for replicate sampling.
- Reproducibility: There must be a method to measure sample reproducibility. Reduce post-sampling error using strategic bagging techniques.

We have often seen data presented with six or more significant digits. This occurs most often with data for grab samples or data for which no QA has been performed. A plethora of significant digits is no substitute for replicate sample data and is a sign that the data is suspect. Data for heterogeneously-distributed analytes on soil in the environment should be represented by no more than two significant digits.

2.5.4 Best practices

Decision Units—DUs will be defined in the sampling plan. The size of the DU, as previously mentioned, should be strongly linked to the typical area in which the receptor will be exposed to the analytes of interest. Do not scale DUs to fit 'hot spots', as sampling for hot spots is unreliable and, in almost every case, unjustified. If finer resolution of the distribution of an analyte in an area is needed, the DU should be sized accordingly.

Error reduction—For data quality to meet almost any goal, active error reduction must take place when sampling a DU. Minimizing error will result in data reproducibility and robustness. The number of increments in the sample ($\approx$100) and the sample mass ($\approx$2 kg) must be sufficient to address FSE and GSE. Both the mass and the number of increments may need to be increased to meet the project-specific SQC.

Representativeness of increments—Each increment must represent an equal area and volume of soil within the DU. Geometrically consistent increments will greatly enhance the ability to obtain reproducible data, so the choice of the sampling tool is critical. Whenever possible, a fixed diameter/fixed depth cylindrical sampling tool should be used. Co-located increments collected during replicate sampling should be avoided as colocation eliminates the randomness of the replicate samples, thus degrading the data.

Reproducibility/QA—Replicate sampling is the best method of determining sample reproducibility. Don't co-locate replicate samples! Sampling outside DUs may be conducted to determine if the DU correctly captures the analyte in a use-specific DU (fuel point, HAZMAT storage facility). Do not overuse significant digits. Two significant digits is about the limit for most environmental work. More gives a false sense of accuracy and will raise suspicion about the data quality. Do not field split samples! That will reintroduce FSE and GSE into the aliquots. Subsampling is best left to the processing lab after proper sample preparation. Often, analyte concentrations are quite low for most samples but some can be quite high. Double-bag samples in the field after collection and before transport. Keep samples cool during storage and shipment.

2.6 Sampling on snow and ice—a special case

Over the past 15 years, CRREL, Envirostat and DRDC have conducted extensive research on energetics' deposition from live-fire training with munitions. This research seeks to characterize the mass deposition and distribution of residues of energetics for various weapon systems to better understand the source terms of military range contamination. The tests for this research have been conducted on snow- and ice-covered ranges to enable the isolation of the test samples from previous activities on the ranges [9, 31, 32]. The sampling methods were derived from those used for the characterization of munitions constituents on soils. There are some differences, however, and these differences may be applicable to sampling of soils.

The major difference between soil sampling for energetics and sampling on snow or ice is the ability to visually discern the approximate area of contamination on the surface of the snow or ice. To verify that the DU area is accurate, one or two 3 m wide annuli are sampled outside the DU to ensure that the demarcated area captures the contamination (figure 2.8) [33]. Another QA procedure typically employed is to sample beneath each increment as a separate sample to determine if the depth of the sample increments is sufficient to capture the contaminant. Both these procedures may be applied to soil sampling for contamination.

A third QA procedure was recently conducted to check the validity of MI sampling. Munitions were detonated on ice and all the visible residues swept up and analyzed as a 'whole population' sample (table 2.3). There was no significant difference between individual plume MI sample results, overlying multiple-shot/single-plume MI sample results, and swept plume 'whole population' sample results, proving that MI sampling is a valid method for characterizing a DU, be it on snow, ice, or, by extension, soil.

Sampling on snow or ice is a very informative method for determining contaminant loading of ranges from training activities. It can be used as a supplement or verification of data obtained using soil sampling on DUs. Verification of very low residue concentrations at a 155 mm firing point and very high residue mass at a shoulder-fired rocket range has been successfully carried out by comparing snow sampling and soil sampling results. A database of residues' deposition rates for training activities can be used to augment the tools needed to define DUs at a site being investigated for contamination [10, 11, 25, 34–36].

Figure 2.8. Example of sampling detonation residues on snow, showing the main area of deposition with a 3 m QA annuli.

Table 2.3. Comparison of various methods to characterize a DU containing detonation residues.

Matrix - Deposition	NTO		DNAN		RDX plus HMX	
	Mass (g)	Detonation efficiency	Mass (g)	Detonation efficiency	Mass (g)	Detonation efficiency
Snow—Individual	0.54	99.87%	0.008	99.997%	0.0077	99.995%
Snow—Overlying	1.15	99.73%	0.013	99.995%	0.0038	99.998%
Swept Ice—Individual	0.72	99.83%	0.017	99.993%	0.0041	99.997%

2.7 Sample processing and analysis

Sample processing and analysis should achieve the following objectives:
- *Preserve the analyte integrity within the sample and subsample*
- *Ensure that each subsample represents the sample from which it is taken*
- *Reduce error in the final aliquot taken for extraction prior to analysis*
- *Result in reproducible analyses that represent the concentration of the field analyte.*

Quality assurance procedures should be conducted throughout the process to ensure error is minimized or not introduced. Replicate subsampling, analytical duplicates and whole-sample extraction are methods to ensure that errors are kept in check. An example of a QA method for one of the steps in the processing of samples is the grinding of silica sand to check for analyte carryover between grinding of different samples. As with sampling, the goal is reproducibility of results for each step of the process.

2.7.1 How to process multi-increment soil samples to determine energetics

Soil sample processing involves the following steps: air-drying, sieving and machine grinding until all the particles in the multi-increment sample are very fine (<75 μm), followed by subsampling and extraction of the analytes with solvent. Reduction in particle size results in the formation of many, many fine energetic particles so that a small analytical soil subsample (≈ 10 g) can have the same proportion of constituents as the processed multi-increment sample. *The ultimate goal of this procedure is to adequately represent the several tons of soil in the field with the few grams of soil that are actually subjected to solvent extraction and analysis in the lab.*

Step 1. Air-dry the entire sample.

Soil samples with energetics should be stored at <4 °C (can be frozen) until they are ready for processing. Samples that are known to have very high energetic concentrations should be isolated from samples that potentially have low or undetectable energetic concentrations to prevent cross contamination.

The first step in processing is to air-dry the soils at room temperature. Drying arrests microbial activity that could biotransform some of the explosives and it enhances the ability to further process the soil. The entire field-moist multi-increment sample is spread onto a clean tray and air-dried away from light. The time required to thoroughly air-dry a sample depends on the relative humidity and the initial soil moisture content. Generally, two to three days are required. Oven-drying IS NOT acceptable due to thermal degradation and/or sublimation of the energetics. Once a soil sample is air-dried, it may be returned to a sample bag and stored at room temperature in the dark.

Step 2. Sieving or manual removal of oversized fraction.

Unvegetated or sparsely vegetated soil samples are forced through a #10 sieve. The #10 sieve has a mesh with 2 mm openings, the size division between course sand and gravel. Most of the energetics are found in the <2 mm size fraction [37]. Soil aggregates and dried vegetation such as moss or grass should be broken apart while sieving the soil. Each size fraction is weighed, the less than 2 mm fraction processed further as described below, and the oversized (>2 mm) fraction is saved for further study if needed to meet the objectives of the investigation.

At some ranges, the mineral soil may be overlain by a thick (>3 cm) organic layer, and the multi-increment samples will be primarily composed of dry, decomposed and/or partially decomposed organic material with little mineral soil. These samples are not easily forced through a sieve, so the oversized fraction consisting of pebble-size or greater rocks and woody sticks are manually removed and the rest of the sample is machine-ground as described below [38].

Step 3. Machine grinding.

The less than 2 mm fraction is ground using a puck mill (LabTech Essa LM2 or equivalent). Other types of grinders do not adequately grind the energetic particles [39].

Soils from impact areas that contain only high explosive residues (i.e., HMX, RDX, TNT) are ground continuously for 60 s [40]. Soils from ranges that contain propellant residues (i.e. firing points, demolition areas, rocket impact ranges) are ground for five 60 s periods with a cooling time between each grind. The extra grinding time is needed to pulverize the propellant fibers.

Grinding reduces the particle size of the course soil to the texture of flour (<75 µm) and it significantly reduces the uncertainty associated with subsampling [41, 42].

Step 4. Subsampling.

The ground sample is spread evenly over a clean surface in an exhaust hood. The thickness of the sample should be about 1 cm. A lab spatula is used to obtain increments of soil through the thickness of the sample. Although the energetic particles are more evenly distributed throughout the ground sample than in the field sample, at least 30 increments from random locations in the ground sample should be combined until a 10 g subsample is obtained.

Step 5. Extraction with solvent.

The energetic compounds are extracted from the soil matrix using acetonitrile. Extraction kinetic studies conducted many years ago using field-contaminated soils from arsenals showed that equilibration time varied with concentration, the soil, the solvent and agitation method. Acetonitrile was superior for extraction of the nitramine explosives (HMX and RDX) and it is compatible with the analytical instruments used to determine the energetics. Equilibration using a sonic bath was achieved rapidly for higher concentrations from sandy soils and slowly for low concentrations and clay soils. An 18 h time was sufficient for most soils. The 18 h extraction is not a necessarily a burden as solvent can be added to the subsamples in the afternoon and extracts prepared for analysis the following morning. More recent studies showed that a platform shaker can be substituted for a sonic bath [41].

2.7.2 Summary of the sources of error

The following short list summarizes the errors that can occur when processing the sample.

- *Heterogeneity of the sample:* Caused by heterogeneously-distributed analyte particle(s).

  Sample heterogeneity is the result of the analyte particle properties in the field that translates to the sample. Nothing in the processing lab will affect the *initial* heterogeneity of the sample, which is unknown.

- *Incorrect sample drying:* Loss of analyte caused by a drying temperature that is too high.

  Most energetic materials have some volatility associated with them. Overheating the sample will drive off some of the analyte, while under-drying will not arrest microbial effects on the sample and will interfere with the proper grinding of the sample.

- *Particle heterogeneity:* Segregation of particles based on size.

  A large variance in particle sizes cause problems with segregation. The largest particles (stones and material over a pre-determined cut-off size such as 2 mm) may be removed prior to grinding, along with non-contributing vegetation, such as leaves and sticks. Particle size reduction (comminution) using a grinder will reduce segregation error as well as more evenly distribute the analyte particles throughout the ground sample mass, making the representative subsampling of the sample much more likely. Use only a puck grinder, and ensure that the sample temperature does not increase to the point of volatilizing the analyte.

- *Inappropriate grinding:* Lack of comminution and loss of analyte due to overgrinding and overheating of the sample and insufficient particle reduction of the sample.

  Improper grinding of the sample will not reduce the heterogeneity of the sample sufficiently to allow reproducible subsampling. The dry sample must be ground to the protocols outlined above to ensure proper particle reduction (comminution) without volatilizing the energetics in the sample. The mass of the sample for each grind should be about half the capacity of the bowl to allow proper grinding without heating of the sample. Sample aliquots that require additional grinding should be allowed to cool between grinds. Ground aliquots can be recombined prior to subsampling.

- *Incorrect subsampling:* Subsample does not represent the sample.

  Comminution of the sample will greatly aid in subsampling accuracy, but there is still a need to ensure that the subsample represents the original sample. This can be achieved through MI sampling, where the ground sample is spread out to a near-uniform thickness and 30 or more increments are taken through the thickness of the material to build the subsample(s). A rotary splitter may also be used to obtain subsamples, but only AFTER the sample has been properly ground.

- *Incorrect extraction procedures:* Process does not capture or insufficiently captures analyte in solvent.

  The extraction solvent must dissolve the analytes from the sample matrix and be compatible with the analytical instruments. Adequate extraction time will vary with the soil matrix (sand versus clay) and with analyte concentrations.

  The following is a source of error not discussed above because it should *never* be conducted with a sample collected in the field.

- *Field splitting of large samples:* Causes a loss of representativeness of the sample because of the heterogeneity of the energetic particles in the collected sample.

Mass reduction in the field to reduce the amount of soil sent to the laboratory is a major source of error and should not be performed. Splitting an unprocessed sample that contains heterogeneously-distributed energetics does not work in the laboratory. It certainly will not work in the field [27, 37].

2.7.3 Analytical processes

Concentrations of energetics are measured using a liquid chromatograph (LC) equipped with a UV detector (dual wavelength or diode array). Identifications are based on retention time, so all measurements must be confirmed, either different chromatographic conditions that result in a different retention order, a mass spectrometer, or another method such as gas chromatography with an electron capture detector (GC-ECD).

The method using the UV detector (US Environmental Protection Agency, SW-846 Method 8330B Nitroaromatics, Nitramines, Nitrate Esters by High Performance Liquid Chromatography (HPLC) [43]) has proven to be extremely rugged over many years (high accuracy and high precision) for the analysis of most soils. A mass spectrometer is needed when analyzing complex matrices that contain many other UV absorbing compounds. For confirmation of analyte identity and to estimate concentrations in samples below the HPLC detection limit, extracts may be analyzed by the GC-ECD method (Method 8095 Nitroaromatics and Nitramines by GC-ECD [44]). The disadvantages of the GC-ECD method are that the detector has a limited calibration range and the GC yields concentration estimates with higher variance than the HPLC.

The following are typical analytical conditions for the analysis of energetics:

HPLC-UV: Extracts are mixed 1/3 v/v with water and filtered (PTFE, 0.45 μm) prior to injection into the HPLC-UV. The output from the UV detector is monitored at two wavelengths: 254 nm for the nitroaromatics and nitramines and 210 nm for NG. The chromatographic separation is achieved on a 15 cm × 3.9 mm (4 μm) NovaPak C-8 column maintained at 28 °C and eluted with 15:85 isopropanol/water (v/v) at 1.4 ml min^{-1}. Concentrations are estimated from peak height measurements compared to calibration standards with concentrations spanning the concentrations in the sample extracts and within the linear range of the detector.

GC-ECD: Acetonitrile extracts are filtered (PTFE, 0.45 μm) to autosampler vials, which are then placed into an autosampler tray. A 1-μL aliquot of each extract is directly injected into the HP 7890A purged packed inlet port (250 °C) containing a deactivated liner. Separation is conducted on a 6 m- × 0.53 mm-ID RTX-TNT (Restek) fused-silica column that has a 1.5 μm-thick film of a proprietary Crossbond phase. The GC oven is temperature-programmed as follows: 100 °C for 2 min, 10 °C min^{-1} ramp to 250 °C. The carrier gas is hydrogen at 1.4 psi inlet pressure. The μECD detector temperature is 290 °C; the make-up gas is nitrogen at 45 ml min^{-1}.

2.8 Conclusion

Live-fire training ranges are critical assets for militaries worldwide. As countries have become more environmentally aware of the impacts of munitions on the range ecosystems, a need has arisen for a method to sample ranges such that the results are scientifically robust, reproducible and legally defensible. Current methods, which do not take into consideration the theory of sampling and make little if any effort towards error reduction in the collection and processing of field samples, are unable to deliver reproducible results. The methods and tools developed for *Multi-Increment*® sampling (MIS) have enabled the reduction of field sampling error, the ability to collect reproducible samples, and obtain consistent subsamples after following the recommended processing procedures. MIS has been successfully applied to many different sampling problems over areas from 1000 ha to square-meter cells on military training ranges. Quality assurance methods have been developed and refined to enable the assessment of the robustness of the samples. The integration of MIS into the US Environmental Protection Agency's Method 8330B has enabled the development of sound and defensible sampling plans in the USA to address a variety of sampling situations [28]. It is important to remember, however, that error will never be completely eliminated in any step of the characterization process. By knowing the source of the possible error and where to concentrate error reduction and quality assurance efforts, a sampling and sample processing plan can be developed that will result in reproducible data and a defensible inference of the mean contaminant concentration in a decision unit. All procedures described in this chapter must be carried out with the proper sampling and processing equipment to ensure that the error associated with each sample will be within acceptable limits for reproducibility, legal defensibility and statistical robustness.

References

[1] Clausen J L, Robb J, Curry D and Korte N 2004 A case study of contaminants on military ranges: Camp Edwards, Massachusetts, USA *Environ. Pollut.* **129** 13–21

[2] Walsh M E, Walsh M R, Collins C M and Racine C H 2014 White phosphorus contamination of an active army training range *Water, Air, Soil Pollut.* **225** 2001

[3] Ramsey C A 2015 Considerations for sampling of water *J. AOAC Int.* **98** 316–20

[4] Racine C H, Walsh M E, Roebuck B D, Collins C M, Calkins D J, Reitsma L R, Buchli P J and Goldfarb G 1992 White phosphorus poisoning of waterfowl in an Alaskan salt marsh *J. Wildl. Dis.* **28** 669–73

[5] Walsh M R, Walsh M E and Voie Ø 2014 Presence and persistence of white phosphorus on military training ranges *Propellants Explos. Pyrotech.* **39** 922–31

[6] Jenkins T F, Grant C L, Brarr G S, Thorne P G, Schumacher P W and Ranney T A 1996 Special Report: Assessment of sampling error associated with the collection and analysis of soil samples at explosives contaminated sites, SR 96-15, U.S. Army Cold Regions Research and Engineering Laboratory, Hanover, USA

[7] Pitard F F 1993 *Pierre Gy's Sampling Theory and Sampling Practice: Heterogeneity, Sampling Correctness, and Statistical Process Control* (Baton Rouge, LA: CRC Press)

[8] Thiboutot S *et al* 2012 Environmental characterization of military training ranges for munitions-related contaminants: understanding and minimizing the impacts of live-fire training *Int. J. Energ. Mater. Chem. Propul.* **11** 17–57

[9] Walsh M R, Walsh M E and Ramsey C A 2012 Measuring energetic contamination deposition rates on snow *Water, Air, Soil Pollut.* **223** 3689–99

[10] Walsh M R, Walsh M E, Taylor S, Ramsey C A, Ringelberg D B, Thiboutot S, Ampleman G and Diaz E 2013 Characterization of PAX-21 insensitive munition detonation residues *Propellants Explos. Pyrotech.* **38** 1–11

[11] Walsh M R, Walsh M E, Ramsey C A, Brochu S, Thiboutot S and Ampleman G 2013 Perchlorate contamination from the detonation of insensitive high-explosive rounds *J. Hazard. Mater.* **262** 228–33

[12] Taylor S, Jenkins T F, Rieck H, Bigl S R, Hewitt A D, Walsh M E and Walsh M R 2011 *MMRP Guidance Document for Soil Sampling of Energetics and Metals* 11-15 ERDC/CRREL (Hanover, NH USA) Technical Report

[13] AAFCO 2015 *Guidance on Obtaining Defensible Samples: GOOD Samples* (Champaign, IL: Association of American Feed Control Officials)

[14] AAFCO 2018 *Guidance on Obtaining Defensible Test Portions: GOOD Test Portions* (Champaign, IL: Association of American Feed Control Officials)

[15] Ramsey C A and Wagner C 2015 Sample quality criteria *J. AOAC Int.* **98** 265–8

[16] Ramsey C A 2015 Considerations for inference to decision units *J. AOAC Int.* **98** 288–94

[17] Ramsey C A 2015 The decision unit—a lot with objectives *TOS Forum* **5**

[18] Ramsey C A 2015 Considerations for sampling contaminants in agricultural soils *J. AOAC Int.* **98** 309–15

[19] Hathaway J E, Rishel J P, Walsh M E, Walsh M R and Taylor S 2015 Explosive particle soil surface dispersion model for detonated military munitions *Environ. Monit. Assess.* **186** 415

[20] Hewitt A D, Jenkins T F, Walsh M E, Walsh M R, Bigl S R and Ramsey C A 2007 *Protocols for Collection of Surface Soil Samples at Military Training and Testing Ranges for the Characterization of Energetic Munitions Constituents* TR-07-10 ERDC/CRREL (Hanover, NH) Technical Report

[21] Walsh M R 2009 *User's Manual for the CRREL Multi-increment Sampling Tool* SR-01-1 ERDC/CRREL (Hanover, NH, USA) Special Report

[22] Hewitt A D and Walsh M E 2003 Technical Report: On-site processing and subsampling of surface soil samples for the analysis of explosives (CRREL)) ERDC/ CRREL TR-03-14. US Army Cold Regions Research and Engineering Laboratory, Hanover, USA

[23] Ramsey C A and Hewitt A D 2005 A methodology for assessing sample representativeness *Environ. Forensics* **6** 71–5

[24] Rasemann W and Dumps I W 2000 Sampling and analysis *Encyclopedia of Analytical Chemistry* ed R A Meyers (Hoboken, NJ: Wiley), pp 2692–719

[25] Walsh M R, Bigl M F, Walsh M E, Wrobel E T, Beal S A and Temple T 2018 Physical simulation of live-fire detonations using command-detonation fuzing *Propellants Explos. Pyrotech.* **43** 602–8

[26] Ramsey C A and Suggs J 2001 Improving laboratory performance through scientific subsampling techniques *Environ. Test. Anal.* **10** 12–6

[27] Walsh M E, Hewitt A D, Jenkins T F, Ramsey C A, Walsh M R, Bigl S R, Collins C M and Chappell M A 2011 Assessing sample processing and sampling uncertainty for energetic residues on military training ranges: Method 8330B *Environmental Chemistry of Explosives and Propellant Compounds in Soils and Marine Systems: Distributed Source Characterization and Remedial Technologies ACS Symposium Series* (Washington, DC: ACS), chapter 5, 91–106

[28] US EPA 2006 *Guidance on Systematic Planning using the Data Quality Objectives Process: EPA QA/G-4* (Washington, DC USA: US Environmental Protection Agency Document EPA 240-B-06-001. USEPA)

[29] Hadley P W and Sedman R M 1992 How hot is that spot? *J. Soil Contam.* **1** 217–25

[30] Hadley P W and Mueller S D 2012 Evaluating 'hot spots' of soil contamination (Redux) *Soil Sediment Contam.* **21** 335–50

[31] Jenkins T F, Walsh M E, Miyares P H, Hewitt A D, Collins N H and Ranney T A 2002 Evaluation of the use of snow-covered ranges to estimate the explosives residues that result from high order detonations of army munitions *Thermochim. Acta* **384** 173–85

[32] Walsh M R, Walsh M E, Ramsey C A and Jenkins T F 2005 *An Examination of Protocols for the Collection of Munitions-derived Residues on Snow-Covered Ice* TR-05-8 ERDC/CRREL (Hanover, NH) Technical Report

[33] Walsh M R, Walsh M E, Ramsey C A, Rachow R J, Zufelt J E, Collins C M, Gelvin A B, Perron N M and Saari S P 2006 *Energetic Residues Deposition from 60-mm and 81-mm Mortars* TR-06-10 ERDC/CRREL (Hanover, NH) Technical Report

[34] Walsh M E, Walsh M R, Taylor S, Douglas T A, Collins C M and Ramsey C A 2011 Accumulation of propellant residues in surface soils of military training range firing points *Int. J. Energ. Mater. Chem. Propul.* **10** 421–35

[35] Walsh M R, Walsh M E, Poulin I, Taylor S and Douglas T A 2011 Energetic residues from the detonation of common US ordnance *Int. J. Energ. Mater. Chem. Propul.* **10** 169–86

[36] Walsh M R, Walsh M E, Ramsey C A, Thiboutot S, Ampleman G, Diaz E and Zufelt J E 2014 Energetic residues from the detonation of IMX-104 insensitive munitions *Propellants Explos. Pyrotech.* **38** 243–50

[37] Walsh M E, Ramsey C A, Collins C M, Hewitt A D, Walsh M R, Bjella K L, Lambert D J and Perron N M 2005 *Collection Methods and Laboratory Processing of Samples from Donnelly Training Area Firing Points, Alaska, 2003* TR-05-6 ERDC/CRREL (Hanover, NH) Technical Report

[38] Walsh M E, Ramsey C A, Taylor S, Hewitt A D, Bjella K and Collins C M 2007 Subsampling variance for 2,4-DNT in firing point soils *Soil Sediment Contam.* **16** 459–72

[39] Penfold L 2008 *Personal Communication: Grinding Methods for Soils Containing Propellants—Methods Comparison (RSD)* (TestAmerica Laboratories, Inc.)

[40] Walsh M E, Ramsey C A and Jenkins T F 2002 The effect of particle size reduction by grinding on subsampling variance for explosives residues in soil *Chemosphere.* **49** 1267–73

[41] Walsh M E and Lambert D J 2006 *Extraction Kinetics of Energetic Compounds from Training Range and Army Ammunition Plant Soils: Platform Shaker versus Sonic Bath Methods* TR-06-6 Technical Report ERDC/CRREL

[42] Walsh M R, Walsh M E, Gagnon K, Hewitt A D and Jenkins T F 2014 Subsampling of soils containing energetics residues *Soil Sediment Contam.* **23** 452–63

[43] US EPA Method 8330B (SW846) 2006 *Nitroaromatics, Nitramines, and Nitrate Esters by High Performance Liquid Chromatography (HPLC), Revision 2. US Environmental Protection Agency Document EPA 240-B-06-001* (Washington, DC: USEPA)

[44] US EPA Method 8095 (SW846) 2007 *Explosives by Gas Chromatography, Revision 0* (Washington, DC: US Environmental Protection Agency)

IOP Publishing

Chapter 3

Hydrologeological characterization of military training ranges and production of maps for land management

Richard Martel and Sylvie Brochu

3.1 Introduction

The training activities of the land forces include the use of a wide variety of weapons that liberate energetic materials, propellants, metals and other compounds in the environment. Research provides knowledge to reduce the environmental impact by optimized practices and the use of less hazardous materials but there is a large heritage of accumulated residues from training activities of the past as well as—even though reduced—new residues from the ongoing military training. The present section of this chapter provides information on appropriate methods for ground-water (GW) monitoring within range and training areas (RTAs). In fact, GW is a media that integrates the behavior of contaminants at the surface and the interaction/transformation of these dissolved contaminants within the vadose zone. Munition residues may migrate off-range mostly via surface water and groundwater, and wells give information about a potential off-range migration of dissolved residues. Wells are tools for long term surveillance of GW quality and flow direction and help in sustainable training.

The past two decades of hydrogeological characterization of Canadian RTAs has shown that based on potential risks of GW contamination by munition residues, activities may be ranked as indicated in table 3.1 [1]. Table 3.2 shows which ranges are likely to be a source zone of munition residues.

GW is rarely directly accessible for measurements or sampling. Its access is limited to outcropping aquifers where GW may drain as natural springs. In most cases, water discharge from aquifers occurs in wetlands and through river bed and

3-1

Table 3.1. Ranking of military activities based on potential impact on groundwater.

Priority	Description[a]	Accumulated munition residues on soil surface[b]
1	**Air-surface bombing(Air Force)**	**ClO$_4$, RDX**, TNT and degradation products
2	**Demolition range**	**RDX, HMX**, TNT, NG, 2,4-DNT, 2,6-DNT, 2-ADNT, 4-ADNT,Cr, Cu, Zn
3	**Experimental testing range**	Depends on activity—all of above
4	**Anti-tank range—impact area**	**ClO$_4$, RDX, HMX**, TNT, NG, 2-ADNT, 4-ADNT, Pb, Cu
5	**Hand grenade range**	**RDX**, TNT,Zn, Cu, Ni, Sb, Cr, Pb
6a	**Artillery range—impact area**	**RDX, HMX,** TNT, NG, 2,4-DNT, 2,6-DNT, 2-ADNT, 4-ADNT, Cr, Cu, Zn, Mo, Ni, Cd
6b	**Artillery range—firing position**	NG and degradation products,2,4-DNT, 2,6-DNT, Cu, Cr, Pb
7	**Anti-tank range—firing position**	**ClO$_4$**, NG
8	Skeet range	PAH, Pb
9	Small arm range	Firing position: NG, 2,4-DNT, Backstop: Pb, Cu, **Sb**, Zn
10	Striffing (Air Force)	Metals
11	Battlerun—impact area	Like artillery impact area, but more diluted
12	Manoeuver range, munitions depot	Depending on activity

[a] In bold: anticipated higher potential impact on groundwater.
[b] In bold: highly mobile residues.

Table 3.2. Most likely source zone of munition residues migrating to/in groundwater.

Ranges	ClO$_4$	RDX	HMX	TNT	Others
(1) Air-surface bombing (Air Force)	X	X	-	X	-
(2) Demolition range	-	X	X	X	2,4-DNT, 2,6-DNT
(3) Experimental testing range	X	X	X	X	depend on activity
(4) Anti-tank range—impact area	X	X	X	X	NG
(5) Grenade range	-	X	-	X	-
(6A) Artillery range—impact area	-	X	X	X	NG, 2,4-DNT, 2,6-DNT
(6B) Artillery range—firing position	-	-	-	-	NG, 2,4-DNT, 2,6-DNT
(7) Anti-tank range—firing position	X	-	-	-	NG

TNT = TNT and transformation products

lake bottoms. So, boreholes have to be drilled and wells installed in order to access GW. This section is subdivided following the necessary steps: (1) well location; (2) preparation of drilling sites and safety procedure; (3) drilling methods and cleaning procedure; (4) well components; (5) well development; (6) water level measurements; (7) well purging and groundwater sampling; (8) analytical methods and quality control; (9) hydraulic conductivity testing (slug tests).

3.1.1 Well location

GW is naturally accessible in springs where the discharging GW may be directly sampled. In all other cases, boreholes have to be drilled and wells installed to sample GW and to measure water level, hydraulic conductivity of the aquifer and to evaluate the GW quality. Generally, the potentially altered aquifer is the first to be encountered when drilling, which is most of the time unconfined. In some cases, the first aquifer encountered is a perched aquifer of limited extent or a confined one below a low permeability unit (clay, silt, till). The following criteria are proposed for well emplacement: (1) locate wells with short polyvinylchloride screens (1 to 1.5 m long) in the upper part of the first aquifer meet in or close to the source zone (the screen length may be of 3 m long in a bedrock aquifer or in sediment aquifer in the pathway of potentially altered GW); (2) in unconfined conditions locate the top of the screen at 0.5 to 1 m below the groundwater level to be sure to get a water sample at any time of the year; (3) define well depth according to distance from the source zone, the GW velocity and recharge rate; (4) locate wells close to the inner side of base boundaries downstream of the source zone or close to sensitive receptors for liability reasons; (5) do not locate wells close to targets in impact areas; (6) locate wells along existing roads and trails for a safer access and well protection; (7) use a transect approach for plume delineation and mass flux estimation downstream of a source zone using several wells installed in different (separate) boreholes (multi-level sampling point) or combined with a direct push method; and (8) use a multiple phases approach with the wells' emplacement improved following knowledge from every phase.

3.1.2 Preparation of drilling sites and safety procedures

The safety procedure includes a safety briefing given by range control personnel to inform all involved workers on the dangers related to the training areas. A level one clearance (visual) is made by the explosive ordnance disposal (EOD) specialists on the road that give access to the drilling site. Then, as shown in figures 3.1 and 3.2, the proofing of the drill sites is performed in the field by the EOD specialists. The proofing is done with a metal detector (e.g. Forester or Vallon MWD1) or a magnetometer (e.g. EM 61 or Vallon EL-1302-D). After the sweeping of the surface of the drilling area, a pre-borehole can be made with a hand auger. The sweeping is done down the hole at every 30 cm to a total depth of 1 m. If potentially dangerous

Figure 3.1. Drilling site preparation: proofing and top soil removal (upper left), pre-hole boring (upper right) and proofing down hole (lower).

Figure 3.2. Different instruments for UXO proofing before borehole drilling.

objects are found at the soil surface or detected into the ground of the drilling site, unexploded ordnance (UXO) avoidance is made by selecting a new drill location beside the initially proposed one. The EOD specialists are present during the entire field work procedure if necessary. The drilling site is prepared by removing the surface soil on 15 cm depth on a surface area of 1 m^2 at the well location before the drilling is performed to avoid any vertical migration of contaminated soil into the borehole and to avoid contact of the drilling bit, augers and rods or soil sampler with surface soil during drilling operations.

3.1.3 Drilling methods and cleaning procedures

The drilling method proposed for well installation in sediment is the hollow stem auger and the air rotary method for well installation in bedrock. Both techniques have the advantage of using no or a minimal amount of water during drilling operations. Injected water may bias water quality of the following GW sampling campaigns if the well is not purged properly. Soil sampling can be made with a 60 cm long spilt spoon or a corer 1.5 m long when using the hollow stem auger, whereas chips are recovered at the surface when using the air rotary technique in bedrock. The solid stem auger can also be used for fast well installation in sediment without the possibility of getting representative soil samples along the drilling profile. The rotosonic drilling method can also be used for a complete recovery of soil samples (corer 3 m long) from the borehole but has the disadvantage of using a lot of water under difficult drilling conditions (deep boreholes through hard and compact geological units). The direct push technique can be used for fast soil sampling at shallow depths (core of 1.5 m long) and 1″ well installation. Sediments can also be defined along the soil profile using a CPT (cone penetration test) probe on a direct push drill. The data collected are put on a classification chart [2, 3] to identify them.

In training areas, the munition residues accumulate on the soil surface. Decontamination of drilling bits, augers and rods between drilling sites is mandatory to avoid cross-contamination. It is recommended to store the drilling equipment in a clean area (e.g. on a clean tarpaulin) to avoid direct contact with the surface soil. Corers and split spoons must also be cleaned between each sampling event. The measuring tape used for well installation must also be cleaned and kept for being in contact with the surface soil particles.

The equipment decontamination has the following sequence: (1) brush to remove soil particles followed by high pressure washing with non-phosphated detergent mixed in tap water; (2) wash with hydrochloric acid 10% (HCl) to dissolve metals; (3) rinse with distilled water; (4) wash with American Chemical Society grade acetone to dissolve energetic materials; and (5) rinse with distilled water. All chemical waste used must be securely stored and eliminated according to appropriate regulations.

3.1.4 Well components

It is recommended to complete all monitoring wells according to [4–6] (figure 3.3 and table 3.3). Observation wells are made of four different components. The PVC screen and riser, the filter sand pack, the sealing material and the protective well casing. The material selected has to have minimal interference with the chemistry of the GW [7]. The PVC screen and riser are usually of schedule 40 (wall thickness of 0.154″) and of 2″ in diameter. The slot size of the PVC screen is designed according to the d_{10} (the diameter in the particle-size distribution curve corresponding to 10% finer) of the filter sand pack used around the screen. The filter pack is a sub round to round silica sand at 99% quartz mineralogy. The d_{30} of this commercially available sand is 4 to 6 times the d_{30} of the sediment along the screen length (sampled during borehole drilling) and has a uniformity coefficient (d_{60}/d_{10}) of 2 to 3. The filter pack in the annular space between the borehole wall and the screen extent is up to 1 m

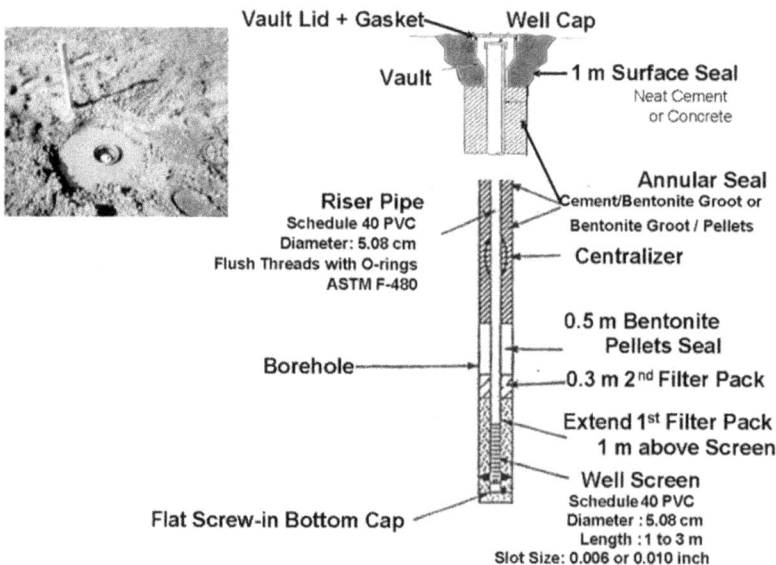

Figure 3.3. Monitoring well flush-to-ground-surface completion.

Table 3.3. Materials used in well installation.

Materials
PVC screen and riser pipe
PVC well screen: slot size (d_{10} of primary filter pack)
1st filter sand around screen and until 1 m above screen: prevents intrusion of sediments in filter; d_{30} = 4 to 6 times the d_{30} of sediments; d_{60}/d_{10} = 2 to 3
2nd filter sand from 1.0 to 1.3 m above the screen: prevents intrusion of grout into 1st filter sand; d_{90} = d_{10} of 1st filter sand; d_{10} = 0.2 to 0.3 of d_{10} of 1st filter sand
Bentonite pellets (above the sand pack)
Grouting above screen (above the bentonite pellets to 1.5 m below surface)
Concrete (from 1.5 m below surface to surface) and well casing

Figure 3.4. Metallic above-ground casing (left) and adapted flush mount casing (right).

above the top of the screen. The sealing material in the annular space between the borehole wall and the PVC riser pipe is bentonite pellets up to 60 cm length above the filter pack followed by a bentonite grout (put in place via a tremie pipe) up to 1.5 m below the surface followed by concrete. As shown in figure 3.4, the metallic protective casing is installed into the concrete and can be above ground or an adapted flush mount can be used. The most critical operation is the screen, the filter sand pack and the bentonite pellets installation where the drillers need to wear clean nitrile gloves and be cautious to avoid contamination by surface soil.

3.1.5 Well development

During drilling operations, augers and drill bits tend to smear clay and other fine soil particles along borehole walls, which alters the hydraulic response of the well. Drilling fluid is also introduced into the well. To minimize the impacts of drilling fluids and debris, well development, which essentially consists in pumping GW out of the well, is accomplished. This operation is usually made once just after well installation. New wells must be developed before the water level measurement and the slug testing to ensure proper hydraulic connection between the well and the

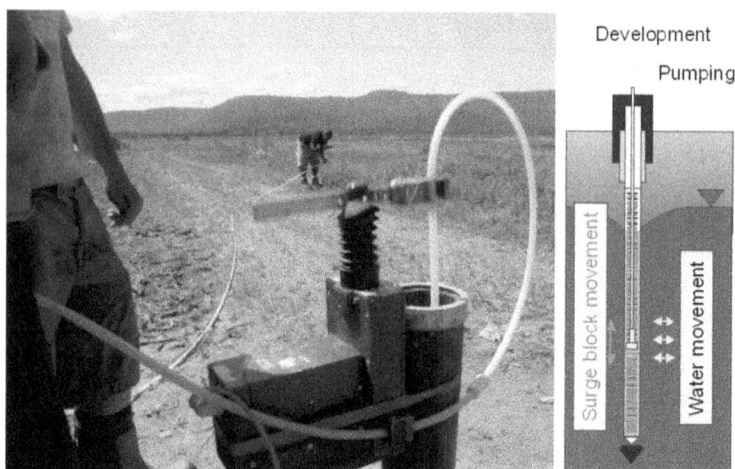

Figure 3.5. Verification of fine particles in GW from a well developed with an inertia pump (see length of tubing for disposal of effluent water).

aquifer. Pumping is accomplished using an inertia pump (Waterra) that is a linearly oscillating foot valve into the screen of the well (figure 3.5). The foot valve directs the water out of the well through dedicated high density polyethylene (HDPE) tubing. A surge block can be installed on the foot valve to promote water stirring into the well and to remove sediments blocking the well screen or the filtering sand pack. Linear motion of the system (tubing/foot valve/surge block) is ensured by a hydrolift actuator. All equipment inside the well is strictly dedicated. The water is pumped in 20 cm intervals until it is clear (without any sediment). This operation is repeated along the screen until the bottom of the well is reached. The pumped water is discarded at least 15 m down-gradient from the well.

Another method of development is the method by pumping/over pumping that consists in rapidly lowering the water level to the top of the screen and letting it go back up. Development of wells where a screen is installed in fine materials, such as silt or clay, is done upon installation of the well. Clean water is circulated in the well's PVC risers, through the screen and the sand filter before sealing the rest of the annular space between the PVC tubing and the borehole walls with bentonite pellets and grout.

3.1.6 Water level measurements

Measured water levels (figure 3.6) are the only large scale method to determine the regional GW flow direction. In order to get reliable measurements, all wells have to be carefully surveyed to determine the exact elevation of the well head above sea level. The measure is done with a measuring tape with a dipole at the tip. The measured depth of water below the top of the well PVC tubing can then be subtracted from the elevation of the well head to obtain the elevation of the water table. At the end of a sampling or a characterization campaign, a water level 'snap shot' is usually done. This means that all water levels are measured within a short period of time (usually less than a few days) to avoid fluctuation of the GW level due

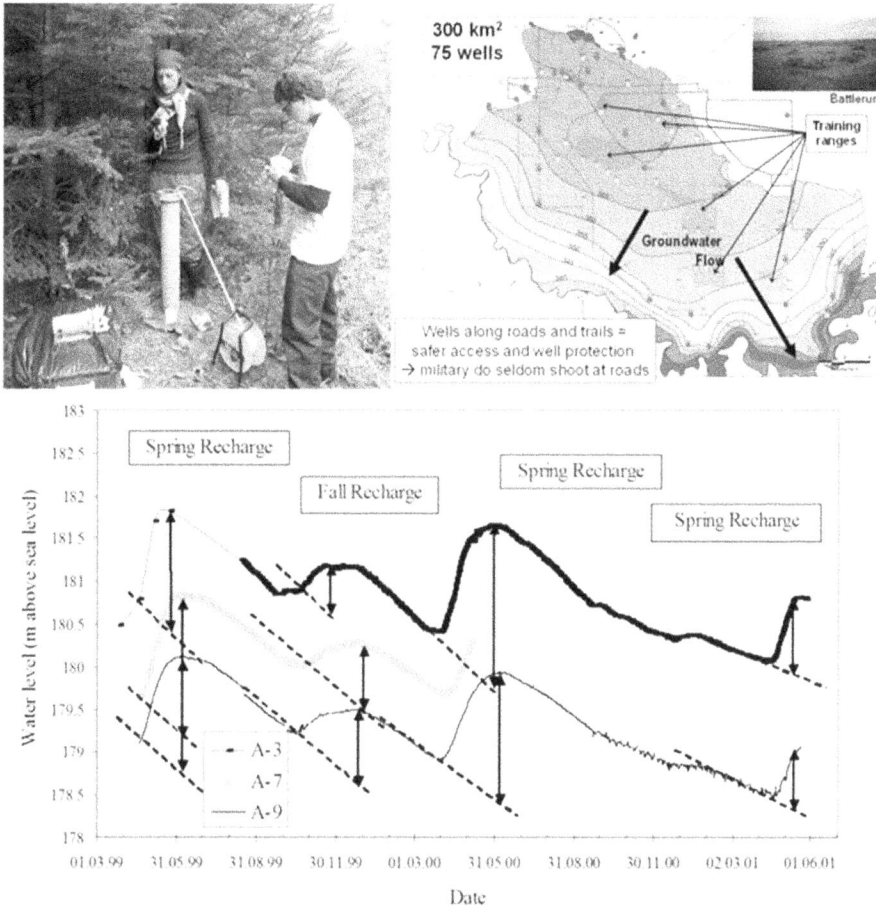

Figure 3.6. Water level measurement (top left) hydraulic head map made by interpolation with the ordinary kriging method at CFB Shilo (top right), water level fluctuations over two years in three wells (bottom).

to climatic events (drought or precipitations). This data set of water levels is then used to elaborate a potentiometric map of the GW surface. It is very important to use water levels measured in the same aquifer to produce the hydraulic head map of this aquifer otherwise different maps need to be produced for each one. The map is drawn by interpolation of individual water level observations. Interpolators can be ordinary kriging, co-located co-kriging, or smooth interpolation. In complicated situations linear interpolation and hand drawing can be more effective. Long term monitoring of the water levels may be performed by putting a level logger into the well that records changes in hydrostatic pressures (the height of the column of water). These data can then be used for the aquifer recharge estimation using software such as GWHAT [8]. The model HELP [9] can also be used for an estimation of the recharge at the larger scale. All equipment used for water level measurements must be dedicated or thoroughly washed between sites (see *Drilling methods and cleaning procedures*).

Figure 3.7. Flow-through cell and multi-parameter-probe to monitor physico-chemical parameters when purging the well with a peristaltic pump on a protected work zone (left), groundwater sampling with peristaltic pump installed on a table (right).

3.1.7 Well purging and groundwater sampling

During water purging and sampling for munition residues in training ranges special precautions must be taken because residues are present mainly in soil at the ground surface (0–5 cm). In order to reduce the risk of contamination of the GW samples by the surface soil contaminants, the work zone can be protected by a dedicated polyethylene sheet (tarpaulin placed around the well or on a table). All the equipment needed for sampling is placed on that sheet and thus isolated from soil particles (figure 3.7). The goal of water sampling is to verify if contamination is present and it generally takes four sampling sessions in the first year to be able to record the fluctuations of the contaminant concentrations. However, due to limited access to training areas, it is often difficult to have four sampling sessions per year. The first session should be detailed to have a complete assessment of the site and the following sessions can be partial assessments to contaminated areas identified in the first session.

To avoid cross-contamination between wells it is recommended to have a dedicated sampling system in each well. If parts of the sampling or measuring equipment (bladder pump, measuring tape) are going to be used at several wells, decontamination procedures (2 to 5 at section 1.3) must be applied between each well location. It is also good practice to proceed from less contaminated areas to more contaminated areas when this information is known. Ideally, the material in contact with water of the well must be made of inert material; Teflon®, Viton® and stainless steel are recommended.

All wells must be purged before sampling to remove the stagnant water that is not representative of the water quality that circulates in the aquifer [10]. When using the low flow technique [11] (figure 3.8) and prior to sampling, the pumped water is directed to a flow-through cell, isolating water from the atmosphere. The low flow technique aims to withdraw GW at a very low rate (typically between 100 and 500 ml min^{-1}) in order to minimize drawdown, usually below a value of 10 cm of hydraulic head. A multi-parameter probe is connected to the flow-through cell allowing measurement of the GW's physico-chemical properties (N.B. the multi-parameter probe needs to be calibrated with the appropriate standard every three days). Measured parameters

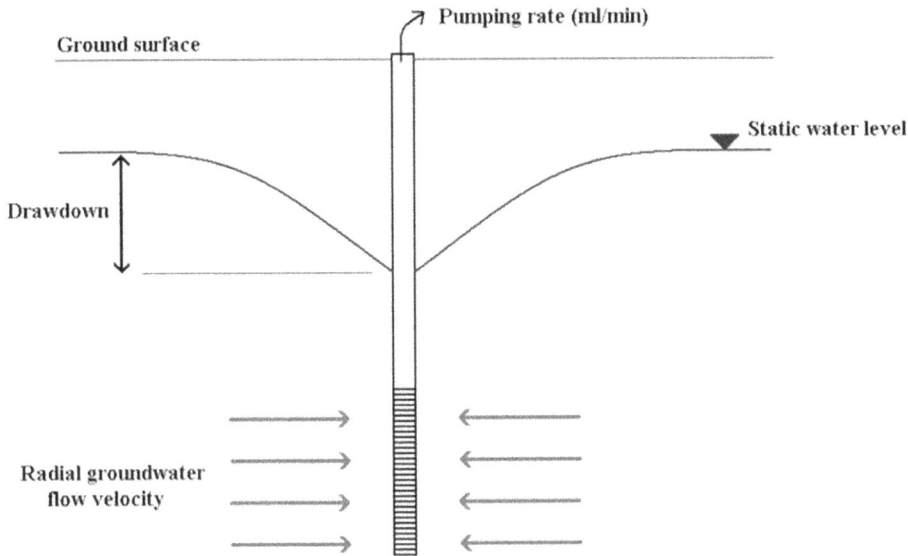

Figure 3.8. Principle of low flow sampling—the drawdown should be less than 10 cm.

are: temperature, electric conductivity, dissolved oxygen, pH and oxydo-reduction potential (ORP). Parameter readings are taken every 5 to 10 min depending on flow rate. Parameter stabilization indicates that the purge is completed and that the pumped water comes directly from the geologic formation. Stabilization is usually achieved when three consecutive values are within 0.1 pH unit for pH, within 3% for electrical conductivity and 10% for dissolved oxygen. At the end of the purge, the flow-through cell is disconnected and water is collected directly from the flexible Viton® tubing of the pump used.

For wells located in low permeability materials (i.e. silt and clay), it is necessary to empty the well's tubing once while refraining from drying the screen to avoid infiltration of air from the aquifer, which would lead to the alteration of the water quality.

The advantages of low flow sampling over high flow sampling methods are: (1) less turbulence generated in the well; (2) there are fewer suspended solids in samples and adsorbed compounds; (3) sampling of naturally flowing suspended particles in the GW (1–1000 nm); (4) samples are more representative of real GW chemical composition; (5) samples for dissolved metals do not need to be filtered; (6) lightweight and dedicated equipment saves field time and eliminates cross-contamination problems; and (7) reduced purge volume compared to other sampling methods, which makes it a very time and labor efficient method saving costs in the end.

Low flow sampling can be accomplished with a peristaltic pump (figure 3.9) that is connected to a dedicated system permanently installed inside every well. The

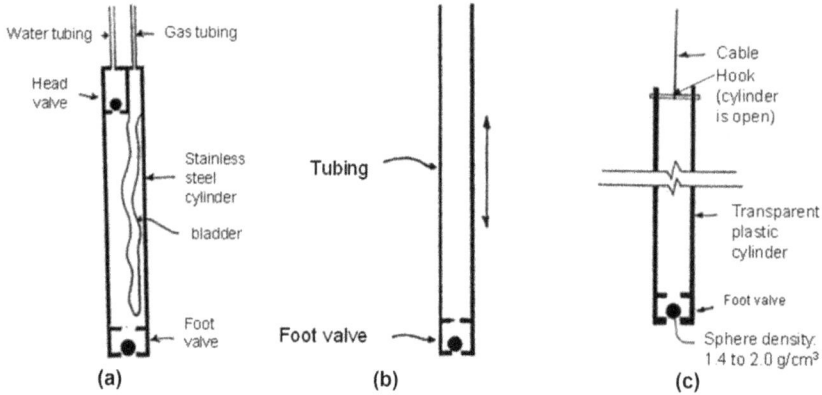

Figure 3.9. Foot valves allowing to lift GW with (a) a bladder pump (low flow), (b) a Waterra pump (high flow) and (c) a bailer (high flow).

system is made of a Teflon® tubing (¼″) connected to a ¼″ flexible Viton® tubing. The Teflon® tubing must be pulled up to place the tubing end at the middle of the well screen in order to sample the upper part of the saturated thickness. It is recommended to proceed to GW sampling with the same constant flow rate and at the same depth of the purging into the well to avoid sampling localized water up in the tubing of the well or down on the bottom of the well that may contain fine particles from the well development. Peristaltic pumps have a maximum suction lift of about 8 m. When the static water level in a well exceeds that depth, a bladder pump is used.

A bladder pump (figure 3.9(a)) consists of a chamber in which a membrane (bladder) can expand to occupy the space inside the chamber thereby expelling the water inside it. A bladder pump can have a diameter as small as 2.5 cm to allow it to enter into narrow wells. A bladder pump reduces degassing and volatilization of the water sample (good for sampling VOCs) because there is no contact between the air used to inflate and deflate the bladder of the pump and the water sample. If not dedicated, the inside and outside of the pump must be thoroughly washed following the cleaning procedures (2–5 in section 1.3). The bottom of the pump should match the desired depth of sampling (mid-screen for a well in sediments and levelled at the fracture's depth in bedrock).

Another option is to make groundwater sampling with the high flow technique using a bailer or an inertia pump (Waterra). When performing high flow sampling, purging 3 to 5 times the volume of water in the well tubing and filtering sand pack prior to sampling is mandatory. Both methods have the inconvenience of aerating the sample and bringing small solid particles (colloids) that do not participate in the natural groundwater flow. Samples collected with those techniques need to be filtered for dissolved metal analysis. A bailer (figure 3.9(c)) is a thin plastic hollow cylinder on which a foot valve is fixed at one extremity. When plunged below the groundwater surface, the valve allows water to get in the cylinder. The valve also prevents water from coming out as it is removed from the well. A bailer can be made of inert material such as Teflon®, stainless steel or a clear polyethylene (disposable after usage). The bailer's diameter can vary within a range of 1 to 10 cm and has no limitation to the depth of the water inside the well. However, it only allows removal of a small volume of water each time. The inertia pump (figure 3.9(b)) is composed of semi-rigid tubing that is fixed to a ball valve at its lower end. When the tubing is lowered into the well, the ball valve opens and lets the water enter. When the tubing is pulled back up, the foot valve closes and the water is brought up towards the surface. It should be used in sampling only on rare occasions, when no other pump is available. It is convenient for sampling wells up to 40 m in depth manually and wells up to 100 m in depth when using a hydrolift. It is light, low maintenance, easy to use and represents good dedicated equipment.

Filtration of the water samples is required for perchlorate analysis in order to avoid sample deterioration by bacteria. It is required for dissolved metals' analysis if a high flow sampling technique is used because metals may be liberated from particles (colloids) after sample acidification. The filtration is done with a disposable syringe and 0.45 μm filter.

3.1.8 Analytical methods and quality control

Collected groundwater samples are put in specific bottles depending on the parameters to be analyzed (figure 3.10) and sent to the laboratory for chemical analyses using the suggested methods presented in table 3.4. Data interpretation involves: comparing detected concentrations to existing water quality guidelines and background values (for metals and anions) and verifying consistency of values between sampling events and neighboring sites. It also includes the quality control/ quality assurance program with field blank (1 per week during campaign), trip blank

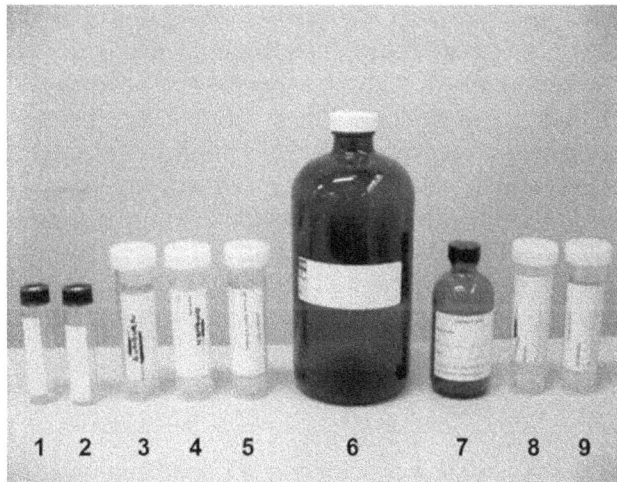

Figure 3.10. Water sample bottles suggested. From left to right: Total organic compound (1, 2); anions (3); alkalinity (4); pH, electrical conductivity, turbidity, color (5); energetic materials (6); nutrients (7); metals (8); perchlorate (9).

Table 3.4. Recommended analytical methods for water analysis.

Parameter group	Analytical method
Metals (traces)	US EPA method 6020A ICP-MS (US EPA, 2007)
Major components	APHA 3120 B-ICP-AESUS EPA (1994b) method 200.7
Anions	APHA 4110 B-ion chromatography
Ammonium	APHA 4500-NH3 D ion selective electrode APHA 4500-NH3 G automated phenate method
Total organic carbon (TOC)	US EPA 415.1 modified
Total alkalinity (as $CaCO_3$)	APHA 2320B titration method
Perchlorate	LC-MS-MS (Backus *et al* [12])
Energetic materials	US EPA method 8330B HPLC (US EPA 1994a) US EPA method 8095 GC-ECD (US EPA 1999)

(1 per week during campaign), field equipment wash blank (1 per sampling campaign), drilling water samples (1 per tank truck), field duplicates or triplicates (10% of all samples) and internal laboratory duplicates and controls.

3.1.9 Hydraulic conductivity testing (slug tests)

Slug tests offer an economical way to determine near-well geological unit hydraulic conductivity. A slug test consists in instantaneously changing the water level in an observation well and monitoring the change in hydraulic head through time until complete recovery of the static water level. The sudden change in hydraulic head is achieved by dropping or by removing a slug into the well. This slug can be a bailer of water or simply a Teflon® cylinder. Compressed air can also be used to make the slug in water above the screen. The well response is monitored with a level logger previously installed into the well that measures the hydrostatic pressure (the height of the column of water). All equipment must be dedicated or thoroughly washed prior to the test according to procedures 2–5 in section 1.3. Various methods exist to interpret the water level data such as Bower and Rice [13] for unconfined aquifers, and Cooper [14] or Hvorslev [15] for confined aquifers. Software such as AQTESOLV allow one to evaluate the hydraulic conductivity using those methods. Three slug tests are recommended per well and the geometric mean of those tests can be used to make the calculation of the groundwater velocity using the Darcy law equation.

3.1.10 Conclusion

Hydrogeological characterization of a RTA in a Canadian military base has shown that activities may be ranked based on their potential risks of GW contamination by munition residues. Wells give information about a potential off-range migration of dissolved contaminants in GW. Wells are tools for long term monitoring of GW quality that help one to manage and sustain training. Different methods were proposed for borehole drilling and GW sampling. The challenge of water sampling in military training ranges is to avoid contamination bias of the samples with surface soil particles that may potentially contain traces of EM, perchlorate or metals. A good GW sampling program involves many campaigns, appropriate analytical methods and a good quality control/quality assurance to produce reliable water quality data.

3.2 Production of maps for land management of range training areas

3.2.1 Introduction

One challenge in assessing the risk of contamination by military training comes from the fact that different types of munitions are used in many activities. The use of practice or live munition also change the amount of residues (energetic materials (EM), metals and propellant) deposited on site. Low order or null detonation can generate residues that accumulate over time in the impact areas whereas propellant residues accumulate at the firing position. The EM content of damaged munitions and propellant can interact with the surrounding soil and can be transported by precipitation/infiltration events. This process can lead to the contamination of soils, surface water and groundwater.

To develop a military training area management tool, information is needed on aquifer vulnerability (vulnerability assessment) and on hazards of munitions residues (contamination risk assessment) to produce a risk map (management tool) (figure 3.11). On many Canadian army bases, aquifer vulnerability maps were made using a method that combines a 3D geologic model in gOcad and the calculated downward advective time (DAT) of infiltration water in the vadose zone (step 1). DAT was calculated throughout the RTA from the soil surface to the water table. A vulnerability classification index was developed from DAT to make those maps. The hazard maps (step 2) were drawn combining a hazard-classification index developed according to the surface area of the range, the quantity of munition used and the dangerousness of the residues (drinking water toxicity and the

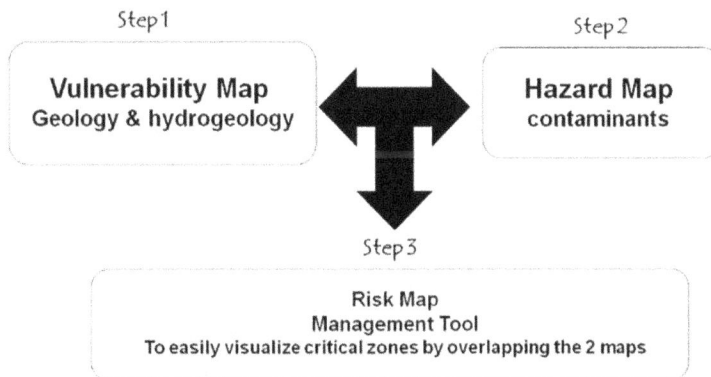

Figure 3.11. Methodology developed to produce a risk map.

environmental persistence) associated with each type of munition used in each range. Finally, the risk maps (step 3) combined (overlay) the vulnerability and the hazard maps. The risk map offers an effective tool to easily visualize the critical zones. Thus, it is possible to make a better land-use management of the actual or future activities by checking their locations in order to minimize the risk of groundwater contamination. It is possible to reduce this risk by completely moving the area of training in a zone of low or very low vulnerability, modifying the type of activities in an area or reducing the amount of munition fired.

The Canadian Force Base (CFB) Shilo was selected to illustrate the methodology developed to produce vulnerability, hazard and risk maps in Canadian RTAs. CFB Shilo occupies an area of 900 km^2 in South-Western Manitoba (Canadian prairie province). It has four small arms firing ranges, one anti-tank range, one grenade range and five multi-purpose battlerun ranges. Surficial deposits consist of deltaic sediments underlying discontinuous aeolian sands; sand dunes exist on the northern and eastern parts of the garrison. The groundwater bearing formation under CFB Shilo is part of the great Assiniboine regional aquifer, which is characterized by relatively homogenous sand. Regional average annual precipitation is 470 mm from which 5% to 15% accounts for aquifer recharge (23 mm to 71 mm).

3.2.2 Aquifer vulnerability

Aquifer vulnerability to contamination is a concept for which several definitions and assessment methods exist [16–23]. In the Canadian approach the method described by Ross *et al* [16] was used, whereby the aquifer vulnerability to contamination is defined as the relative ease with which dissolved contaminants can reach the upper boundary of an aquifer by downward advective non-retarded and nonreactive transport (worst-case scenario) following the introduction of a contaminant at or near the land surface. It is a conservative definition of aquifer vulnerability since all the processes that could potentially limit the impact of contamination, such as adsorption and dispersion, are not considered. Following this definition, vulnerability is an intrinsic characteristic of the natural environment, which is independent of contaminant type. This concept provides insights on the potential of a natural setting to act as a downward migration route for a contaminant to reach an underlying aquifer. The vulnerability of an aquifer to contamination is interpreted from estimates of groundwater DAT along a vertical travel distance from the land surface down to the water table of the first aquifer met. Therefore, the fate and transport of contaminants, once they have reached the aquifer, are not taken into account in this approach.

As described by Ross *et al* [16, 24], the method makes use of a 3D geologic (stratigraphic) model constructed using the software gOcad 2.0.8 that allows for a relative regional estimate of aquifer vulnerability to downward transport of dissolved and persistent contaminants based on groundwater DAT. Such an approach is a simplification of the actual complexity of the natural system and assumptions have to be made. The main assumptions are: (1) The relative vulnerability of an aquifer can be obtained by estimating DAT using geologic and hydrogeologic information; (2) factors that may change over time such as land use or seasonal effects are not considered; (3) contaminant behavior is the same as water (worst-case scenario); (4) contaminants are released at the land surface; (5) ground-water flow is vertical along the entire length considered for DAT estimation; and (6) saturated properties are considered because it represents the fastest DAT (worst-case scenario).

Hydrogeologic properties (infiltration and porosity) are integrated into the 3D grid of the geological model to make the estimation of DAT from the surface to the water table of the first intercepted aquifer. The one-dimensional DAT through an unsaturated layered system can be approximated by the following equation (3.1):

$$DAT = \frac{1}{q}\sum_{i=1}^{n}m_i\theta_i \qquad (3.1)$$

where q (m s^{-1}) is the groundwater recharge rate [8, 9], while m_i (m) and θ_i (mL cm^{-3}) are, respectively, the thickness and volumetric water contents of geological layers (n) at every location where the calculation is applied [25–29]. The sum of m_i is limited by the travel distance $D(m)$, which is the thickness of the unsaturated zone. Also, when assuming water saturation through D, θ is replaced in equation (3.1) by the effective porosity n_{eff} (cm^3 cm^{-3}).

The input parameters (porosity, infiltration) are randomly chosen within their estimated range for each geological unit to generate multiple DAT estimates [16, 24]. The parameters are assigned to each geological unit and are considered constant throughout the study area for each iteration. A total of 250 iterations were run with the script in the model gOcad providing a DAT for each 105 m × 105 m cell. The average scenario represents the mean DAT for each cell, whereas the optimistic scenario represents the 25th percentile longest DAT value and the pessimistic scenario, the 25th percentile shortest DAT value for each cell. The aquifer vulnerability map is an intrinsic vulnerability assessment meaning that it is independent of the contaminant type and indicates areas most sensitive to contamination based on the physical properties of the ground. Only surface area inside of the RTA was evaluated for aquifer vulnerability. The vulnerability index from very high to very low is related to DAT that varies from ≤0.5 year to ≥100 years (figure 3.12).

For CFB Shilo, the Assiniboine Delta Aquifer has a mean infiltration rate of 50 mm year^{-1} (23 to 68 mm/year) calculated with water level fluctuation in one observation well over a period of 17 years [30, 31]. The recharge corresponds to 15% of precipitation. The effective porosity was estimated at 0.25 (cm^3 cm^{-3}), which is typical of a medium sand and in agreement with the literature [29, 32]. Figure 3.12 presents the aquifer vulnerability index maps for CFB Shilo (average scenario). On the CFB Shilo map the vulnerability is very high to the north-east because it is associated with a very shallow water table that eventually flows into the Epinette Creek. Very low vulnerability in the southern part of the RTA is associated with a deep water table combined with very low infiltration rate.

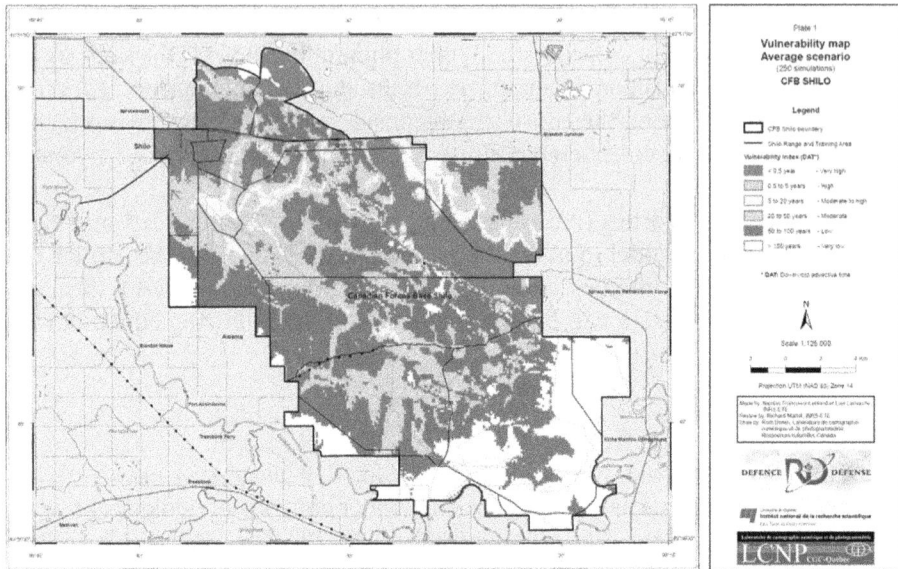

Figure 3.12. Vulnerability map of CFB Shilo RTAs (average scenario).

3.2.3 Hazard

Environmental contamination associated with military activities and munition used can occur in many type of ranges (demolition, grenade, anti-tank, air/ground, mortar/artillery, small arms (see section 3.1)). Each type of activity generates different potential residues. The use of practice or live munition also changes the amount of residues deposited on range. There are two broad classes of munition residues that can be deposited in the field: energetic materials (i.e. TNT, RDX, HMX, perchlorate), and metals (Sb, Pb, Cu, Zn, etc) [33]. Only energetic materials (EMs) have been taken into account in our approach since metals are prone to complex interactions with the soil matrix and dissolved metals in water are considered much less mobile than propellants and EMs. The hazard map is built using a hazard index, which is the multiplication of the frequency, the dangerousness and the surface area indexes [34] [equation (3.2)].

$$\text{Hazard index} = \text{Frequency index} \times \text{dangerousness index} \times \text{surface area index} \tag{3.2}$$

Frequency index

The frequency is the quantity of residues deposited per year for each EM per range. To evaluate the frequency, the quantity of munition fired on ranges has to be known. Firing statistics have been obtained either from range control personnel of Canadian army base that provide a list of munition names and the number of rounds fired per range over a period of 10 years, or from the Canadian Force Range Integration System (CFRIS) database. Also the composition and the dud rate of munitions need to be known. The munition composition and the mass of each EM in specific munition was obtained from Canadian Forces Technical Orders (CFTO). Broad class and generalizations had to be used in some munitions under trade secrets. The dud rate was obtained from the literature. The anti-tank munition can have a dud rate of up to 40% [35]. Munitions of artilleries and mortars normally have rates of failure between 3% and 8% [36]. Other important information needed is the proportion of residues remaining after a low order detonation. This information was obtained from the literature. Canadian and US deposition rate studies on snow [37–44] generated data for many munition types. At firing positions, unburned propellant deposited behind or in front of the weapon have a deposition rate that vary in the literature with the munition type [35–55]. Finally, the calculation of the frequency in impact areas is different from firing positions. The mass of EM residues deposited on impact areas is the product of the number of munitions fired, the mass of each EM (in the main explosive charge) in that munition, the dud rate and the proportion of remaining residues for low order detonation (from deposition tests on snow). The mass of EM residues deposited on firing position is the product of the number of munitions fired, the mass of propellant in each munition and the proportion of remaining residues after firing (from deposition tests on snow).

The frequency index takes into account the munition used and deposition rate for each type of munition in order to give a deposited quantity for each evaluated EM. The evaluated EM are TNT, RDX, HMX, 2-4-DNT, PETN, NG and perchlorate. The decision matrix (table 3.5) then gives a score from 0 to 5 for each of the seven energetic materials. The yearly average mass is used for frequency index estimation based on the ten-year activities record.

Dangerousness index
Dangerousness of a compound depends on danger to humans and other organisms and on the behavior in the environment, soil and aquifers. Dangerousness was evaluated for each EM potentially released by training activities based on the toxicity and the environmental persistence. The toxicity of EMs was established from the water quality guidelines for Canadian Forces [56]. The parameters considered for the environmental persistence were the mean solubility and natural attenuation factors such as adsorption on soil, biodegradation and photodegradation. Each energetic material has been given a dangerousness level for each criteria from very low (1) to very high (5) according to table 3.6. The weighted criteria values were then added and

Table 3.5. Frequency index per range per energetic material per year.

0	0 (g)
1	<100 (g)
2	100–1000 (g)
3	1000–5000 (g)
4	5000–10 000 (g)
5	>10 000 (g)

Table 3.6. Environmental dangerousness criteria.

Dangerousness index	Criteria	Drinking water toxicity (μg l^{-1})	Mean solubility (mg l^{-1})	Biodegradation	Photodegradation	Adsorption depends on soil (l kg^{-1})
1	Very low	no criteria	<10	aerobic and anaerobic < 15 days	<1 day	Very high: >100
2	Low	100–300	10–100	aerobic and anaerobic 15 days–30 days	1–10 days	High: 30–100
3	Moderate	30–100	100–1000	aerobic and anaerobic 30 days–0 days	10–60 days	Moderate: 5–30
4	High	10–30	1000–10 000	>90 days	>60 days	Low: 1–5
5	Very high	<10	>10 000	anaerobic or strict aerobic	limited	Very low: <1

the dangerousness index calculated for the most common EMs found in RTAs (TNT, RDX, HMX, PETN, 2–4-DNT, NG and perchlorate) (table 3.7) according to the following formulae: (Drinking Water Toxicity + Environmental Persistence) ÷ 2.

The most dangerous EMs considering the criteria evaluated is thus perchlorate, followed by 2,4-DNT and RDX. TNT and HMX have a medium relative dangerousness whereas NG and PETN are considered of low and very low dangerousness, respectively.

Surface area index

The surface area index is inversely proportional to the surface area of the range studied. In fact, hazard is higher in a small area because more rounds are fired per unit surface area and dilution is less present. Firing positions have very small deposition areas (order of 100 m^2) located behind or in front of them, whereas an impact area may extend to millions of square meters (table 3.8).

Hazard index

The hazard index assesses potential contamination by using weighted parameters in decision matrices of residue deposition frequency, environmental dangerousness of each type of EM and of the surface area where munitions are used. For each of the EMs evaluated by this method and for a given range, the three above parameters are multiplied together (equation (3.2)). Because each index has a maximum value of 5, the hazard value calculated cannot be more than 125. Since this method is dependent on the availability of data, zero should not be considered as an absence of hazard. Also the hazard index that varies from 0 to 5 (vector file) is attributed on a map showing the whole training area with the corresponding colors according to table 3.9.

Six different EMs (TNT, HMX, RDX, 2,4-DNT, NG and perchlorate) were evaluated for CFB Shilo. PETN was not evaluated because no data on dangerousness were available when the maps were made. Table 3.10 gives maximum values for each EM in a worst-case scenario (high residues deposition, dangerousness and small surface area of deposition).

The six hazard values are added together for a potential total value of 0 to 550, and a global relative hazard index is attributed according to table 3.9. For example, according to tables 3.5–3.9, perchlorate (environmental dangerousness index of 5) (table 3.7) with deposition quantities over 10 kg (frequency index of 5) (table 3.5) in a small range with a surface area equal or less to 50 000 m^2 (surface area index of 5) (table 3.8) gives a hazard value of 125, which is sufficient to give a global hazard index of 5 (table 3.9). If more than one EM is used on a range, then all the hazard values are added to give the total hazard value, which is converted to the hazard index according to table 3.9. Figure 3.13 presents the hazard map for CFB Shilo.

Globally, the hazard is very low to moderate throughout the RTAs with the exception of the hand grenade range and the rocket launcher position that show a very high hazard.

Table 3.7. Dangerousness index for energetic materials.

EM	Toxicity (drinking water) Weight 1/1	Environmental persistence				Sum Weight 4/4	Dangerousness index $\frac{(Toxicity + Persistence)}{2}$	Relative dangerousness
		Mean solubility Weight 1/4	Biodegradation Weight 1/4	Photodegradation Weight 1/4	Adsorption Weight 1/4			
TNT	5	2	2	1	3	2	3	Medium
RDX	5	2	5	3	4	4	4	High
HMX	3	1	4	4	3	3	3	Medium
PETN	1	2	5	5	1	1	1	Very low
2,4-DNT	5	3	3	1	2	2	4	High
NG	2	4	1	2	4	3	2	Low
Perchlorate	5	5	5	5	5	5	5	Very high

Table 3.8. Surface area index.

1	> 6 000 000 (m^2)
2	1 000 001–6 000 000 (m^2)
3	100 001–1 000 000 (m^2)
4	50 001–100 000 (m^2)
5	< or = 50 000 (m^2)

Table 3.9. Global hazard index.

Range	Hazard value	Index
=	0	0
1	25	1
26	50	2
51	75	3
76	100	4
>	100	5

Table 3.10. Potential hazard index per energetic material.

Energetic material	Maximum hazard value
TNT	100
RDX	100
HMX	75
2,4-DNT	100
NG	50
Perchlorate	125
Total	550

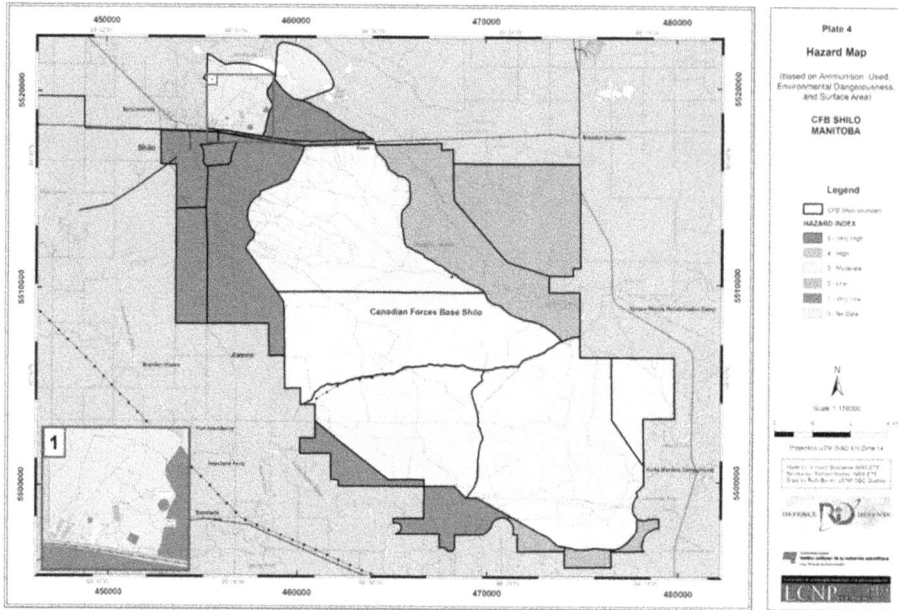

Figure 3.13. Hazard map of CFB Shilo RTAs.

3.2.4 Risk

The risk map is the management tool. It represents a synthesis of both hazard and aquifer vulnerability maps. The global hazard index and vulnerability index values are added and averaged to produce a single value that is then classified from 0 to 5. Figure 3.14 shows the risk map of CFB Shilo.

The risk map of CFB Shilo shows that the groundwater contamination risk is globally very low to moderate throughout the RTAs except for the hand grenade range and rocket launcher position that have a moderate risk. Since those ranges with high hazard index are located in a very low vulnerability area, moving those ranges to other locations will not decrease the risk, so they are considered at the best position.

3.2.5 Vulnerability of potential receptors

The risk map does not give the vulnerability of potential receptors (e.g. wetlands, lakes, rivers, springs and wells (private, livestock, irrigation, municipal and industrial)) that are located inside the RTAs or outside but close to the borders. To protect those receptors, it may be necessary to delineate protection zones in the RTAs or priority zones for prevention and land management (figure 3.15). The identification of those zones can be based on the total time of travel of EM between ranges and the receptors. The protocol that was applied at CFB Shilo was as follows (figure 3.16): (1) locate the potential receptors inside and outside of the RTAs within the hydrogeological watershed; (2) evaluate the advective horizontal traveling time

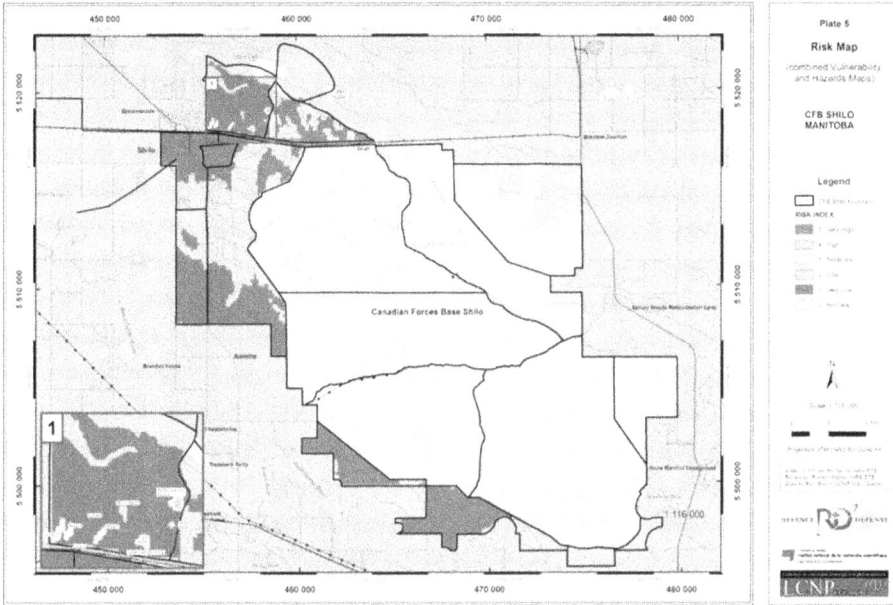

Figure 3.14. Risk map of CFB Shilo RTAs.

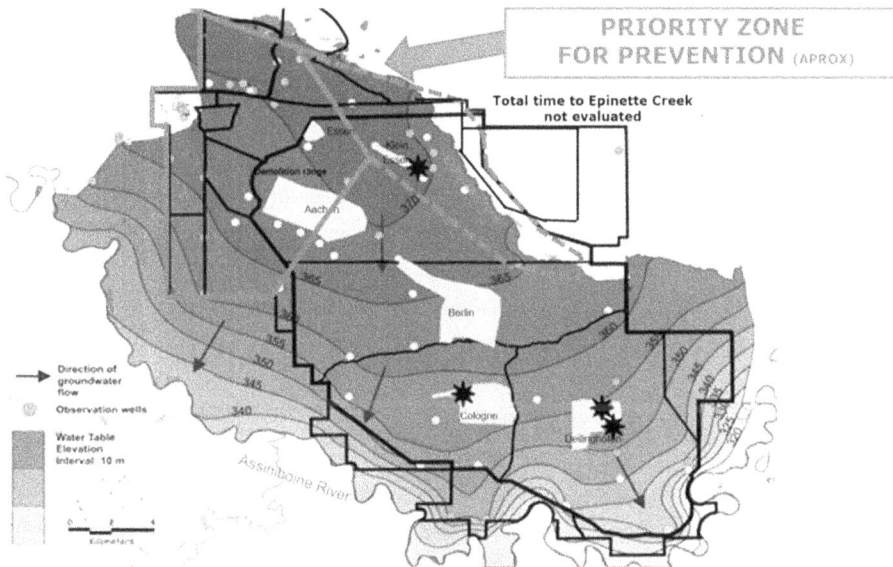

Figure 3.15. Priority zones for receptor protection, prevention and land management at CFB Shilo.

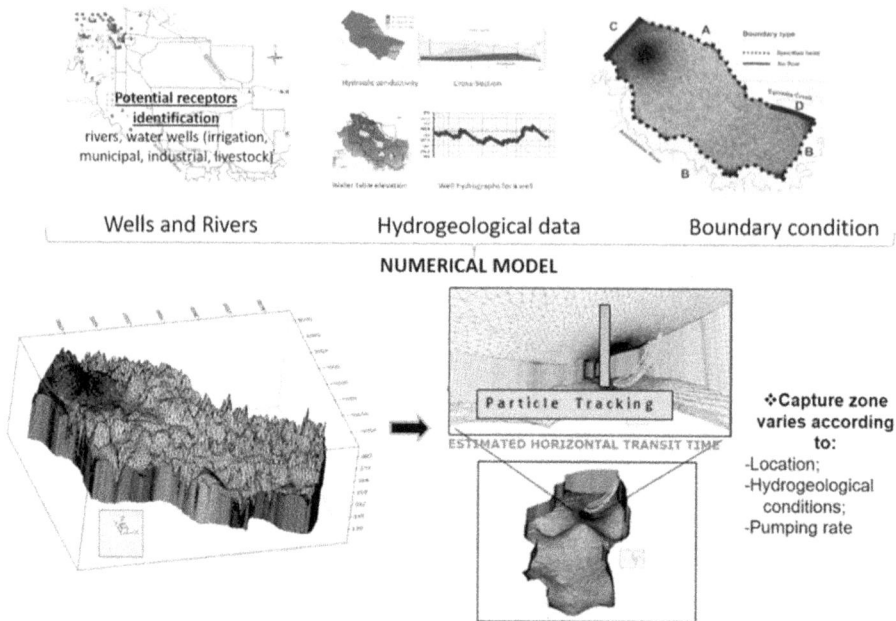

Figure 3.16. Protocol for the delineation of protection zones in the RTAs at CFB Shilo.

of groundwater between the training ranges and the receptors using a groundwater flow numerical model (Feflow v 6.0. advection in steady state, no degradation, no adsorption, no dispersion) with a domain that extends outside of the RTAs where the potential receptors are located; (3) use all hydrogeological information available (hydraulic head, hydraulic conductivity, stratigraphy, pumping rate, recharge) in the numerical model to calculate the groundwater horizontal traveling time; (4) in the groundwater flow model use the advective backward particle tracking isochrones markers to delineate the capture zone of a well or a well field to see if a range may be located inside the capture zones; (5) add the traveling time calculated by the groundwater flow model to the DAT estimated in gOcad to make the total time between a range and the receptor.

3.2.6 Conclusions

The aquifer vulnerability map can serve as a management tool for activity relocation. The hazard estimation is highly dependent on the quality of data such as statistics on munitions used provided by range control personnel and composition of munitions in data banks. The risk map is presently a good method to identify the risk of groundwater contamination from the current use of RTAs. It may be used to avoid immediate danger. Vulnerability of receptors may be used to indicate ranges that need to be managed in priority.

References

[1] Martel R and Gabriel U 2015 Sampling and characterization protocol for groundwater, surface water in the land forces military training areas *Munitions Related Contamination Source Characterization, Fate and Transport*, NATO, Brussels pp 7-1–7-21

[2] Fellenius B H and Eslami A 2000 Soil profile interpreted from CPTu data *Year 2000 Geotechnics, Geotechnical Engineering Conf. (Asian Institute of Technology, Bangkok, Thailand, November 27–30)* p 18

[3] Robertson P K, Campanella R G, Gillespie D and Greig J 1986 Use of Piezometer Cone Data *In-Situ'86 Specialty Conf. (Blacksburg, VA, June 23–25 1986)* Geotechnical Publication (GSP) 6, ASCE, Reston, VA pp 1263–1280

[4] ASTM 2016 *ASTM-D5092-02 Standard Practice for Design and Installation of Ground Water Monitoring Wells in Aquifers* (West Conshocken, PA: ASTM International)

[5] Driscoll F 1986 *Groundwater & Wells* 2nd edn (Amarillo, TX: Johnson Filtration Systems Inc.), p 1089

[6] Nielson D M 1991 *Practical Handbook of Ground-Water Monitoring* (Chelsea, MI: Lewis)

[7] Byrnes M E 1994 *Field Sampling Methods for Remedial Investigations* (Boca Raton, FL: Lewis), p 254

[8] GWHAT (Groundwater Hydrograph Analysis Toolbox) https://gwhat.readthedocs.io/en/gwhat-0.2.5/groundwater_recharge.html

[9] Schroeder P R, Aziz N M, Lloyd C M and Zappi P A 1994 *The Hydrologic Evaluation of Landfill Performance (HELP) Model: User's Guide for Version 3,'EPA/600/R-94/168a* (Washington, DC: U.S. Environmental Protection Agency Office of Research and Development)

[10] CCME 2011 Protocols Manual for water quality sampling in Canada PN 1461 http://www.ccme.ca/publications/list_publications.html#link2

[11] Puls R W and Barcelona M J 1995 Low-flow (minimal drawdown) ground-water sampling procedures, United States Environmental Protection Agency Office of Research and Development Office of Solid Waste and Emergency Response EPA/540/S-95/504, December 1995

[12] Backus S M, Klawuun P, Brown S S, D'sa I, Sharp S, Surette C and Williams D J 2005 Determination of perchlorate in selected surface waters in the Great Lakes Basin by HPLC/MS/MS *Chemosphere* **61** 834–43

[13] Bouwer H and Rice R C 1976 A slug test method for determining hydraulic conductivity of unconfined aquifers with completely or partially penetrating wells *Water Resour. Res.* **12** 423–28

[14] Cooper H H, Bredehoeft J D and Papadopulos S S 1967 Response of a finite-diameter well to an instantaneous charge of water *Water Resour. Res.* **3** 263–69

[15] Hvorslev M 1951 Time Lag and Soil Permeability in Ground-Water Observations, Waterways Exper. Sta. Corps of Engrs, U.S. Army, Vicksburg. http://www.csus.edu/indiv/h/hornert/geol_210_summer_2012/week%203%20reading s/hvorslev%201951.pdf

[16] Ross M, Martel R, Lefebvre R, Parent M and Savard M 2004 Assessing rock aquifer vulnerability using downward advective times from a 3D model of surficial geology: a case study from the St. Lawrence Lowlands *Canada: Geofis. Int.* **43** 591–602

[17] National Research Council 1993 *Groundwater Vulnerability Assessment: Predicting Relative Contamination Potential Under Conditions of Uncertainty* (Washington, DC: National Academy Press), p 204

[18] U.S. Environmental Protection Agency 1993 *A review of methods for assessing aquifer sensitivity and groundwater vulnerability to pesticide contamination* (Washington, DC: U.S. Environmental Protection Agency) (EPA/813/R-93/002) p 147

[19] Civita M, Forti P, Marini P, Micheli L, Piccini L and Pranzini G Carta della Vulnerabilità degli acquiferi carsici delle Alpi Apuane *Atti del 1. Convegno nazionale sulla protezione e gestione delle acque sotterranee : metodologie, tecnologie e obiettivi (Modena, 20-21–22 Settembre 1990)* [S.l.] : [s.n.], v. 2 pp 465–468 [in italian]

[20] Gogu R C and Dassargues A 2000 Current trends and future challenges in groundwater vulnerability assessment using overlay and index methods *Environ. Geol.* **39** 549–59

[21] Vrba J and Zaporozec A 1994 *Guidebook on Mapping Groundwater Vulnerability* (International Contributions to Hydrogeology 16) (Hannover: H Heise)

[22] Frind E O and Molson J W 2002 Well vulnerability mapping: a new approach to wellhead protection and assessing bedrock aquifer vulnerability in St. Lawrence Lowland, Canada Water: theory to practice *3rd Joint IAH-CGS Groundwater Conf.* Session 15: 707–712

[23] Frind E O, Molson J W and Rudolph D L 2006 Well vulnerability: a quantitative approach for source water protection *Ground Water* **44** 732–42

[24] Ross M, Parent M and Lefebvre R 2005 3D geologic framework models for regional hydrogeology and land-use management: a case study from a Quaternary basin of southwestern Quebec, Canada *Hydrol. J.* **13** 690–707

[25] Kalinski R J, Kelly W E, Bogardi I, Ehrman R L and Yaniamoto P D 1994 Correlation between DRASTIC vulnerabilities and incidents of VOC contamination of municipal wells in Nebraska *Ground Water* **32** 31–4

[26] Wösten J H M, Bannink M H, De Gruijter J J and Bouma J 1986 A procedure to identify different groups of hydraulic-conductivity and moisture-retention curves for soil horizons *J. Hydrol.* **86** 133–45

[27] Laden E M 1989 Screening of groundwater contaminants by travel time distributions *J. Environ. Eng.* **115** 497

[28] Cherkauer D S and Ansari S A 2005 Estimating ground water recharge from topography, hydrogeology, and land cover *Ground Water* **43** 102–12

[29] Freeze R A and Cherry J A 1979 *Groundwater* (Englewood Cliffs, NJ: Prentice Hall)

[30] Ampleman G 2003 *Evaluation of the Impacts of Live Training at CFB Shilo (Final Report)* TR 2003-066 Defence R&D Canada-Valcartier 216 Technical Report

[31] Gauthier C 2005 Caractérisation environnementale et hydrogéologique de l'écoulement à la base des forces canadiennes (BFC) Shilo, Manitoba, Canada. Unisersité du Québec—Institut National de Recherche Scientifique: Quebec

[32] De Marsily G 1986 *Quantitative hydrogeology* (Fontainebleau: Paris School of Mines)

[33] Martel R, Bordeleau G, Ait-Ssi L, Ross M, Comeau G, Lewis J, Ampleman G and Thiboutot S 2007 *Groundwater and surface water study for potential contamination by energetic materials, metals and related compounds at the Cold Lake Air Weapons Range (CLAWR)* Final Report Québec, INRS - Eau, Terre & Environnement (Rapport de recherche; 746)

[34] Parent G, Martel R, and Ross M 2007 Land use management of Canadian forces bases using aquifer vulnerability and risk maps *8th Joint CGS/IAH-CNC Groundwater Conf. (Ottawa, Canada, 21–24 October 2007)* p 8

[35] Thiboutot S, Ampleman G, Marois A, Gagnon A and Gilbert D 2010 *Energetic Residues Deposition from 66-mm Antitank Rockets* TR 10-13 chapter 6ERDC/CRREL

[36] Lewis J, Martel R, Thiboutot S, Ampleman G and Trépanier L 2009 Quantifying the transport of energetic materials in unsaturated sediments from cracked unexploded ordnance, Québec, Canada *J. Environ. Qual.* **38** 1–8

[37] Thiboutot S 2017 Scientific Methodology for the Precise Determination of Post-Detonation Explosive's Footprint, Québec, National Defense Canada (DRDC-RDDC-2017-R112)

[38] Diaz E and Thiboutot S 2016 Deposition rates from blow-in-place of different donor charges: comparison of composition C-4 and shaped charges *Propell. Explos.Pyrotech.* **42** 90–7

[39] Thiboutot S, Brousseau P and Ampleman G 2015 Deposition of PETN following the detonation of Seismoplast plastic explosive *Propell. Explos. Pyrotech.* **40** 329–32

[40] Walsh M R, Walsh M E, Ramsey C A, Zufelt J, Thiboutot S, Ampleman G and Zunino L 2014 Energetic residues from the detonation of IMX-104 insensitive munitions *Propell. Explos. Pyrotech.* **39** 243–50

[41] Walsh M R, Walsh M E, Ramsey C A, Brochu S, Thiboutot S and Ampleman G 2013 Perchlorate contamination from the detonation of insensitive high-explosive rounds *J. Hazardous Mat.* **262** 228–33

[42] Walsh M R, Walsh M A, Taylor S, Ramsey C A, Ringelberg D B, Thiboutot S, Ampleman G and Diaz E 2013 Characterization of PAX-21 insensitive munition detonation residues *Propell. Explos. Pyrotech.* **38** 399–409

[43] Walsh M R, Walsh M E, Ampleman G, Thiboutot S and Walker D 2006 Comparison of explosive residues from the blow-in-place detonation of 155-mm high-explosive projectiles, Hanover, NH, US Army Corps of Engineers, Engineer Research and Development Center, Cold Regions Research and Engineering Laboratory ERDC/CRREL TR-06-13, June, pp 1–40

[44] Ampleman G, Thiboutot S, Marois A, Gagnon A, Gilbert D, Walsh M R, Walsh M E and Woods P 2010 *Study of the Propellant Residues Emitted During the Live Firing of 105-mm Leopard Tank Squash-Head Practice Rounds at CFB Valcartier, Canada* TR 10-13 Chapter 4 ERDC

[45] Thiboutot S, Ampleman G, Gagnon A, Marois A, Martel R and Bordeleau G 2010 Persistence and Fate of Nitroglycerin in Legacy Antitank Range, Québec, National Defense Canada. (DRDC TR-10-059)

[46] Martel R, Lange S, Coté S, Ampleman G and Thiboutot S 2010 *Fate and Behaviour of Energetic Material Residues in the Unsaturated Zone: Sand Columns and Dissolution Tests* R-1161 INRS

[47] Bordeleau G, Martel R, Ampleman G, Thiboutot S and Poulin I 2012 The fate and transport of nitroglycerin in soils and groundwater at active and legacy anti-tank firing positions *J. Contam. Hydrol.* **142–43** 11–21

[48] Walsh M R, Walsh M E, Thiboutot S, Ampleman G and Bryant J 2010 *Propellant Residues Deposition from Firing of AT4 Rockets* TR-09-13 ERDC/CRREL

[49] Thiboutot S, Ampleman G, Marois A, Gagnon A and Gilbert D 2009 *Nitroglycerine Deposition from M-72 Antitank Rocket Firing* TR-2009-003 (Unclassified) DRDC (February 2009)

[50] Thiboutot S, Ampleman G, Lapointe M C, Brochu S, Brassard M, Stowe R, Farinaccio R, Gagnon A, Marois A and Gamache T 2008 Study of the Dispersion of Ammonium Perchlorate Following the Static Firing of MK-58 Rocket Motors, Québec, National Defense Canada (DRDC TR 2008-240)

[51] Thiboutot S, Ampleman G, Marois A, Gagnon A, Gilbert D, Tanguay V and Poulin I 2008 *Energetic Residues Deposition from 84-mm Carl Gustav Antitank Live Firing (in report Characterization and Fate of Gun and Rocket Propellant Residues on Testing and Training Ranges)* TR-08-1 chapter 4 US Army Engineer Research and Development Center Cold Regions Research and Engineering Laboratory Technical Report ERDC (January 2008)

[52] Ampleman G, Thiboutot S, Marois A, Gagnon A and Gilbert D 2008 *Study of Propellant Residues Emitted During 105-mm Leopard Tank Live Firing and Sampling of Demolition Ranges at CFB Gagetown (in report Characterization and Fate of Gun and Rocket Propellant Residues on Testing and Training Ranges)* US Army Engineer Research and Development Center TR-08-1 Cold Regions Research and Engineering Laboratory Technical Cold Regions Research and Engineering Laboratory Technical Report ERDC (January 2008)

[53] Jenkins T F 2008 *Characterization and Fate of Gun and Rocket Propellant Residues on Testing and Training Ranges* TR-08-01 US Army Engineer Research and Development Center Cold Regions Research and Engineering Laboratory Technical Report ERDC TR-08-1 (January 2008)

[54] Ampleman G, Thiboutot S, Marois A, Gagnon A and Gilbert D 2008 Evaluation of the Propellant Residues Emitted During 105-mm Leopard Tank Live Firing and Sampling of Demolition Ranges at CFB Gagetown, Canada, Québec, National Defense Canada (DRDC TR 2007-515)

[55] Thiboutot S, Ampleman G, Marois A, Gagnon A, Gilbert D, Tanguay V and Poulin I 2007 Deposition of Gun Propellant Residues from 84-mm Carl Gustav Rocket Firing, Québec, National Defense Canada (DRDC TR 2007-408)

[56] Robidoux P Y, Lachance B, Didillon L, Dion F O and Sunahara G I 2006 Development of ecological and human health preliminary soil quality guidelines for energetic materials to ensure training sustainability of Canadian Forces. Final Report (revised) (Report no. 45936) Ottawa, National Research Council

IOP Publishing

Global Approaches to Environmental Management on Military Training Ranges

Tracey J Temple and Melissa K Ladyman

Chapter 4

Analysis of explosives in the environment

N Mai, J-F Pons, D McAteer and P P Gill

Detection and analysis are important at all stages in the life-cycle of an explosive, from manufacture through storage to deployment; the actual technique employed for a particular job being dependent upon the type of information required. Detection techniques are used in a variety of situations (e.g. forensic laboratories, public buildings, manufacturing plants) to establish the presence of energetic materials. No single technique is suitable for all situations or, indeed, is capable of detecting every type of explosive. A range therefore exists that encompasses techniques differing in terms of cost, scale, sensitivity, specificity and ease of use. Analysis techniques for explosives are capable of providing detailed qualitative and quantitative information. Thus, analysis can establish both the identity and the concentration of a component. However, in order to yield this sort of information, analysis techniques invariably require a certain amount of sample preparation. Interference from the water or soil matrix will generally require the additional steps of sample extraction and chromatography to separate the explosive of interest from the background matrices. This chapter will discuss the main techniques used for the analysis of explosives in soils and water; focusing in the areas of sample preparation, chromatography and detection. Sample preparation will focus upon solvent extraction methodologies such as sonication, Soxhlet extraction and accelerated solvent extraction. In addition, techniques such as solid phase extraction and solid phase micro-extraction will also be discussed. Thin layer chromatography, high performance liquid chromatography and gas chromatography methods from the literature will be evaluated for explosive analysis, together with detection techniques such as infra-red, Raman and mass spectrometry. The number of techniques available for the detection of explosives is vast and this chapter does not seek to give a comprehensive description of each of them. What it does attempt to do is provide a broad introduction to the major techniques, their applications and advantages, at a level which does not assume a detailed technical knowledge.

doi:10.1088/978-0-7503-1605-7ch4

4.1 Sample preparation

The sample preparation and extraction is an important step in assessing the degree of contamination of soil or water. It is important to fully extract the potential contaminant prior to its analysis and quantification. Several extraction methodologies are available, ranging from a simple solvent stirring; solvent shaking or sonication assisted solvent extraction, to more complex techniques such as Soxhlet extraction (SE) or accelerated solvent extraction (ASE).

These solvent extraction techniques rely on the higher solubility or affinity of the contaminant in the extraction solvent than in the soil or in water (table 4.1) [1, 2]. The most prevalent methods are described below with a discussion on their relative uses and limitations. With respect to their toxicity and extraction power, acetone, acetonitrile and methanol are the solvents of choice for the extraction of explosives from soil samples.

Shaking and mixing soil samples with a solvent followed by a filtration, is the simplest method for solid–liquid extraction. These techniques can be viewed as a cost effective compromise for the extraction as described for TNT and its degradation products [24]. To achieve a high extraction efficiency or good recovery, several extractions are often required and large amounts of solvents may be necessary.

Microwave assisted [25] and sonicated assisted [26, 27] extractions (MAE and SAE, respectively) can be used in combination with stirring or shaking. The solubility of the solutes increases when using microwaves by heating solvents with dielectric constant. Sonication on the other hand, helps break down small particles allowing a greater extraction through higher contact surface area. However, neither of these techniques has shown a real advantage compared to ASE or SE, which are more efficient methods.

SE uses a percolator (boiler and reflux) to continuously circulate fresh solvent through the sample. The solvent vapour travels up a distillation arm and condenses above the sample. The solvent drips back down into the sample chamber. When the chamber is almost full it empties back into the distillation flask using a syphon. This technique is frequently viewed as the most effective. However, extraction times are long (often over 24 h) and can use large amounts of solvents.

Developed in the mid-1990s, ASE is one of the most recent extraction techniques currently employed [28, 29]. A solid sample is extracted under pressure and heat [30]. Under such conditions, solvents can be used above their boiling points, leading to a higher efficiency of extraction. Typical extraction only uses a fraction of the solvent required by standard methods, and can be conducted within 30 min. Although it is possible to achieve high temperatures, for extractions of energetic materials, it is recommended to work below 100 °C. In addition, current equipment allows for multiple samples to be analysed automatically. ASE extraction of a range of explosives and their degradation products shows comparable results to SE [29, 31] and often superior results when compared to supercritical fluid extraction (SFE) and MAE [25, 30].

Table 4.1. Solubility of explosives in common solvents (g l⁻¹) at room temperature and atmospheric pressure, unless stated. * Neutral pH, unless stated.

Type	Solvent explosive	Water (pH)*	Acetone	Methanol	Acetonitrile	Ethanol	Ethyl acetate	Toluene	Chloroform	DMSO	Other	References
Nitroaromatics	DNAN	0.276										[3, 4]
	HNS 30 °C	0.014	0.64	0.03	1.18					13		[5, 6]
	LLM-105											[7]
	NTO	13 (1) 13–16.6 (6) 92 (14)	16.8						<0.2			[8–10]
	Picric acid	11.7	1580	240		84	466	150				[11]
	TATB	insoluble			0.03					0.06	76 (sulphuric acid)	[12, 13]
	Tetryl	0.075	948			6.9	443	98	2.6			[11]
	TNT	0.11–0.15	1378			15.3	809	635	127			[11, 14]
Nitramines	CL20	<0.05	748–1196			7.72	45				0.03 (DCM)	[15, 16]
	HMX,	0.005	35		25					522		[5, 12, 17]
	NQ	3.8–4.4										[11]
	RDX	0.05–0.060	90–110	2.91	55	1.29	32.15	0.23	0.07	375	392 (DMF)	[4, 5, 11, 18]
Nitrate esters	NG	1.95	Fully miscible	55.5		430–540	Fully miscible		Fully miscible			[19–21]
	PETN	0.021	314	5.82		2.61	156	4.97	2.83		8 (DCM)	[11, 22, 23]

SFE [25] is mostly used with supercritical CO_2, sometimes modified with methanol or ethanol co-solvents. Supercritical CO_2 (above 75 bar) passes through the sample and dissolves the analytes [32]. The cost of the equipment and the requirement for a highly pressurised system means that this technique is only infrequently used.

Although it is acknowledged that SE is the more universal method, its slow process and the high requirement of solvent makes it inconvenient for rapid response and field work. The use of ASE shows speed and necessitates low amounts of solvent, making this technique a real asset; however, its high equipment costs may hinder its deployment for field analysis. In contrast, the more traditional shaking or stirring assisted by microwave or sonication extraction methods are simple to use and may be sufficiently efficient for a rapid estimation of contamination (table 4.2).

Heavily contaminated water or soil samples can be difficult to directly analyse. A direct analysis would lead to a complex readout. It is possible to purify the samples using solid phase extraction (SPE) cartridges. These catch and release cartridges are specific for the analyte of interest and follow a simple procedure: conditioning of the column followed by the sample loading, washing and finally releasing the analyte. The purified sample is then analysed by any of the techniques described in the section below. This method was shown to be versatile with nitramines, nitrate esters and nitroaromatics in soils and wastewaters, using different SPE cartridges [33, 34]; with detection concentrations as low as sub microgram per litre with polymeric divinylbenzene-based sorbents reported.

Solid phase micro-extraction (SPME) fibres have been developed as a much simpler method. Direct sorption of the analyte from the sample (solid, soil or gas) is desorbed directly upstream of a gas or liquid chromatograph [35–37]. SPME has been successfully applied to contaminated water samples, allowing the detection of traces of nitramines and nitroaromatics explosives as low as 10 ppb, comparable to the SPE results (~1 ppb) [38]. Different types of fibres are available with diverse polarities, such as the nonpolar polydimethylsiloxane and the polar polyacrylate or polyethyleneglycol, and will therefore have a range of different affinities with different explosives.

Table 4.2. Extraction method summary [24–32].

	Shaking/ Stirring	MAE/SAE	SE	ASE	CO_2 SFE
Extraction time	Up to 24 h	Up to 1 h	24–48 h	Up to 30 min	Up to 1 h
Sample size	1–50 g	Up to 10 g	10–50 g	Up to 25 g	>1 g
Equipment cost	Low	Low	Low	High	High
Advantages	Easy to use	Easy to use	High recovery	High recovery and fast	Low solvent usage
Disadvantages	Large amounts of solvents	Limited added value	Very long extraction time	Expensive equipment	Expensive equipment

4.2 Detection and chemical analysis of explosives

The detection of explosives in the environment is an issue since most of the explosives and their metabolites are toxic. To minimise consequence to humans it is important to analyse soil or groundwater where there is a potential risk of contamination, e.g. military firing ranges. Many new techniques, available to sensitively detect explosives even at trace level, are emerging in the literature but this chapter will focus upon the most common applications.

4.2.1 Spectroscopic and spectrometric techniques

Fourier transform infra-red and Raman spectroscopy

Infra-red (IR) spectroscopy is one of the most widely used spectroscopic techniques in energetic materials analysis. When a molecule absorbs energy in the IR region of the electromagnetic spectrum, excitation of its vibrational energy levels occurs. Particular groups of atoms can therefore be identified from their characteristic absorption frequencies. For example, the IR spectrum of TNT contains various absorption bands at wavelengths (or frequencies) characteristic of the nitro ($-NO_2$) group. Similar bands are found in the IR spectra of all molecules containing this group but the precise position depends upon the molecular environment of the $-NO_2$ group. The overall IR spectrum of a molecule provides a 'fingerprint' of that molecule.

Fourier transform IR and Raman spectroscopy are complementary analytical techniques based on vibrational spectrometry. Both techniques allow the identification of compounds through the absorption or the scattering of energy, which induces the vibration of covalent bonds. In particular, the modes of excitation that are studied in IR spectroscopy are the symmetric and asymmetric stretching, bending and twisting vibrations of covalent bonds leading to a change in the dipole moment. In Raman spectroscopy, the vibrations leading to a change in polarizability are detected.

The common functionalities observed in explosive compounds such as aliphatic and aromatic nitro, azido and C–H groups are all visible by IR and Raman (table 4.3) [39]. As a result, these techniques have been widely published in the characterisation of energetic compounds.

Successful detections of ppb levels of explosives using Raman and IR spectroscopy have been reported [40–43].

Ion mobility spectrometry

Ion mobility spectrometry (IMS) is a sensitive, fast and potentially low cost technique for the detection and measurement of compounds. IMS operates by ionising molecules and measuring the time taken to travel along an electric field counter to an inert gas stream (drift tube). The time taken to travel through this drift tube is dependent on the charge, mass and cross-sectional area of the ion.

This technique is coupled to an electron capture detector. In this device electrons are produced from a radioactive β emitter (usually nickel-63), which forms the

Table 4.3. Frequencies of the main explosophore groups [39].

Functional group		Frequency (cm^{-1})	Intensity[a]	
			IR	Raman
–NO$_2$	Nitro aliphatic	1555–1530	s	m-w
		1385–1360	vs	s
	Nitroaromatic	1580–1485	s	m-w
		1370–1315	s	s
–O–NO$_2$	Nitroester	1660–1615	s	v
		1300–1250	s	s
–N–NO$_2$	Nitramine	1630–1530	s	v
		1315–1260	s	v
–N=O	Nitroso aliphatic	1625–1540	s	v
–N=O	Nitroso aromatic	1525–1485	s	s
N–N=O	Nitrosamine	1460–1435	s	s
		1150–1025	s	s-m
–N$^+$–O$^-$	N-oxide	1380–1150	m-s	m
–N=N	Diazonium	2300–2130	m-s	m-s
–N$_3$	Azido	2170–2080	vs-s	m-s

[a] vs: very strong; s: strong; m: medium; w: weak; vw: very weak; v: variable.

cathode of an ionisation chamber. Within the chamber electrons collide with the inert carrier gas molecules and thereby lose much of their energy, becoming in the process 'thermal' electrons. These thermal electrons then migrate to the anode under a fixed voltage, thus producing a constant (baseline) current between the electrodes. When an electron-capturing gas or vapour is introduced into the ionisation chamber it causes a decrease in the baseline current due to the interaction of the thermal electrons with the electron-capturing molecules. The extent of the decrease in current is an indication of the concentration of the electron-capturing molecules present. An electron capture detector is very sensitive towards compounds with a high affinity for electrons. These include compounds containing highly electronegative groups such as –F and –NO$_2$. An IMS operates like a separation technique such as gas or liquid chromatography but with reduced operational requirement as no vacuum or sample preparation is required.

Mass spectrometry
Mass spectrometry (MS) is a powerful technique for qualitative and quantitative analysis of a wide range of molecules. It involves the introduction of a sample in the ionisation source where it is converted into ions that are accelerated and separated in the mass analyser according to their mass-to-charge ratio. The ions are then detected and a mass spectrum is produced. The data generated determines the molecular weight and/or structural information allowing identification and quantification of the molecules analysed.

Sampling system: The introduction of the sample in the mass spectrometer depends on the type of samples and the method of ionisation used. Samples can be introduced directly into the instrument using a solid probe or an inlet for gases, liquids or solids. Introduction of the sample can also be done with an on-line chromatography column, e.g. gas chromatography (GC) or high performance liquid chromatography (HPLC) or another type of separation column, e.g. capillary electrophoresis (CE). The separation of the compounds prior to their analysis in MS reduces the complex matrix interference allowing a lower limit of detection and an improved identification.

Main ionisation methods used in explosives detection: For MS analysis the sample molecules must be ionised. This step is potentially one of the most important steps in the mass spectrometer system. Many ionisation methods are available depending on the polarity, stability and size of the compounds analysed.

Electron ionisation: Electron ionisation (EI) is normally used to ionise and detect molecules in a gas phase. They are therefore nonpolar, low molecular weight, volatile and thermally labile organic compounds. Once in the gas phase, the analytes enter the EI source. A beam of high-energy electrons charges the molecule under vacuum by removing electrons to form a positively charged molecular ion, which is detected. Since excess energy is transferred to the system, the ions undergo a further degree of fragmentation depending on the stability of the charged molecule. The fragmentation into smaller ions is also monitored. This ionisation technique is called 'hard' ionisation because of the large number of fragments generated. Fragmentation provides structural information of the parent compound with the molecular ion not always observed in the spectrum.

The EI-MS technique is very often hyphenated with GC as the samples are already in gas phase when they elute from the gas column. Detection of explosives using GC-EI-MS has been extensively investigated. Trace nitroaromatics, nitramines and nitrate esters have been detected in water at concentrations of 5–100 ppb [44]. One of the drawbacks of using an EI source with a low level of nitrate esters is that due to their extensive fragmentation to nitronium and nitrate ions, the spectra produced are not unique, complicating their identification, as no structural information is available.

Chemical ionisation: Chemical ionisation (CI) is also used for the detection of nonpolar low molecular weight, volatile and thermally labile organic compounds as they require vaporisation. A reagent gas (methane or ammonia) is added to the ion chamber and is ionised by a beam of electrons. In positive CI mode (PCI), the ionised reagent gas reacts with more gas molecules to form a carbocation, which in turn protonates the analyte molecules under vacuum. The protonated analyte molecule doesn't retain enough residual energy to fragment further. In negative CI mode (NCI), the negative ions are formed by the capture of thermal low-energy electron by the analyte molecules. NCI is selective and sensitive for electron-capturing molecules like nitro-containing explosive compounds, which reduces the problem of matrix interferences. This ionisation method is called 'soft' ionisation as it greatly reduces fragmentation and tends to preserve the protonated molecular ion, which identifies the molar mass of the analyte.

While PCI-MS is mainly used for aromatic explosives and their derivatives, NCI-MS is more commonly used for the analysis of common high explosives. Analysis in NCI-MS of molecules with high electron affinities such as nitroaromatics will result in the detection of deprotonated molecular ions and a peak with a loss of OH [45], while molecules with low electron affinity such as nitramines or nitrate esters will partially decompose as they cannot stabilise the thermal electron [46]. CI can complement EI by providing more information on the structure of the explosives and when used in negative mode offers a better selectivity.

Atmospheric pressure ionisation: Both ionisation sources EI and CI are maintained at high vacuum to facilitate the formation of ions from the analytes in the gas phase. New atmospheric pressure ionisation (API) sources have been developed to broaden the range of molecules previously analysed in MS and include non-volatile, polar and thermally stable compounds. In the last decade, many ionisation sources operating at atmospheric pressure have been developed. The two predominant ones are *electrospray ionisation* (ESI) and *atmospheric pressure chemical ionisation* (APCI); both are well suited for flow-injection and LC/MS techniques. ESI and APCI can be complementary methods

In the ESI source, the solvated analyte passes through a highly charged electrospray needle to form charged aerosol droplets. The solvent is evaporated, leaving only positive or negative ions ready to enter the mass analyser. This ionisation technique is a soft ionisation method producing mainly the protonated or deprotonated molecular ion with fewer fragmentations. Stable adducts can be observed in the presence of additives or contaminants and assist in the identification of the sample.

ESI-MS is a very sensitive ionisation method for the detection of high explosives [47–49]. When used in negative mode, the sensitivity is low for aromatics but adduct signals of nitramines and nitrate esters intensify in the presence of additives like ammonium salts. When ammonium nitrate is used as a mobile phase, a peak corresponding to the molecular ion adduct with NO_3^- is observed in the negative mode [48] and when the mobile phase contained ammonium formate, the nitramines are detected as the molecular ion adduct with HCO_2^- [50]. The relative intensity and stability of various adducts of neutral high explosives like EGDN, NG, TNT, PETN, RDX and HMX with chloride, formate, acetate and nitrate were assessed and compared to assist with the identification of explosives tested [51].

The APCI is comparable to CI although the ionisation occurs at atmospheric pressure and the reactant gas is replaced by the solvent. The solvated sample is first nebulised to form a spray, which is then rapidly vaporised before being ionised when subjected to a corona discharge. As opposed to ESI, APCI can be used with a low polarity solvent with high flow rate (>1 ml min^{-1}) to analyse less polar and lower molecular mass molecules. APCI is a soft ionisation method resulting predominantly in the formation of a deprotonated molecular ion in negative mode with very little fragmentation. APCI is used for the detection of nonpolar or slightly polar species. It is compatible with the solvents used in normal phase HPLC (see section 4.2.2).

Thermally labile samples could decompose in the heated nebuliser of the source and should be avoided.

APCI-MS is a simple and powerful detection tool to detect semi-volatile and slightly polar high explosives. Various experimental conditions in combination with HPLC have been developed and validated for explosives and their degradation products [47, 52, 53]. The analysis of explosives is preferable in negative mode because of the presence of electron-withdrawing nitro-groups [54].

Atmospheric pressure photoionisation (APPI) is a new API technique that has been reported for low to moderately polar compounds. The molecular ion is formed by photoionisation (absorption of a high-energy photon by the molecule). Explosives with high electron affinity, like nitroaromatics, form the molecular ion via electron capture while low electron affinity explosives in the presence of a halogenated solvent additive produce the molecular ion with halide anions. Explosives in water have been analysed using negative ion APPI [55] though no signal was observed for the nitramines RDX and HMX when dissolved in pure toluene. Adduct ions were obtained when 1% (v/v) methylene chloride was added to the solvent.

Ambient ionisation: Ambient ionisation processes have been developed to reduce the sample pre-treatment before MS analysis and to avoid using a separation technique resulting in a decrease of analysis time without compromise on the sensitivity. A variety of surfaces or matrices are exposed to an ionising gas or aerosol under ambient conditions. These simple, rapid and low cost techniques have recently emerged with more than 30 methods published in the last ten years [56]. One such technique is desorption electrospray ionisation (DESI), which generates ions by directing a charged solvent spray onto the contaminated non-conducting surface. Another established technique is direct analysis in real time (DART), which uses an ionised gas rather than a charged solvent spray to produce the ions. Other quite promising ambient MS methods have been described for explosive detection but they are not discussed in this chapter.

Due to its versatility, sensitivity and rapid response time, the application of DESI-MS for explosive detection is growing rapidly. It has been applied for direct analysis of nitrated explosive traces with minimal sample preparation. Fast analysis of TNT, RDX and HMX in groundwater have been performed with the DESI-MS method using methanol/water/hydrochloric acid as the spray solvent [57]. The molecular radical anion was detected for TNT while RDX and HMX generated chloride adduct species with a limit of detection (LoD) of 1 μg l^{-1} for TNT and 10 μg l^{-1} for RDX. The type and polarity of explosives studied with the nature of the surface investigated will have an influence on the efficiency of the technique. On average, the mass detection limit of explosives is in the picogram range.

DART is a fast non-destructive technique which has been successfully applied to the direct analysis of energetic material on surfaces, avoiding any sample loss. Fifteen common explosives were deposited on five solid surfaces with various physical properties and tested with DART [58]. All the MS spectra, 75 combinations in total, were a perfect match of the explosives tested. Analyses of traces (nanogram levels) of inorganic salts [59] were performed with a DART-MS coupled with a resistive Joule heating thermal desorption (JHTD). A JHTD-DART-MS method in negative mode was sucessfully applied to desorb low vapour pressure explosives like PETN and non-volatile salts like nitrite, nitrate, chlorate and perchlorate without excess degradation, decomposition and fragmentation.

Main mass analysers and detectors used in explosives detection: After ionisation, the ions are separated in a mass analyser according to their mass-to-charge ratio before being detected. The main mass analysers used for explosives analysis are quadrupole (Q), ion-trap (IT) and time-of-flight (ToF). The selection of mass analysers depends on the applications and the performance required.

In a classical sector mass spectrometer, the ions are accelerated and deflected by a magnetic field with high resolution and sensitivity. This analyser with excellent quantitative performance can be used for accurate mass measurement but it is not well suited for a pulsed ionisation method like matrix assisted laser desorption/ionisation (MALDI). The Q mass spectrometer has an array of four interposed rods with combined direct current and radiofrequency (RF) potentials controlling the ions' path. Only selected mass-to-charge ratio ions will have a stable trajectory towards the detector. This low cost analyser is relatively small and can be miniaturised but it has a limited resolution. An IT mass analyser holds the ions in a sphere using a superimposed RF, which is adjusted to allow selective release of ions from the sphere according to their mass. It's a powerful instrument with a very high mass resolution and high sensitivity but trapping the ion for too long could result in changes in the mass spectrum. ToF mass spectrometers use analysers that measure the time it takes for ions of the same initial energy to travel down a field-free tube to the detectors. Lighter ions travel faster. It is the fastest of the four analysers described and has the highest practical mass range of all the MS analysers.

Explosives have been identified and quantified with Q [60, 61], IT [52] and ToF [62] mass spectrometers.

Once separated, the ions need to be detected by being converted into a measurable electrical signal. The most common means of ion detection is by an electron multiplier tube that consists of a series of dynodes amplifying the current generated by the ions when they strike the surface of each dynode to produce more electrons. It is generally applied with Q, IT and sector analysers. Microchannel plate detectors, which contain small parallel electron multipliers array, are used with ToF.

Tandem MS: With the development of API and ambient ionisation techniques, MS has proven to be a very versatile analytical technique allowing the analysis of an enormous variety of compounds with different chemical characteristics in different matrices. These soft ionisations methods generally generate spectra with a molecular mass and only a few fragments that don't always allow structure elucidation, especially when analysing a mixture. To overcome this drawback and to achieve better selectivity and sensitivity for quantitative analysis, tandem MS (also known as MS/MS) has been developed. It involves multiple steps where firstly the selected parent ions obtained by ionisation undergo a chemical reaction that results in their dissociation. The second stage involves the mass analysis of the product ions generated. There are two types of MS/MS instrument: *tandem-in-space* or *tandem-in-time*. Tandem-in-space instruments use two or more separated analysers connected in series. The successive analysers, usually trapping analysers, could be identical, e.g. triple quadrupole (QqQ) or hybrid, e.g. Q/ToF, ToF/ToF. In tandem-in-time instruments, the sequential steps are performed in the same device, usually beam-type analysers, e.g. IT, orbitrap.

Many papers have summarised the potentialities of MS/MS to identify common military-grade organic nitrated explosives. Various ionisation sources have been applied in positive or negative mode. Trace level of explosives such as TNT and its mono- and diamino metabolites, RDX, HMX, NG, tetryl and PETN from water samples were detected by direct injection in a LC/ESI (PI or NI)-QqQ [49, 63], while LC-NI-APCI-QqQ enables the detection of nitroaromatic explosives in water samples [64]. A GC/NCI-Q-IT analysis was performed with nine explosive compounds including EGDN, 4-NT, NG, 2,6-DNT, 2,4-DNT, TNT, PETN, RDX, Tetryl solubilised in acetonitrile [65]. The analyses of TNT and its degradation products in environmental samples were carried out using a GC-PI-EI-QqQ [24].

Conclusion: MS is a sensitive and selective technique which has been widely used in the detection of explosives, especially when hyphenated with a separation technique. The choice of MS will depend on a number of parameters such as resolution, sensitivity, etc, but the sample matrix or fragmentation pattern will also need to be considered.

4.2.2 Chromatographic techniques

Most of the explosives found in the environment are usually present in complex sample matrices, which makes their detection more challenging. Chromatography is normally used to separate the components from a mixture before being detected with spectroscopic/spectrometric techniques.

Chromatography, a qualitative or quantitative analytical technique, is widely used for the separation of organic explosives mixtures. Chromatography covers a large range of different techniques but they all follow the same general principle, which is based on the differential partitioning of the components between a mobile and a stationary phase, because of differences in their solubility, polarity, vapour pressure, or molecular size (hydrodynamic volume). The success of the separation process relies on the type and degree of interactions between the analytes and the two phases. The four main mechanisms of separation are partitioning, adsorption, size exclusion and ion exchange.

The classification of chromatographic techniques depends on the type of mobile phase and stationary phase used, as summarised in figures 4.1 and 4.2.

Gas chromatography
GC is a very sensitive and selective technique that separates gases and volatile organic mixtures. This versatile technique is applicable to analyse a widespread of gaseous, liquid or solid samples. The sample injection can be liquid or as a vapour

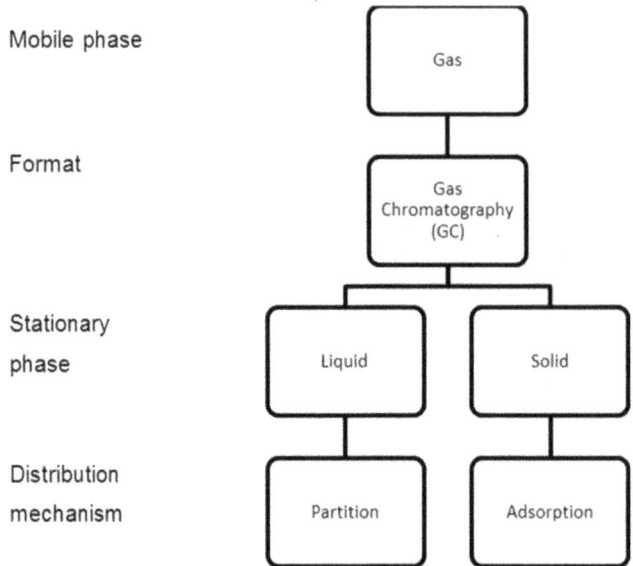

Figure 4.1. Principal chromatography techniques using a gas as a mobile phase.

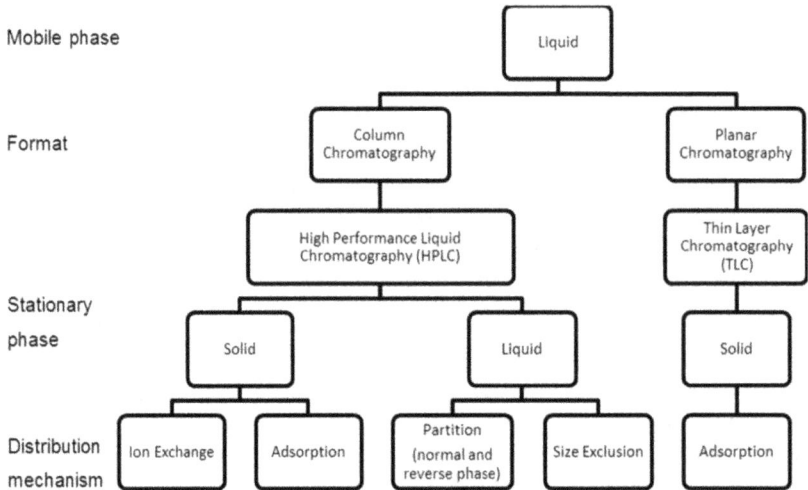

Figure 4.2. Principal chromatography techniques using a liquid mobile phase.

injection, e.g. headspace or SPME. The sample is first volatilised in the inlet and transported by an inert carrier gas through a heated column, either packed or hollow capillary column, which separates the sample into its constituent components. The analytes elute from the column into the detector where a signal is sent to the data system, which generates a chromatogram. A solid or liquid stationary phase is used in the column.

The separation of the compounds is based both on their volatility and their different distribution between the stationary/mobile phases and is influenced by the flow rate of the carrier gas, the column characteristics (stationary phase chemistry, column length, internal diameter, temperature limits) and the column temperature.

For GC analysis the organic compounds need to vaporise and need to be stable when exposed to high temperatures. This is not applicable to many explosives. The thermal breakdown of the energetic materials occurring in the hot inlet will result in poor chromatography and sensitivity. However, explosives and in particular thermal labile nitrate esters have been routinely analysed by GC over the years. On-column injectors or deactivated inlet liners [66] minimise the contact between analytes and metal components in the injection port. Together with the use of short capillary columns with high carrier gas flow rate, or the use of temperature-programmed injection have made the analyses of thermally labile explosives possible [44, 67–69].

The resolution of the separation can be improved by selecting a stationary phase matching the polarity of the analytes, with a preference for the least-polar phase available to avoid column bleed observed with more polar phases. Nonpolar fused silica capillary columns are typically used for the explosives analysis. The columns are coated with a thin liquid film of diphenyl dimethylpolysiloxane derivatives supported on an inert solid [66]. Other stationary phases are also available [67, 70, 71]. The short column (5–15 m) can be used with a narrow-bore (< 0.25 mm) and thin stationary phase (< 0.25 µm) [72] or with a wide-bore (0.53 mm) and thick stationary phase (1.25 µm) [68]. The injector is often maintained below 180 °C. Components of new insensitive munitions like NTO or NQ cannot be analysed by GC due to their low vapour pressure and thermal lability [73, 74]. Using the same GC conditions for other thermally labile or non-volatile explosives such as HMX, resulted in a broad peak for NQ and no peak for NTO [74].

Following the separation in the GC column, another key factor is the detection of the explosive components. The range of detectors available in GC is wider than in other separation techniques because of the low interference or background level of the carrier gas used. Some detectors are universal and will respond to any compounds eluting from the column while some are selective and will respond only to specific molecular, elemental or physical properties. Universal GC detectors provide only retention time data that depends on the specific chemical properties of the analytes. The compounds are identified by comparison of their retention times with the retention times of standards injected in the same conditions. For a definitive identification, an additional injection of the same solution using another column with a different stationary phase is required, which is time consuming. Further confirmation or unambiguous identification can be accomplished by using an identification method, e.g. GCMS, which records the retention time and also the mass spectrum of each component. Table 4.4 summarises the main detectors and their characteristics for the analysis of explosives.

Table 4.4. Specifications of common GC detectors for explosives analysis.

Detector name	Selectivity	Sensitivity	Application to explosives
Flame ionisation detector	Compounds with C–H bonds universal	Poor sensitivity for nitramines and nitrate esters. Better results for nitroaromatic compounds	Limited applications for detection of organic explosives [21]. More suitable for explosives containing hydrocarbons/oils
Nitrogen-phosphorus detector	Nitrogen and phosphorous containing compounds	Low picogram range	Used for nitramine and nitroaromatic explosives [75] but peak might be overlapped by other nitrogen containing contaminants. Insensitive to nitrate esters
Electron capture detector	Selective for electronegative group	Low nanogram range for nitrate group	Non-destructive sensitive detector suitable for the detection of nitroaromatics, nitramines, and nitrate esters and their by-products or their transformation products [66–68]
Thermal energy analyser (chemiluminescence detector)	High selectivity for nitro or nitroso, containing molecules	Low picogram range	Contaminations can deteriorate the detector [21] Chemiluminescence detector for nitroaromatics [76], nitrate esters [77] and nitramines
Mass spectrometer Electron ionisation detector	Any compound that produces fragments within the selected mass range	Low nanogram range	Can positively identify compounds Suitable detector for nitroaromatics and nitramines [44, 78, 79]. The identification of specific nitrate esters is more difficult as the molecular ion is not present and the mass spectra are very similar with the abundance of only low mass fragments

Detector / method	Compounds	Detection level	Comments
Mass spectrometer Chemical ionisation (negative or positive mode)	Any compound that produces fragments within the selected mass range	Low nanogram range	Can be used as a complementary method to EI Suitable detector for nitroaromatics, nitrate esters and nitramines [60, 65, 80]
Tandem mass spectrometer		Low nanogram range	Can improve selectivity or be used as a confirmatory test [78]
Thermal conductivity detector	All compounds	Low nanogram range	Non-destructive detector suitable for permanent gases Rather low sensitivity and a great dependence on temperature and gas flow rate [81]
UV and vacuum ultra violet spectroscopy	Good selectivity for light adsorbing compound	Low nanogram range	Non-destructive UV detector used to study TNT, DNT, EGDN, NG and PETN [82]. Nitrate esters and their degradation products were analysed by vacuum ultra violet [83]
Fourier transform infra-red detector		Low nanogram range	Non-destructive detector suitable for the analysis of nitramines, nitrate esters and nitroaromatics [84]

Liquid chromatography techniques

Liquid chromatography (LC) is a widely used chromatographic technique that separates soluble compounds into its constituent components using a liquid mobile phase. LC can be performed using either a planar or column technique and the compounds are separated according to their polarity, electrical charge and molecular size depending on the choice of the column. Planar chromatography, e.g. thin layer chromatography, coats the stationary phase onto a plate. In column techniques, e.g. HPLC, the stationary phase is packed or coated into a narrow tube.

Thin layer chromatography: Thin layer chromatography (TLC) is a rapid, easy to carry out and low cost screening planar chromatographic method to detect common organic compounds with minimal sample preparation. In TLC the analytes are spotted on a thin layer of sorbent material (stationary phase) supported on an inert base. The mobile phase, containing a minimal amount of solvent or mixture of solvents, diffuses along the plate by capillary action and the analytes travel up the plate at different rates depending on their interactions with the mobile and stationary phases, achieving separation. Various chemical or spectroscopic detection methods are used to visualise the separated components. Each compound is then detected by comparison of its retardation factor (R_f) value (ratio of the travelled distance by the substance from the migration distance of the mobile phase) and the R_f value of a reference standard compound analysed on the same plate. While TLC is a presumptive test, confirmation of the chemical composition of the analytes can be obtained when the separated TLC spots are further analysed using alternative techniques like MS, IMS or Raman spectroscopy.

High performance thin layer chromatography (HPTLC) plates are pre-coated with smaller particle size and narrower size distribution sorbent material, and ultra-thin layer chromatography (UTLC) with a thinner layer of adsorbent material resulting in a shorter R_f and development time. These plates are described as being more sensitive than the conventional plates (low microgram to high nanogram) allowing trace amount of material to be detected. The separated analytes can also be quantified using a densitometric (optical density) scanning method with different wavelengths in absorbance or fluorescence mode.

Since the early 1960s, TLC has been successfully used to detect organic explosive compounds using normal phase silica-gel TLC plates with various solvent systems. Six eluent systems (chloroform, carbon-tetrachloride/dichloroethane, benzene/hexane, xylene/hexane, benzene, hexane/acetone and chloroform/acetone) were tested with a mixture of RDX, PETN, TNT, NG, tetryl and NC and a good separation for all the explosive compounds with the exception of PETN and NG was achieved in all the mobile phases studied [85]. NC didn't migrate in any of the solvent systems investigated. Other binary solvent systems were investigated using less toxic solvents and the best separation for nitroaromatics was achieved with petroleum/isopropanol (4:1 ratio), whereas petroleum ether/acetone (1:1 ratio) was more efficient for separating nitramines and nitrate esters [86]. Another 38 binary solvent systems were compared for a mixture of TNT, RDX, HMX and tetryl. Petroleum ether/ acetone (2:1 or 7:3 ratio) and hexane/acetone (2:1 ratio) were found to be the most effective [87]. Two-dimensional TLC has also be carried out to separate and identify high explosives using a binary solvent system made of petroleum/acetone (3:1 ratio) for the first separation and petroleum/ethyl acetate (3:1 ratio) for the second [88]. A tertiary solvent system composed of 8:4:3 v/v mixture of toluene, hexanes and acetone was also successfully used to separate nine military explosives [89]. Many solvent systems have been extensively studied to separate a wide range of military explosive compounds. However, the separation of all classes of explosives using a single eluent system is currently unachievable.

Various types of TLC plates have been investigated to separate nitroaromatics, nitramines and nitrate esters. In 1997, a study comparing TLC plates (fluorescence vs. nonfluorescence, and pre-concentration zone vs. no pre-concentration zone, HPTLC vs. TLC) concluded that conventional TLC plates with pre-concentration zones were recommended for better separation resolution of explosives compounds [86]. Research using HPTLC modified with a cyano propyl group (CN-HPTLC) to separate TNT, RDX, PETN, NG, EGDN and NC reported that CN-HPTLC was more effective than conventional silica TLC [90]. The best eluent system to separate the six explosives was found to be toluene.

After separation by TLC, the sample spots can be visualised by irradiating UV light or by spraying chromogenic reagents on the plate. Nitroaromatic and nitramine compounds are observed under 254 nm UV radiation while the nitrate esters are detected with 254 nm UV light after treatment with 1% (w/w) diphenylamine in hexane [89]. Other visualisation reagents have been used such as the modified Griess reagent for nitramines and nitrate esters [91] or ethylene diamine in DMSO solution for nitroaromatics [92].

For a formal identification of explosives, IMS has been coupled with TLC, using laser desorption to successfully quantify explosives (TNT, RDX, PETN) on the surface of the plate without any extraction [93]. Electrospray ionisation mass spectrometry (ESI/MS) has been applied to scan and characterise compounds on TLC plates [94]. However, they have yet to be applied to the detection of explosives.

TLC is such a simple, sensitive screening method to analyse explosives that it can be used in the field. Lawrence Livermore National Laboratory developed a miniaturised, field-portable detection system based on TLC technologies. It can be used to detect and identify common military explosives (i.e. TNT, tetryl, picric acid, Explosive D, RDX and HMX) in environmental samples but also propellants (through propellant stabilisers), peroxide explosives, nitrates and inorganic oxidiser precursors [89].

High performance liquid chromatography: HPLC is a form of column chromatography that separates dissolved chemical mixtures by elution through a column holding a tightly packed stationary phase. The choice of the stationary phase is governed by the characteristics of the analytes to be analysed. The different modes of HPLC separation rely on the different types of interactions between the analytes and the stationary phase. The three main mechanisms of separation are based on the polarity (normal or reversed phase chromatography), the electrical charge (e.g. ion exchange chromatography) and the size (e.g. size exclusion chromatography) of the chemical compounds to be analysed.

The process is similar to GC except the compounds are transported by a liquid mobile phase. The instrument, called a chromatograph, is composed of an injector port to introduce the sample into the continuously flowing mobile phase, a high-pressure pump to deliver the mobile phase, a column to separate the mixture and a detector to identify the eluting compounds. The resulting detector responses are recorded on a chromatogram. The composition of the mobile phase can influence the separation. The separation can be carried out by *isocratic elution* when the solvent composition remains constant throughout the analysis or *gradient elution* when the composition of the solvent changes during the run toward a better desorbing mobile phase. Gradient can provide better resolution for complex samples. Unlike GC, this technique is carried out at room temperature, which allows thermally labile or non-volatile compounds to be easily analysed.

Reversed and normal phase chromatography: The most popular HPLC separation mode based predominantly on polarity is reversed phase HPLC (RP-HPLC), which consists of a mobile phase more polar than the stationary phase. In this system, the polar analytes elute before the nonpolar analytes. Normal phase HPLC (NP-HPLC), which is used for water-insoluble samples, consists of a polar stationary phase with a less polar mobile phase containing 100% of organic solvents. The nonpolar analytes will elute before the polar analytes. RP-HPLC allows direct injection of water samples. NP-HPLC requires a solvent extraction from water as only a nonpolar mobile phase can be used with the stationary phase. For a reversed and normal phase, separation will be achieved if the polarities of the mobile phase, the stationary phase and the analytes are different. For analytes that are difficult to separate, such as isomers, a porous graphitic carbon (PGC) stationary phase can be used as it can retain polar molecules and separate isomers.

For the last 30 years, RP-HPLC has been used routinely for the analysis of explosives. The mobile phase is conventionally made of a mixture of acetonitrile/water or methanol/water with the high surface area stationary phase containing a modified alkyl chain (e.g. C8 n-octyl silane or C18 n-octadecylsilane) bonded to a silica surface. PGC is a good substitute for RP C18 that poorly retains the most polar explosives and doesn't separate some regio-isomers. In 2005, a single analysis of 21 common explosives, including their by-products, was successfully carried out using a PGC column and a gradient elution with acetonitrile/methanol/water [52]. The addition of toluene in the mobile phase improved the peak shape and eluted the most strongly retained compounds.

Columns packed with small fully porous particles or small core–shell particles (core superficially wrapped with a porous shell) are gaining popularity as they result in sharper peak shapes. This can increase efficiency compared to traditional columns. With the core–shell technology, 14 explosives (nitroaromatics mainly and nitramines) were chromatographically separated in 3 min reducing dramatically the time of analysis and therefore the quantity of solvent necessary [95]. Simultaneous analysis of 17 explosives from the US EPA 8330 method including nitrates esters, nitramines and nitroaromatics was carried out in 31 min using the same type of column [96]. Three constituents of the new insensitive munitions, DNAN, NTO and NQ, have been separated by RP-HPLC using a BEH amide stationary phase [74], while the four constituents DNAN, NTO, NQ and RDX were separated with hydroRP [97] or C18 columns [29] and acidified mobile phase gradient.

Particle size of the stationary phase plays a crucial role in the separation efficiency of the column. A decrease in the particle size of the packing material results in an increase of the separation efficiency. Particle size as small as sub 2 μm are being used in columns. However, smaller particle sizes increase the operating pressure of the systems. A new technology, ultra-high performance liquid chromatography (UPLC) was developed to allow pressure above 6 000 PSI and reduce analysis time.

Detectors: HPLC detectors might be less sensitive that the GC detectors but they are successful for a broader range of applications. A wide variety of HPLC detectors are integrated within the HPLC system such as the ultra violet/visible (UV–vis) detectors or hyphenated techniques like a mass spectrometer LC-MS [98]. The main detectors used in the analysis of explosives [99] are listed in table 4.5.

Ion exchange chromatography: Ion exchange chromatography is a specific and sensitive (range of ppm) chromatographic method to separate organic and inorganic ions in solution. The stationary phase contains a resin with immobilised charged functional groups, which can be either positive (anion exchanger) or negative (cation exchanger), and exchangeable counter-ions. The mobile phase is a solution of salt in water that acts as a buffer to maintain the pH throughout the analysis and to change the eluting power depending on the type of salt. When the sample reaches the column, the neutral analytes and ions of the same charge as the stationary phase will elute without any retention whilst the ions of opposite charge to the stationary phase will be retained. The separation mode is based on the degrees of interactions between the analytes and the opposite charged stationary phase and can be controlled by modifying the pH and other solution parameters. An extensive range of stationary phases and eluents makes this technique versatile. The main detectors used in IC include conductivity, UV/vis, and amperometric detectors. When hyphenated with a MS or MS-MS detector, IC becomes a very sensitive and selective method.

IC is well suited to analyse organic and inorganic ions often present in pyrotechnics, propellants and some insensitive high explosive formulations. Several IC methods with suppressed conductivity detection have been developed [104] and approved by US EPA [105, 106] to determine low levels of perchlorate in water samples. A new improved method, using two-dimensional IC, has been applied to detect the level of perchlorate in drinking waters as low as 55 ng l^{-1} [107, 108]. When coupled with a MS or MS-MS spectrometer, IC can be used to identify both cationic metals such as aluminium, barium, potassium, magnesium, ammonium, lead, etc, and anionic oxidisers such as chlorate, perchlorate, fulminate, nitrate, etc. This can lead to identification of the explosive being detected as well as its quantification, with the limit of detection ranging from low μg l^{-1} to pg l^{-1} [109].

Table 4.5. Specifications of common HPLC detectors for explosives analysis.

Name	Selectivity	Sensitivity (minimum detectable quantity)	Application to explosives
Absorbance detectors Ultra violet/visible and photo diode array (PDA)	Good selectivity for light adsorbing compound	Picogram range	Used with reverse phase application. Easy to use and low cost detector for nitro explosives Fourteen explosives included in the US EPA Method 8330 were detected by RP-HPLC-PDA [100] DNAN, NTO, NQ and RDX were separated simultaneously [29, 74, 97]
Electrochemical detector	Good selectivity for compounds that can be oxidised	Picogram range	Used for the detection of nitrated explosives [21]
Thermal energy analyser	Specific to compounds that decompose to form products that can be derivatised with a fluorophore	Picogram range. Better sensitivity for nitramine and nitrate esters than nitroaromatics [21]	Not as popular as a UV detector Used with normal phase only (no water) [21] TEA detection has been used to analyse explosives, such as RDX, HMX, NG and PETN [101]
Mass spectrometer with: ESI APCI, negative or positive mode APPI		Picogram range	Different API methods have been apply to detect a variety of conventional explosives: ESI-MS [47] APCI [34, 52, 102] NQ, NTO, RDX and DNAN in water were separated and detected by HPLC–UV–ESI-MS [97] and by LC-NI-APCI-MS [74] but NTO was detected in positive APCI Level of perchlorates in water, soils and solid wastes have also been analysed by LC-MS [103]
Tandem mass spectrometer		Picogram range	Trace level of TNT, RDX and HMX in soil have been analysed by LC-MS-MS [49] [63] and by direct injection analysis

4.3 Conclusion

This chapter summarises the current state of the analysis of explosive contaminated samples, whether soil or water. The range of techniques for extraction, the use of pre-concentrators and the extensive analysis methods of these samples, offer the operator a comprehensive set of tools to fully qualify the samples of interest. An ever-developing area, the current field techniques are becoming more accurate (ppb to ppt), easier to use and with a quicker response.

References

[1] Taylor S, Park E, Bullion K and Dontsova K 2015 Dissolution of three insensitive munitions formulations *Chemosphere.* **119** 342–8

[2] Meyer R, Köhler J and Homburg A 2007 *Explosives* (New York: Wiley-VCH)

[3] Boddu V M, Abburi K, Maloney S W and Damavarapu R 2008 Thermophysical properties of an insensitive munitions compound, 2,4-dinitroanisole *J. Chem. Eng. Data* **53** 1120–5

[4] Phelan J, Romero J, Barnett J L and Parker D R 2002 Solubility and Dissolution Kinetics of Composition B Explosive in Water, SAND Report SAND2002-2420. Sandia National Laboratories, Albuquerque, NM

[5] Singh B and Singh H 1981 Synthesis of 2,2',4,4',6,6' hexanitrostilbene *Def. Sci. J.* **31** 305–8

[6] Sitzmann M E and Foti S C 1975 Solubilities of explosives–dimethylformamide as general solvent for explosives *J. Chem. Eng. Data* **20** 53–5

[7] Pesce-Rodriguez R A, Munson C, Giri L and Blaudeau L B 2018 Experimental determination of physical properties of 1-Propyl-2-Nitroguanidine (PrNQ) (Report No. ARL-TN-0927) US Army Research Laboratory Aberdeen Proving Ground, USA

[8] Tennant M, Chew S C, Krämer T, Mai N, McAteer D and Pons J-F 2019 Practical colorimetry of 3-nitro-1,2,4-triazol-5-one *Propell. Explos. Pyrotech.* **44** 198–202

[9] Spear R J, Louey C N and Wolfson M G 1989 *A Preliminary Assessment of NTO as an Insensitive High Explosive* MRL-TR-89-18 MRL MRL Technical Report

[10] Smith M W and Cliff M D 1999 NTO based explosive formulations: a technology review (Report No. DSTO-TR-0796). Weapons Systems Division, Aeronautical and Maritime Research Laboratory. Defence Science and Technology, Australian Department of Defence, Salisbury, South Australia

[11] Tomlinson W R and Sheffield O E 1971 *Properties of Explosives of Military Interest* 706-177 AMCP, Washington DC, USA Report

[12] Sitzmann M E, Foti S and Misener C C 1973 *Solubilities of High-Explosives: Removal of High Explosive Fillers from Munitions by Chemical Dissolution* (Maryland: White Oak)

[13] Selig W 1977 Estimation of the Solubility of 1,3,5-Triamino-2,4,6-Trinitrobenzene (TATB) in Various Solvents (Report No. UCID-17412). California Univ., Livermore (USA). Lawrence Livermore Lab.

[14] Ro K S *et al* 1996 Solubility of 2,4,6-trinitrotoluene (TNT) in water *J. Chem. Eng. Data* **41** 758–61

[15] von Holtz E, Ornellas D, Foltz M F and Clarkson J E 1994 The solubility of ε-CL-20 in selected materials *Propell. Explos. Pyrotech.* **19** 206–12

[16] von Holtz E, Ornellas D, Foltz M F and Clarkson J E 1992 The Solubility of e-CL-20 in selected materials *Energetic Materials for Munitions Conf. (China Lake, CA, September 22–24,*

1992) Order Number DE93 009304. NTIS issue 199319; subject category 79A - Ammunition, Explosives, & Pyrotechnics; 99F - Physical & Theoretical Chemistry

[17] Glover D J and Hoffsommer J C 1973 Thin-layer chromatographic analysis of HMX in water *Bull. Environ. Contam. Toxicol.* **10** 302–4

[18] Kang H, Kim H, Lee C H, Ahn I S and Lee K D 2017 Extraction-based recovery of RDX from obsolete composition B *J. Ind. Eng. Chem.* **56** 394–8

[19] Brannon J and Pennington J 2002 *Environmental Fate and Transport Process Descriptors for Explosives* (Vicksburg, MA: Engineer Research and Development Center)

[20] Naoum P 1928 *Nitroglycerine and Nitroglycerine Explosives* vol 1 (Baltimore, MD: Williams and Wilkins)

[21] Yinon J and Zitrin S 1993 *Modern Methods and Applications in Analysis of Explosives* (Chichester: Wiley)

[22] Roberts R N and Dinegar R H 1958 Solubility of pentaerythritol tetranitrate *J. Phys. Chem.* **62** 1009–11

[23] Pesce-rodriguez R 2010 Solubility of RDX, PETN, and boric acid in methylene chloride (Report No. ARL-TN-0401) US Army Research Lab., Aberdeen Proving Ground

[24] Dawidziuk B, Nawała J, Dziedzic D, Gordon D and Popiel S 2018 Development, validation and comparison of three methods of sample preparation used for identification and quantification of 2,4,6-trinitrotoluene and products of its degradation in sediments by GC-MS/MS *Anal. Methods* **10** 5188–96

[25] Ungrádová I, Šimek Z, Vávrová M, Stoupalová M and Mravcová L 2013 Comparison of extraction techniques for the isolation of explosives and their degradation products from soil *Int. J. Environ. Anal. Chem.* **93** 984–98

[26] Fryš O, Česla P, Bajerová P, Adam M and Ventura K 2012 Optimization of focused ultrasonic extraction of propellant components determined by gas chromatography/mass spectrometry *Talanta* **99** 316–22

[27] Williford C W and Bricka R M 1999 Extraction of TNT from aggregate soil fractions *J. Hazard. Mater.* **66** 1–13

[28] Richter B E, Jones B A, Ezzell J L, Porter N L, Avdalovic N L and Pohl C 1996 Accelerated solvent extraction: a technique for sample preparation *Anal. Chem.* **68** 1033–9

[29] Temple T, Ladyman M K, Mai N, Galante E, Ricamora M, Shirazi R and Coulon F 2018 Investigation into the environmental fate of the combined Insensitive High Explosive constituents 2,4-dinitroanisole (DNAN), 1-nitroguanidine (NQ) and nitrotriazolone (NTO) in soil *Sci. Total Environ.* **625** 1264–71

[30] Giergielewicz-Mozajska H, Dabrowski L and Namieśnik J 2001 Accelerated solvent extraction (ASE) in the analysis of environmental solid samples - Some aspects of theory and practice *Crit. Rev. Anal. Chem.* **31** 149–65

[31] Schantz M M 2006 Pressurized liquid extraction in environmental analysis *Anal. Bioanal. Chem.* **386** 1043–7

[32] Pourmortazavi S M and Hajimirsadeghi S S 2005 Application of supercritical carbon dioxide in energetic materials processes: A review *Ind. Eng. Chem. Res.* **44** 6523–33

[33] Thomas J L, Donnelly C C, Lloyd E W, Mothershead R F and Miller M L 2018 Development and validation of a solid phase extraction sample cleanup procedure for the recovery of trace levels of nitro-organic explosives in soil *Forensic Sci. Int.* **284** 65–77

[34] Rapp-Wright H *et al* 2017 Suspect screening and quantification of trace organic explosives in wastewater using solid phase extraction and liquid chromatography-high resolution accurate mass spectrometry *J. Hazard. Mater.* **329** 11–21

[35] Gaurav, Kaur V, Kumar A, Malik A K and Rai P K 2007 SPME-HPLC: a new approach to the analysis of explosives *J. Hazard. Mater.* **147** 691–7

[36] Gaurav, Malik A K and Rai P K 2008 Solid phase microextraction-high performance liquid chromatographic determination of octahydro-1,3,5,7-tetranitro-1,3,5,7-tetrazocine (HMX) and hexahydro-1,3,5-trinitro-1,3,5-triazine (RDX) in the presence of sodium dodecyl sulfate surfactant *J. Sep. Sci.* **31** 2173–81

[37] Gaurav, Malik A K and Rai P K 2009 Development of a new SPME-HPLC-UV method for the analysis of nitro explosives on reverse phase amide column and application to analysis of aqueous samples *J. Hazard. Mater.* **172** 1652–58

[38] Monteil-Rivera F, Beaulieu C, Deschamps S, Paquet L and Hawari J 2004 Determination of explosives in environmental water samples by solid-phase microextraction-liquid chromatography *J. Chromatogr.* A **1048** 213–21

[39] Socrates G 2004 *Infrared and Raman Characteristic Group Frequencies* (New York: Wiley)

[40] Ghosh M, Wang L and Asher S A 2012 Deep-ultraviolet resonance raman excitation profiles of NH_4NO_3, PETN, TNT, HMX, and RDX *Appl. Spectrosc.* **66** 1013–21

[41] Tuschel D D, Mikhonin A V, Lemoff B E and Asher S A 2010 Deep ultraviolet resonance Raman spectroscopy of explosives in *AIP Conf. Proc.* **1267** 869–70

[42] Hwang J *et al* 2013 Fast and sensitive recognition of various explosive compounds using Raman spectroscopy and principal component analysis *J. Mol. Struct.* **1039** 130–6

[43] López-López M and García-Ruiz C 2014 Infrared and Raman spectroscopy techniques applied to identification of explosives *TrAC - Tren. Anal. Chem.* **54** 36–44

[44] Yinon J 1996 Trace analysis of explosives in water by gas chromatography-mass spectrometry with a temperature-programmed injector *J. Chromatogr.* A **742** 205–9

[45] Yinon J 1980 Analysis of explosives by negative ion chemical ionisation mass spectrometry *J. Forensic Sci.* **25** 401–7

[46] Badjagbo K and Sauvé S 2012 Mass spectrometry for trace analysis of explosives in water *Crit. Rev. Anal. Chem.* **42** 257–71

[47] Zhao X and Yinon J 2002 Identification of nitrate ester explosives by liquid chromatography-electrospray ionization and atmospheric pressure chemical ionization mass spectrometry *J. Chromatogr.* A **977** 59–68

[48] Yinon J, McClellan J E and Yost R A 1997 Electrospray ionization tandem mass spectrometry collision-induced dissociation study of explosives in an ion trap mass spectrometer *Rapid Commun. Mass Spectrom.* **11** 1961–70

[49] Ochsenbein U, Zeh M and Berset J D 2008 Comparing solid phase extraction and direct injection for the analysis of ultra-trace levels of relevant explosives in lake water and tributaries using liquid chromatography-electrospray tandem mass spectrometry *Chemosphere.* **72** 974–80

[50] Cassada D A, Monson S J, Snow D D and Spalding R F 1999 Sensitive determination of RDX, nitroso-RDX metabolites, and other munitions in ground water by solid-phase extraction and isotope dilution liquid chromatography-atmospheric pressure chemical ionization mass spectrometry *J. Chromatogr.* A **844** 87–95

[51] Mathis J A and McCord B R 2005 The analysis of high explosives by liquid chromatography/electrospray ionization mass spectrometry: multiplexed detection of negative ion adducts *Rapid Commun. Mass Spectrom.* **19** 99–104

[52] Holmgren E, Carlsson H, Goede P and Crescenzi C 2005 Determination and characterization of organic explosives using porous graphitic carbon and liquid chromatography-atmospheric pressure chemical ionization mass spectrometry *J. Chromatogr. A* **1099** 127–35

[53] Xu X, Van De Craats A M and De Bruyn P C A M 2004 Highly sensitive screening method for nitroaromatic, nitramine and nitrate ester explosives by high performance liquid chromatography—atmospheric pressure ionization—mass spectrometry (HPLC-API-MS) *J. Forensic Sci.* **49** 1171–80

[54] Schreiber A, Efer J and Engewald W 2000 Application of spectral libraries for high-performance liquid chromatography–atmospheric pressure ionisation mass spectrometry to the analysis of pesticide and explosive residues in environmental samples *J. Chromatogr. A* **869** 411–25

[55] Song L, Wellman A D, Yao H and Bartmess J E 2007 Negative ion-atmospheric pressure photoionization: electron capture, dissociative electron capture, proton transfer, and anion attachment *J. Am. Soc. Mass. Spectrom.* **18** 1789–98

[56] Li L-H, Hsieh H-Y and Hsu C-C 2017 Clinical application of ambient ionization mass spectrometry *Mass Spectrom.* **6** 1–12

[57] Mulligan C C, MacMillan D K, Noll R J and Cooks R G 2007 Fast analysis of high-energy compounds and agricultural chemicals in water with desorption electrospray ionization mass spectrometry *Rapid Commun. Mass Spectrom.* **21** 3729–36

[58] Nilles J M, Connell T R, Stokes S T and Dupont Durst H 2010 Explosives detection using direct analysis in real time (DART) mass spectrometry *Propell. Explos. Pyrotech.* **35** 446–51

[59] Forbes T P, Sisco E, Staymates M and Gillen G 2017 DART-MS analysis of inorganic explosives using high temperature thermal desorption *Anal. Methods* **9** 4988–96

[60] Robarge T, Phillips E and Conoley M 2007 *Analysis of Explosives by Chemical Ionization GC-MS* (Austin, TX: Thermo Electron Corporation)

[61] Xu X, Van De Craats A M and De Bruyn P C A M 2004 Highly sensitive screening method for nitroaromatic, nitramine and nitrate ester explosives by high performance liquid chromatography—atmospheric pressure ionization—mass spectrometry (HPLC-API-MS) *J. Forensic Sci.* **49** 1171–80

[62] Delgado T, Alcántara J F, Vadillo J M and Laserna J J 2013 Condensed-phase laser ionization time-of-flight mass spectrometry of highly energetic nitro-aromatic compounds *Rapid Commun. Mass Spectrom.* **27** 1807–13

[63] Avci G F Y, Anilanmert B and Cengiz S 2017 Rapid and simple analysis of trace levels of three explosives in soil by liquid chromatography—tandem mass spectrometry *Acta Chromatogr.* **29** 45–56

[64] Schramm S, Léonço D, Hubert C, Tabet J C and Bridoux M 2015 Development and validation of an isotope dilution ultra-high performance liquid chromatography tandem mass spectrometry method for the reliable quantification of 1,3,5-Triamino-2,4,6-trinitrobenzene (TATB) and 14 other explosives and their degradation pro *Talanta* **143** 271–8

[65] Collin O L, Zimmermann C M and Jackson G P 2009 Fast gas chromatography negative chemical ionization tandem mass spectrometry of explosive compounds using dynamic collision-induced dissociation *Int. J. Mass spectrom.* **279** 93–9

[66] US Environmental Protection Agency 2007 Method 8095, Nitroaromatics and nitramines by GC/ECD [Online]. Available: http://epa.gov

[67] Waddell R, Dale D E, Monagle M and Smith S A 2005 Determination of nitroaromatic and nitramine explosives from a PTFE wipe using thermal desorption-gas chromatography with electron-capture detection *J. Chromatogr. A* **1062** 125–31

[68] Walsh M E 2001 Determination of nitroaromatic, nitramine, and nitrate ester explosives in soil by gas chromatography and an electron capture detector *Talanta* **54** 427–38

[69] Leppert J, Härtel M, Klapötke T M and Boeker P 2018 Hyperfast flow-field thermal gradient GC/MS of explosives with reduced elution temperatures *Anal. Chem.* **90** 8404–11

[70] Aldrich S GC Analysis of Explosives on SLB®–5ms (12 m × 0.18 mm I.D., 0.30 μm), Fast GC Analysis [Online]. Available: https://sigmaaldrich.com

[71] Restek Explosives Rtx-200 [Online]. Available: https://restek.com

[72] Collin O L, Niegel C, DeRhodes K E, McCord B R and Jackson G P 2006 Fast gas chromatography of explosive compounds using a pulsed-discharge electron capture detector *J. Forensic Sci.* **51** 815–8

[73] Liu Z R, Wu C Y, Kong Y H, Yin C M and Xie J J 1989 Investigation of the thermal stability of nitroguanidine below its melting point *Thermochim. Acta* **146** 115–23

[74] Walsh M E 2016 Analytical methods for detonation residues of insensitive munitions *J. Energ. Mater.* **34** 76–91

[75] Hewitt A D and Jenkins T F 1999 *On-Site Method for Measuring Nitroaromatic and Nitramine Explosives in Soil and Groundwater Using GC-NPD: Feasibility Study* (Hanover, NH: US Army Corps of Engineers, Cold Regions Research & Engineering Laboratory)

[76] Lafleur A L and Mills K M 1981 Trace level determination of selected nitroaromatic compounds by gas chromatography with pyrolysis/chemiluminescent detection *Anal. Chem.* **53** 1202–5

[77] Bowerbank C R, Smith P A, Fetterolf D D and Lee M L 2000 Solvating gas chromatography with chemiluminescence detection of nitroglycerine and other explosives *J. Chromatogr. A* **902** 413–9

[78] Perr J M, Furton K G and Almirall J R 2005 Gas chromatography positive chemical ionization and tandem mass spectrometry for the analysis of organic high explosives *Talanta* **67** 430–6

[79] Kolla P 1994 Gas chromatography, liquid chromatography and ion chromatography adapted to the trace analysis of explosives *J. Chromatogr. A* **674** 309–18

[80] Yinon J 1980 Analysis of explosives by negative ion chemical ionization mass spectrometry *J. Forensic Sci.* **25** 401–7

[81] Yinon J and Zitrin S 1981 *The Analysis of Explosives* (Oxford: Pergamon)

[82] Andrasko J, Lagesson-Andrasko L, Dahlén J and Jonsson B H 2017 Analysis of explosives by GC-UV *J. Forensic Sci.* **62** 1022–7

[83] Cruse C A and Goodpaster J V 2019 Generating highly specific spectra and identifying thermal decomposition products via gas chromatography/vacuum ultraviolet spectroscopy (GC/VUV): application to nitrate ester explosives *Talanta* **195** 580–6

[84] Sharma S P and Lahiri S C 2005 Characterization and identification of explosives and explosive residues using GC-MS, an FTIR microscope, and HPTLC *J. Energ. Mater.* **23** 239–64

[85] Parker R G, Mcowen J M and Cherolis J A 1975 Analysis of explosives and explosive residues. Part 2: thin-layer chromatography *J. Forensic Sci.* **20** 254–6

[86] Nam S-I 1997 *On-Site Analysis of Explosives in Soil: Evaluation of Thin Layer Chromatography for Confirmation of Analyte Identity* CRREL-SR-97-21 US Army Corps of Engineers, Cold Regions Research and Engineering Lab., Hanover, NH

[87] Lutonská T, Kobliha Z and Skaličan Z 2015 The qualitative analysis of the selected explosives using thin-layer chromatography *Adv. Military Tech* **10** 111–7

[88] Bagnato L and Grasso G 1986 Two-dimensional thin-layer chromatography for the separation and identification of nitro derivatives in explosives *J. Chromatogr.* A **357** 440–4

[89] Satcher J H *et al* 2012 *Portable thin layer chromatography for field detection of explosives and propellants Chemical, Biological, Radiological, Nuclear, and Explosives (CBRNE) Sensing XIII Proc. SPIE* **8358** 83580Z

[90] Borusiewicz R and Lerg S 2010 TLC as a screening technique in the analysis of explosives – development of a new analytical procedure *Probl. Forensic Sci.* **81** 114–26

[91] Singh B, Verma M, Vasudeva S K and Shera B M L 1982 Analysis of explosive mixtures by thin layer chromatography *Def. Sci. J.* **32** 151–5

[92] Glover D J and Kayser E G 1968 Quantitative spectrophotometric analysis of polynitroaromatic compounds by reaction with ethylenediamine *Anal. Chem.* **40** 2055–8

[93] Ilbeigi V, Sabo M, Valadbeigi Y, Matejcik S and Tabrizchi M 2016 Laser desorption-ion mobility spectrometry as a useful tool for imaging of thin layer chromatography surface *J. Chromatogr.* A **1459** 145–51

[94] Cheng S C, Bhat S M, Lee C W and Shiea J 2019 Simple interface for scanning chemical compounds on developed thin layer chromatography plates using electrospray ionization mass spectrometry *Anal. Chim. Acta* **1049** 1–9

[95] DeStefano J J, Langlois T J and Kirkland J J 2008 Characteristics of superficially-porous silica particles for fast HPLC some performance comparisons with sub-2-μm particles *J. Chromatogr. Sci.* **46** 254–60

[96] Aqeel Z and Lomas S 2017 Analysis of Explosives Found in EPA 8330A and 8330B Using a Kinetex® 2.6 μm Polar C18 Phenomenex Column Torrance, CA 90501 technical note TN 52260517_W

[97] Russell A L, Seiter J M, Coleman J G, Winstead B and Bednar A J 2014 Analysis of munitions constituents in IMX formulations by HPLC and HPLC-MS *Talanta* **128** 524–30

[98] Swartz M 2010 HPLC detectors: a brief review *J. Liq. Chromatogr. Relat. Technol.* **33** 1130–50

[99] Gaurav D, Malik A K and Rai P K 2007 High-performance liquid chromatographic methods for the analysis of explosives *Crit. Rev. Anal. Chem.* **37** 227–68

[100] Borch T and Gerlach R 2004 Use of reversed-phase high-performance liquid chromatography-diode array detection for complete separation of 2,4,6-trinitrotoluene metabolites and EPA Method 8330 explosives: influence of temperature and an ion-pair reagent *J. Chromatogr.* A **1022** 83–94

[101] Lafleur A L and Morriseau B D 1980 Identification of explosives at trace levels by high performance liquid chromatography with a nitrosyl-specific detector *Anal. Chem.* **52** 1313–8

[102] Xu X, Koeberg M, Kuijpers C J and Kok E 2014 Development and validation of highly selective screening and confirmatory methods for the qualitative forensic analysis of organic explosive compounds with high performance liquid chromatography coupled with (photo-diode array and) LTQ ion trap/Orbitrap ma *Sci. Justice* **54** 3–21

[103] US Environmental Protection Agency Method 6850, Perchlorate in water, soils and solid wastes using high performance liquid chromatography/electrospray ionization/mass spectrometry (HPLC/ESI/MS OR HPLC/ESI/MS/MS) 2007. [Online]. Available: http://epa.gov

[104] Seiler M A, Jensen D, Neist U, Deister U K and Schmitz F 2016 Validation data for the determination of perchlorate in water using ion chromatography with suppressed conductivity detection *Environ. Sci. Eur.* **28** 1

[105] US Environmental Protection Agency Method 314.0, Determination of Perchlorate in Drinking Water Using Ion 1999. [Online]. Available: http://epa.gov

[106] US Environmental Protection Agency 2005 Method 314.1, Determination of perchlorate in drinking water using inline column concentration/matrix elimination ion chromatography with suppressed conductivity detection [Online] Available: http://epa.gov

[107] Wagner H P *et al* 2007 Selective method for the analysis of perchlorate in drinking waters at nanogram per liter levels, using two-dimensional ion chromatography with suppressed conductivity detection *J. Chromatogr.* A **1155** 15–21

[108] US Environmental Protection Agency 2008 Method 314.2, Determination of perchlorate in drinking water using two-dimensional ion chromatography with suppressed conductivity detection [Online]. Available: http://epa.gov

[109] Barron L and Gilchrist E 2014 Ion chromatography-mass spectrometry: a review of recent technologies and applications in forensic and environmental explosives analysis *Anal. Chim. Acta* **806** 27–54

IOP Publishing

Global Approaches to Environmental Management on Military Training Ranges

Tracey J Temple and Melissa K Ladyman

Chapter 5

Environmental management of military ranges with the support of a life-cycle assessment approach

C Ferreira, F Freire and J Ribeiro

5.1 Introduction

The awareness of the general population about the consequences of warfare activities is normally associated with casualties and long lasting devastation. Nevertheless, the majority of munitions are disposed of or used in training actions, and these military activities also need to be addressed due to their negative effects on the environment and human health. The specific case of the environmental problems on military ranges is of great importance, as training facilities are essential to ensure that the Armed Forces are at their finest state of promptness and operability. The contamination of these ranges with energetic material and heavy metals from the use of ammunitions can pose a real problem to the operability of the Armed Forces by forcing the relocation of facilities to remediate the sites or even closure of the facilities. In fact, the Massachusetts Military Reservation (US) was closed in 2001 due to the evidence of surface and groundwater contamination [1].

Training facilities are not abundant, so the environmental management of military ranges is relevant to maintaining the operability of the Armed Forces. The monitoring of contamination and the assessment of its effects on ecosystems and human health facilitate an understanding of the evolution of the contamination associated with live-firing training. Furthermore, management of military ranges will cause a decrease of the cost associated with site remediation or relocation of the facilities.

Motivation to characterise and quantify the emissions (and subsequent level of contamination) in military ranges has been increasing. The main residues associated with the use of ammunition in military ranges are energetic materials and heavy metals [2, 3]. Therefore, studies have been focusing on characterising the emissions and deposition rates from the firing and detonation of different types of explosives [4–8], including the particular case of sensitive ammunition [9–11]. For the case of heavy metals, the major substances found in military ranges are lead, antimony, copper and zinc [12].

The aforesaid studies provide knowledge about how much a range is contaminated from previous activities; however, it is not possible to ascertain which contaminant is worrisome and can pose a higher hazard, and which type of activities were responsible for those contaminations. Moreover, the characterisation or monitoring of military ranges does not prevent future contamination. The employment of life-cycle assessment (LCA) methodology to military ranges can provide an appropriate tool to overcome these limitations by quantifying the environmental consequences associated with military range contamination. With LCA studies it is possible to quantify the impacts and identify what is contributing to those impacts; thus, it enables range managers to decrease or mitigate those impacts. Furthermore, LCA studies can be applied to predict impacts. LCA can be applied from the conception of products until their disposal, giving the opportunity to assess the potential impacts from the use of military ordnances in military ranges to avoid future contamination.

It is important to move forward and use the data from the characterisation and quantification of the emissions to determine the potential toxicological impacts, and identify what is contributing to those impacts. The information obtained from LCA studies is complementary to the information gathered with other environment management tools that will help range managers understand in more detail the consequences from the utilisation of military products in their training facilities, and attempt to avoid future contamination.

This chapter intends to shed some light on the advantages of using the LCA methodology to quantitatively assess the environmental and toxicological impacts associated with the use of munitions in military ranges. The employment of life-cycle impact assessment models can help to facilitate the prediction of the behaviour on the environment and the inherent toxicity effects of energetic materials (or other substances), as an alternative to experimental studies that are time consuming and costly. The employment of models that estimate physicochemical and toxicity properties—EPi suite (Estimation Programs Interface, developed to estimate the physical/chemical property and environmental fate) or QSAR (Quantitative structure–activity relationship)—coupled with other models that rely on that information, and average environmental characteristics, to determine the fate and exposure to the substances, is a way to facilitate the determination of the substance behaviour on the environment. USEtox is the main example of a life-cycle impact assessment model that can provide the information required to assess the potential toxicological impacts, which is described in more detail in section 5.2.2.

5.2 Life-cycle assessment methodology

Life-cycle assessment methodology follows the life-cycle principle in which the potential environmental and toxicological impacts of a product (or system) is quantitatively assessed throughout its life-cycle [13]. This means that LCA studies encompass all the life-cycle phases, including the design, procurement, manufacturing, usage, disposal and transport between the different phases. This type of analysis is called cradle-to-grave, in which LCA handles those processes as a chain of subsystems that exchange inputs and outputs [13]. LCA is governed by the standards ISO 14040 (Environmental management—Life Cycle Assessment—Principles and Framework), and ISO 14044 (Environmental management—Life Cycle Assessment—Requirements and Guidelines) that specifies the main principles and requirements of the methodology, including a description of its phases.

Life-cycle assessment methodology, as described in the ISO standards, has four interconnected phases: the goal and scope where the main objective and purpose of the study is defined, including the identification of the product system boundaries and a functional unit; the inventory analysis (LCI) in which the inputs and outputs necessary to meet the goals of the study are collected and compiled; the life-cycle impact assessment that converts the inventory data into environmental impact values according to various methods; and the interpretation where the results are discussed and conclusions are drawn providing recommendations to decision makers.

The results obtained by an LCA study can be used for different purposes, from informing decision makers in industry, government or non-government organisations to marketing or eco-labelling schemes. The major advantage of performing an LCA study is the opportunity to assess, quantitatively, what the environmental and toxicological impacts of products are and identify what is contributing to those impacts. With this information it is possible to identify opportunities to improve the environmental performance of products, identify which life-cycle phase presents a higher contribution to the impacts; and compare different products or technologies with the same function.

The feasibility and inherent advantages of employing LCA studies to ammunition was demonstrated in four studies covering all the life-cycle phases: two studies that assessed the impacts associated with the production and use of small calibre ammunition [14] and large calibre ammunition [15]; and two studies that focused on the end-of-life—demilitarisation of military ammunition in a static kiln [16], and recycling of energetic material from military ammunition for the production of civil explosives [17]. This methodology was also employed to assess the environmental and toxicological impacts for the production of a civil explosive [18]. The findings from these LCA studies will allow shooting range managers, ammunition producers or procurement officers to become more aware of the main environmental impacts of ammunition as well as defining strategies to manage or mitigate ammunition burdens and carry out tailored modifications to decrease the impacts.

5.2.1 Barriers for assessing the toxicological impacts on military ranges with the life-cycle assessment methodology

One of the disadvantages of the LCA methodology is that usually it neither addresses site-specific impacts nor impacts dynamic with time. Traditional LCA studies use average data to specific regions to calculate the environmental impacts with global effects (e.g. climate change, abiotic depletion). The determination of site-specific impacts is neglected due to the complexity of the site-specific models, which makes them difficult to implement, and a lack of available data for the sites evaluated [19, 20]. Since the toxicological impacts are strongly influenced by site-specific characteristics, it is a challenge to evaluate those impacts in LCA studies.

The complexity of assessing the toxicological impacts is due to the dependence on emission patterns, chemical properties, location and various other parameters [20, 21]. Some studies have been conducted to verify the importance of including site-specific characteristics in the calculation of toxicological impacts. The LCA studies that included site-specific information concluded that the conventional models (non-spatial) underestimate or overestimate the toxicological impacts by some orders of magnitude for some chemicals [20, 22–25]. Consequently, the inclusion of higher spatial resolution could potentially reduce uncertainty in the calculation of toxico-logical impacts [23].

The aforementioned studies show that it is possible to include site-specific characteristics in LCA studies to assess toxicological impacts. However, these studies are hindered due to the exigence arising from the need to adapt the life-cycle models to include site-specific data (when this data is available), causing these studies to be time consuming and costly.

5.2.2 USEtox method

The use of site-specific models to assess the toxicological impacts in LCA studies is vital, and the development of the USEtox model is a step forward to achieve it. The USEtox model estimates the intake fraction based on the physicochemical and toxicity properties in combination with environment characteristics of different emission compartments (indoor, urban air, rural air, freshwater, sea water, natural soil and agricultural soil) [26]. Consequently, the USEtox method converts the physicochemical and toxicity properties with the environment characteristics into potential toxicological impacts for ecosystems and human health.

As mentioned before, USEtox provides the possibility to select where the emission will occur. The selection of an emission compartment is necessary to determine the partition of a substance to different media (e.g. air, soil, water). Each one of these media has its own environment characteristics that will determine the behaviour of a substance. USEtox is not a site-specific model, but permits one to calculate the toxicological impacts for eight different continents and seventeen sub-continental regions (e.g. Europe, North America, Africa, Middle East, etc) of the world considering the different environmental conditions (wind, human population, rain, etc). This possibility can potentially decrease the uncertainty referent to the influence of different environment scenarios on the toxicological impacts.

The toxicological impacts on human health are calculated for cancer and non-cancer effects and its unit expresses, as described by Rosenbaum *et al* [26], 'the estimated increase in morbidity cases in the total human population per unit mass of a chemical emitted'—cases kg^{-1}–; while for ecotoxicity the units are expressed in 'potentially affected fraction of species integrated over time and volume per unit mass of a chemical emitted'—PAF m^3 day kg^{-1}. The units for the toxicological impacts for human health and ecosystems can also be converted to disability-adjusted life years (DALY kg^{-1}) and potentially disappeared fraction of species integrated over time and volume per unit mass of a chemical emitted (PDF m^3 day kg^{-1}).

5.3 Life-cycle assessment of the use of ammunition in military ranges

This chapter intends to demonstrate the feasibility and the type of information obtained when LCA methodology is employed to quantitatively assess the environmental impacts associated with the use of military products in military ranges. For that purpose, a generic large calibre ammunition (155 mm) was selected as an example to attain the aforementioned objectives. The assessment that will be shown was started on the NATO task group RTO-TR-AVT-179—Design for Disposal of Present and Future Munitions and Application of Greener Munition Technology, and contributed to the preparation of a scientific paper [15]—in which it is possible to obtain more information regarding environmental and toxicological impacts associated with the production and use of this munition. Among the main findings of that study, the importance of the production phase of the ammunition for the environmental impact categories is shown (e.g. global warming, acidification); whilst the impacts associated with the use phase of the ammunition are significantly higher for the toxicological impact categories. The main contributors to the human toxicity impacts are the emissions of metals (such as cadmium, chromium, zinc and lead), originating from the detonation of the warhead. Due to a nugatory emission of heavy metals, the influence of the emissions from the propellant combustion (point-firing emissions) for the toxicological impacts categories are lower than the impacts from the detonation emission.

The use phase is explored in more detail, as (i) the use of military munitions is the life-cycle phase of ammunition that matters for military ranges, and (ii) the potential toxicological impacts are most significant for this phase. The main objective is to assess the toxicological impacts of the use of a generic large calibre ammunition considering different environment characteristics in order to comprehend the influence of the site-specific conditions to those impacts. With this assessment, the main contributors to the toxicological impacts are also identified. It is also intended to quantify the toxicological impacts in military ranges associated with unexploded ordnances (UXO).

5.3.1 Description of the generic munition and inventory

The generic 155 mm ammunition was built considering the main constituents of a large calibre ammunition: warhead—a steel casing (35.5 kg), a driving band of copper (0.5 kg), a PETN booster (0.02 kg), and an explosive charge of composition B (8.5 kg) –; fuse—body of aluminium (0.45 kg), safety features made of brass (0.45 kg), electronic parts (0.1 kg) and a RDX detonator (0.002 kg) –; propellant charge—triple base powder (9.5 kg), a black powder primer (0.055 kg), a boron-potassium igniter (0.03 kg), and lead (0.085 kg) as a de-coppering agent –; and a steel container (22 kg).

The inventory associated with the use of the generic ammunition considered the emissions from the point of firing (combustion of propellant) and detonation. The firing emissions were measured from a range of weapons fired in an enclosed facility operated by the Aberdeen Test Center, and the details about the tests carried out are described in Onasch *et al* [27]. The capture of gaseous products and particulates was performed with a sampling probe located either fixed on a tripod or handheld during firing to obtain measurements at the desired locations (muzzle, breech, breathing zone). Then, the emissions were measured in a mobile laboratory stationed within 30 m of the weapons systems. Muzzle plume emissions were sampled downwind for the 155 mm ammunition using the sampling tripod mentioned above, in which twenty individual shots were fired. The gaseous and fine particulate matter in the air were measured, but the energetic material that can originate from the incomplete burning of the propellant was not sampled. The firing emissions associated with the combustion of the propellant for the generic 155 mm ammunition are shown in table 5.1.

Table 5.1. Emissions associated with the point of fire of a large calibre (155 mm) ammunition (Source: Onasch *et al* [27]).

Emissions (kg per ammunition)	
Carbon dioxide	2.74×10^{00}
Carbon monoxide	4.64×10^{00}
Nitrogen monoxide	3.90×10^{-04}
Nitrogen dioxide	3.14×10^{-04}
Formaldehyde	8.63×10^{-05}
Acetonitrile	2.80×10^{-04}
Acetaldehyde	3.20×10^{-05}
Benzene	9.03×10^{-04}
Ethylbenzene	2.19×10^{-04}
Toluene	2.79×10^{-04}
Styrene	4.87×10^{-05}
Hydrogen cyanide	3.78×10^{-02}
Acrylonitrile	2.29×10^{-05}
Acetic acid	1.44×10^{-04}
Ethyl acetate	8.98×10^{-05}
Furan (total)	8.98×10^{-05}
Black carbon	1.55×10^{-03}
Sulphate	3.04×10^{-04}
Nitrate	1.22×10^{-04}
Ammonium	9.69×10^{-05}
PAH	2.85×10^{-06}
Lead	4.30×10^{-07}

The detonation emissions were based on the report from US Army Environmental Command [28] that comprehends air emission factors for three types of calibre munitions (81 mm; 105 mm and 120 mm). The aforementioned report was selected because it has the major representation of substances measured and the sampling procedures carried out are reliable and consistent in accordance with EPA (Environmental Protection Agency) test methods or sound methodology, which were reviewed by the agency Emission Measurement Center (US) [28]. Tables 5.2 and 5.3 show the detonation emissions for the generic ammunition. Nitroglycerine was the only energetic material detected as the concentrations for other energetic materials (e.g. TNT, RDX) were probably under the detection limits.

5.3.2 Impact assessment

The toxicological impact on human health (cancer and non-cancer effects) and ecosystems can be calculated by employing the USEtox model. Beyond the default scenario used by the USEtox model, four different landscapes (Europe; USA + Southern Canada; Japan + Korea peninsula; and Africa + Middle East) were selected to understand the potential influence of the site-specific conditions to the toxicological impact. These diverse landscapes were chosen to represent a wide variety of site-specific characteristics. Tables 5.4 and 5.5 show the values of the main parameters that define the different environment characteristics used by the USEtox model to calculate the toxicological impacts. The characteristics from the different locales show some significant variations. The area of land is different for all locations as expected; the Japan + Korea peninsula presents a higher rain rate and lower irrigation than the other locations. Regarding the intake rates[1], Europe and USA + Southern Canada have a similar consumption pattern, in which the only exception is that Europe presents a higher consumption of below-ground products. Africa + Middle East and the Japan + Korea peninsula show a similar intake rate for above- and below-ground products, but with a lower consumption of meat and fish from freshwater for Africa + Middle East. Table 5.6 shows the values of the environment characteristics that are equal for all the landscapes in the USEtox model.

Toxicological impacts associated with the firing emissions
This section describes the toxicological impacts associated with the point of fire emissions that were calculated using the USEtox model. The impacts were calculated for a direct emission into the air or the natural soil. The analysis of the results obtained with the USEtox model also permits one to identify what is contributing to the intake fraction and the effect factor in order to understand the variations for the different locations. It is important to mention that the variations of the impact for the different locations are principally dependent on the intake factor,

---

[1] Intake rate provides the amount of substance that is ingested or inhaled by humans exposed to a certain concentration of those substances.

Table 5.2. Organic emissions associated with the detonation of a large calibre (155 mm) ammunition (source: US Army Environmental Command [28]).

Emissions (kg per ammunition)	
Carbon dioxide	1.91×10^{-01}
Carbon monoxide	4.54×10^{-03}
Nitrogen oxides	1.32×10^{-02}
Sulphur dioxide	3.54×10^{-04}
Acetaldehyde	3.04×10^{-05}
Acetonitrile	2.68×10^{-06}
Acetophenone	1.09×10^{-06}
Ammonia	1.50×10^{-05}
Benzene	3.54×10^{-05}
Beryllium	9.53×10^{-08}
Carbon disulphide	3.68×10^{-06}
Chloromethane	2.18×10^{-06}
Ethylbenzene	2.45×10^{-06}
Ethene	9.99×10^{-05}
Formaldehyde	1.45×10^{-05}
Methylene chloride	1.68×10^{-05}
2-methylnaphthalene	3.09×10^{-07}
Naphathalene	2.72×10^{-06}
Nitroglycerin	5.45×10^{-06}
Phenol	4.09×10^{-07}
Phosphorus	3.86×10^{-05}
Propinaldehyde	1.50×10^{-05}
Propylene	1.95×10^{-05}
Toluene	1.04×10^{-05}
Xylene	9.08×10^{-07}
Acetylene	2.95×10^{-03}
Benzaldehyde	8.17×10^{-06}
2-butenal	2.41×10^{-06}
1-butene	3.81×10^{-06}
Cis-2-butene	1.09×10^{-06}
Trans-2-butene	1.23×10^{-06}
Diethylphthalate	9.53×10^{-07}
Dodecane	1.95×10^{-06}
Ethane	3.18×10^{-05}
Hexaldehyde	5.90×10^{-06}
Methyl ethyl ketone	4.99×10^{-06}
1-propyne	1.54×10^{-06}
Valeraldehyde	8.17×10^{-06}
Dioxin (total)	6.81×10^{-12}
Furan (total)	7.26×10^{-06}

Table 5.3. Metallic emissions associated with the detonation of a large calibre (155 mm) ammunition (Source: US Army Environmental Command [28]).

Emissions (kg per ammunition)	
Antimony	1.50×10^{-05}
Arsenic	3.45×10^{-07}
Barium	1.32×10^{-05}
Cadmium	2.32×10^{-04}
Magnesium	1.14×10^{-01}
Manganese	6.36×10^{-05}
Lead	3.31×10^{-06}
Chromium	7.72×10^{-06}
Cobalt	2.86×10^{-06}
Copper	6.36×10^{-06}
Zinc	1.18×10^{-04}

as the effect factor is not dependent of the site-specific conditions; however, the effect factor can also influence the potential toxicological impact, as will be demonstrated.

Figure 5.1 shows a comparison between the toxicological impacts on human health, for the emission compartment continental air, associated with the point-firing emissions for different landscapes, and which substances are contributing to those impacts. Different locations present different toxicological impacts for cancer effects: the same emission has a higher impact on Africa + Middle East and a lower impact if released in the USA + Southern Canada. The variations observed for the toxicological impact are associated with the behaviour of the substances on the different landscapes. Benzene, formaldehyde and furan are substances that contribute to the impact for all locations, but the contribution varies with the location, which influences the impact observed.

For the case of non-cancer effects, lead dominates the impact for all locations, as this substance shows a higher effect factor (approximately two orders of magnitude higher) than the other substances emitted from ammunition firing. Therefore, the slight differences for the different landscapes are referent to the different intake factor of lead. The conclusions that can be drawn for the emission compartment natural soil are similar to the interpretation of the results presented before.

Figure 5.2 shows a comparison between the toxicological impacts on ecosystems associated with the point of fire emissions for different landscapes, and which substances are contributing to those impacts (for the emission compartments continental air and natural soil). The emissions to air result in similar ecotoxicological impacts, with a slightly higher impact for an emission occurring in the USA + Southern Canada and Africa + Middle East. Formaldehyde has a high contribution to the total impact in almost all the landscapes. However, lead shows a higher influence on the total impact for the locations with higher ecotoxicological impacts.

Table 5.4. Continental-scale characteristics of the environment properties for different landscapes in the USEtox model.

Location	Properties—continental scale									
	Area land (km^2)	Area sea (km^2)	Area freshwater	Area natural soil	Wind speed over mixing height $(m\ s^{-1})$	Rain rate $(mm\ yr^{-1})$	Depth freshwater (m)	Fraction run off	Fraction infiltration	Irrigation (km^3)
Default USEtox	$9.01 \times 10^{+06}$	$9.87 \times 10^{+05}$	3.00×10^{-02}	4.85×10^{-01}	6.65×10^{00}	$7.00 \times 10^{+02}$	2.50×10^{00}	2.50×10^{-01}	2.50×10^{-01}	$4.21 \times 10^{+02}$
USA + Southern Canada	$1.45 \times 10^{+07}$	$1.84 \times 10^{+06}$	3.44×10^{-02}	8.66×10^{-01}	6.95×10^{00}	$7.07 \times 10^{+02}$	$2.00 \times 10^{+01}$	3.66×10^{-01}	2.69×10^{-01}	$1.97 \times 10^{+02}$
Europe	$8.57 \times 10^{+07}$	$1.70 \times 10^{+06}$	1.57×10^{-02}	8.84×10^{-01}	6.77×10^{00}	$5.53 \times 10^{+02}$	1.54×10^{-01}	1.72×10^{-01}	2.69×10^{-01}	$1.15 \times 10^{+02}$
Japan + Korean peninsula	$5.98 \times 10^{+05}$	$4.25 \times 10^{+05}$	4.36×10^{-02}	8.56×10^{-01}	8.33×10^{00}	$2.37 \times 10^{+03}$	$1.30 \times 10^{+01}$	2.72×10^{-01}	2.69×10^{-01}	$7.72 \times 10^{+01}$
Africa + Middle East	$3.43 \times 10^{+07}$	$1.59 \times 10^{+06}$	1.97×10^{-02}	8.80×10^{-01}	4.26×10^{00}	$6.58 \times 10^{+02}$	$4.60 \times 10^{+01}$	1.84×10^{-01}	2.69×10^{-01}	$2.72 \times 10^{+02}$

Table 5.5. Intake rates considering various vectors for different landscapes in the USEtox model.

Location	Production-based intake rates (continental scale)					
	Above-ground produce (kg/(day.capita))	Below-ground produce (kg/(day.capita))	Meat (kg/ (day.capita))	Dairy products (kg/(day.capita))	Fish freshwater (kg/(day.capita))	Fish coastal marine water (kg/(day.capita))
Default USEtox	1.36×10^{00}	1.12×10^{00}	9.49×10^{-02}	2.37×10^{-01}	1.13×10^{-02}	3.60×10^{-02}
USA + Southern Canada	4.28×10^{00}	9.64×10^{-01}	3.48×10^{-01}	6.88×10^{-01}	3.30×10^{-03}	4.27×10^{-02}
Europe	2.34×10^{00}	1.35×10^{00}	1.88×10^{-01}	7.98×10^{-01}	3.30×10^{-03}	1.81×10^{-02}
Japan + Korean peninsula	9.38×10^{-01}	4.37×10^{-01}	8.71×10^{-02}	1.94×10^{-01}	2.71×10^{-02}	5.66×10^{-02}
Africa + Middle East	5.74×10^{-01}	8.70×10^{-01}	3.47×10^{-02}	8.53×10^{-02}	5.96×10^{-03}	1.53×10^{-02}

Table 5.6. Environment parameters that are equal for all landscapes in the USEtox model.

Parameters	Values	Unit
Temperature	$1.20 \times 10^{+01}$	°C
Surface wind speed	3.00×10^{00}	m s^{-1}
Soil erosion	3.00×10^{-02}	mm yr^{-1}
Human breathing rate	$1.30 \times 10^{+01}$	m^3/person day
Water ingestion	1.40×10^{00}	l/person day

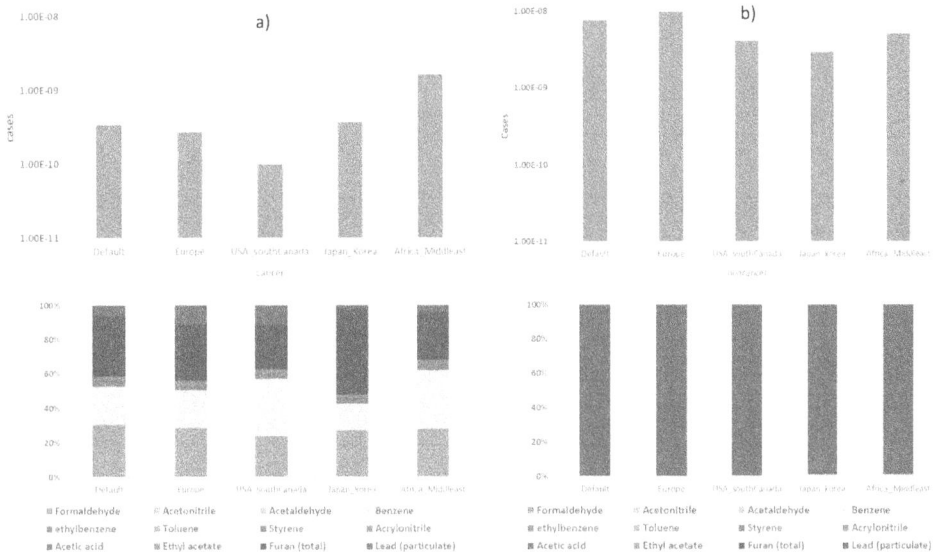

Figure 5.1. Impact comparison of the point-firing emissions for different landscapes on human health with (a) cancer effects, and (b) non-cancer effects for an emission into continental air.

For an emission into natural soil, the Japan + Korea peninsula becomes the location with higher impact, due to the behaviour of benzene, which shows a higher contribution to the total ecotoxicological impact. For instance, the default location presents approximately an impact of one order of magnitude lower than the Japan + Korea peninsula because benzene has a lower influence for the total impact—in fact, formaldehyde is the substance that is contributing more to the total ecotoxicological impact for the default location, as formaldehyde shows a higher persistence than benzene (higher fate factor) for the conditions in the default location. The main reason for the higher influence of benzene than formaldehyde to the total impact for the other landscapes is due to the higher emissions of benzene, as the intake fraction and the effect factor of these two substances for the natural soil compartment are very similar.

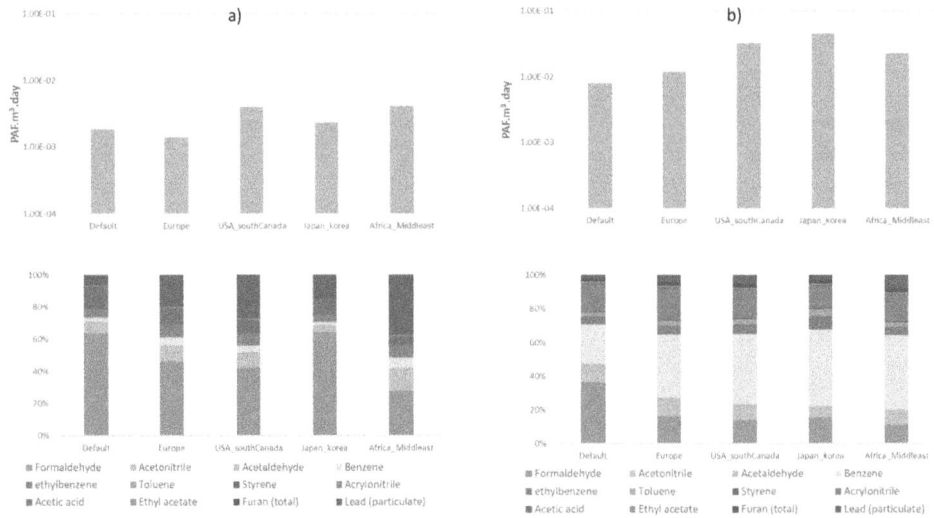

Figure 5.2. Impact comparison of the point-firing emissions on ecosystems for different locations: emission into (a) continental air, and (b) natural soil.

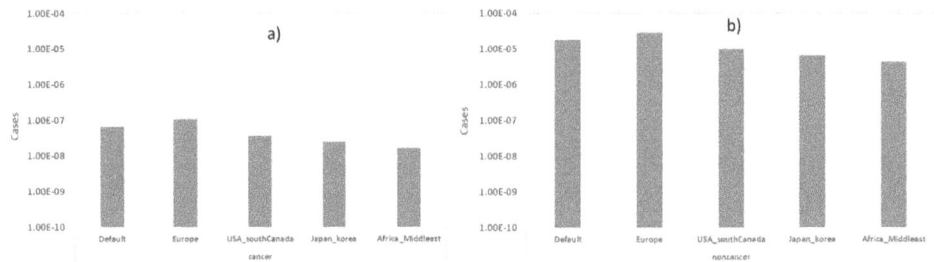

Figure 5.3. Impact comparison of the detonation emissions on human health for different landscapes with (a) cancer effects, and (b) non-cancer effects for an emission into continental air.

Toxicological impacts associated with the detonation emissions

Figure 5.3 shows a comparison between the toxicological impacts on human health associated with the detonation emissions for different landscapes. The toxicological impact is similar for the different locations, with slight variations with minimum magnitude. The total impacts are dominated by the emissions of cadmium in all landscapes, and for both cancer and non-cancer impact categories. The variations observed are referent to the different behaviour of cadmium for the specific locations. The intake fraction of cadmium is higher for an emission in Europe due to a higher contribution of the above-ground products that dominate the intake fraction. For an emission into Africa + Middle East the intake fraction of cadmium, beyond the above-ground products, is also influenced by the consumption of fish (20%) and the ingestion of water (20%) leading to a lower intake factor and, consequently, a lower toxicological impact. The results obtained from the assessment carried out for an emission into natural soil are similar to the continental air, so the interpretation of the results and the conclusions drawn are the same.

Figure 5.4 shows a comparison between the toxicological impacts on ecosystems associated with the detonation emissions for different landscapes. The detonation emissions into Africa + Middle East show higher impacts for both emission compartments, followed by the USA + Southern Canada location. Cadmium and copper are substances that have a higher contribution to the total toxicological impact, so the variations observed are influenced by the different behaviour of these two substances. As an example, the conditions for the default location create a lower impact due to a lower fate factor (lower persistence) of copper and cadmium—(e.g. air compartment: Cu = 4.61 days; Cd = 31.4 days)—than for the conditions of Africa + Middle East—(e.g. air compartment: Cu = 65.2 days; Cd = 552 days).

The analysis of the results associated with the point-firing and detonation emissions permits us to demonstrate that an environment's characteristics have an influence on the calculation of the toxicological impact. Therefore, generic recommendations to avoid the impacts on shooting ranges can lead to misleading solutions, as the environment characteristics of specific locations can influence the behaviour of the substance on the environment and, consequently, the toxicological impact on human health and ecosystems. It is important to carry out tailored LCA studies for specific locations, (eco)regions, or continents to ascertain in detail the potential toxicological impacts of the use of military munitions in shooting ranges, and present recommendations to those locations in particular.

Beyond the influence of environment characteristics to the impact, this type of analysis also helps decision makers and range managers to understand what substances are contributing to the impact in order to take action and present recommendations to decrease or avoid the burden on human health and ecosystems. The interpretation of the results obtained with the USEtox model also allows us to understand which factor (intake or fate factor) has a greater effect on the toxicological impact.

It is important to mention that the USEtox model shows some limitations towards completely comprehending the influence of an environment's characteristics to the calculation of toxicological impacts. Some characteristics are still the same for different landscapes—specifically the average temperature that also influences a substance's physicochemical parameters, type of soil, humidity, etc—that can influence the conclusions shown before. Therefore, research into the influence of

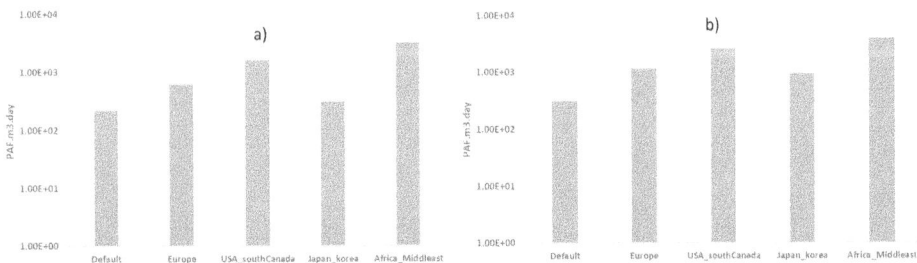

Figure 5.4. Impact comparison of the detonation emissions on ecosystems for different locations for an emission into (a) continental air, and (b) natural soil.

these characteristics to the toxicological impacts is important in order to include this information in the analysis to improve the calculation of site-specific impacts with the USEtox model.

Toxicological impacts associated with UXO

Data referent to the contamination of military ranges to a specific ammunition is difficult to find. However, since the main objective of this assessment is to illustrate the feasibility of the USEtox model to calculate the impacts associated with UXO, some assumptions had to be carried out: the entire warhead is assumed to degrade into the soil over time; therefore, the components and the resulting degradation products from the warhead are considered for emissions into soil. For instance, the components of steel (e.g. iron, magnesium, vanadium, etc) are assumed as an emission into soil. The same approach for the inert and energetic materials from the warhead was carried out. Table 5.7 shows the emissions into natural soil associated with the degradation of generic 155 mm ammunition.

UXO are munitions that do not function, since the ammunition for some reason does not detonate. Therefore, the environmental burdens associated with UXO need to be assessed in combination with ammunition that functions as expected. Therefore, three scenarios are considered for this analysis: in the first scenario the munition detonates as intended (the impacts are equal to the ones shown in this section for the default location in USEtox for an emission into natural soil); in the second and third scenario a rate of 5% and 10% of UXO is considered, respectively.

Figure 5.5 shows the toxicological impacts for human health (cancer and non-cancer effects) considering the three scenarios mentioned above. The toxicological impacts on human health with cancer effects has increased three orders of magnitude for the scenarios with UXO in comparison with the first scenario. The main

Table 5.7. Emissions into natural soil based on the degradation products from a generic 155 mm ammunition warhead.

Components	Degradation products	Amount per ammunition (kg)
Steel	Iron	33.90
	Chromium	0.76
	Nickel	0.36
	Manganese	0.39
	Molybdenum	0.012
	Niobium	0.023
	Tungsten	0.030
	Vanadium	0.021
Driving band	Copper	0.50
Composition B	2,4-DNT	1.70
	TNT	1.70
	RDX	5.10
booster	PETN	0.02

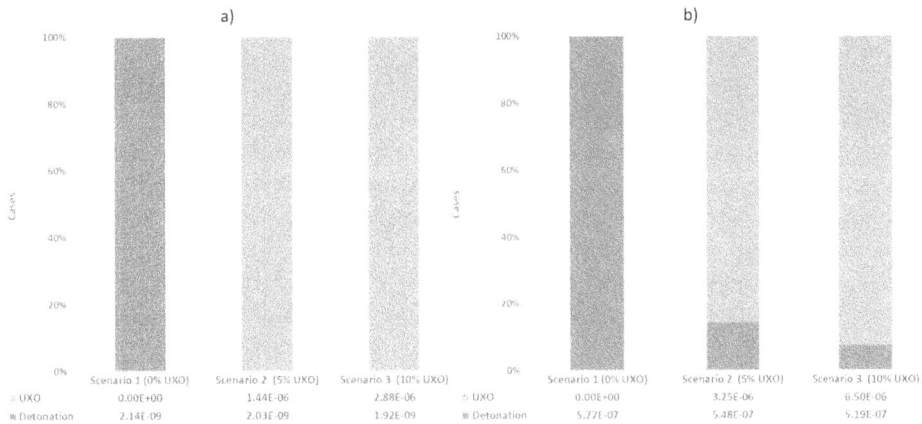

Figure 5.5. Impact comparison of the toxicological impacts on human health for cancer (a) and non-cancer effects (b), considering three scenarios (scenario 1: 0% UXO; scenario 2: 5% UXO; scenario 3: 10% UXO).

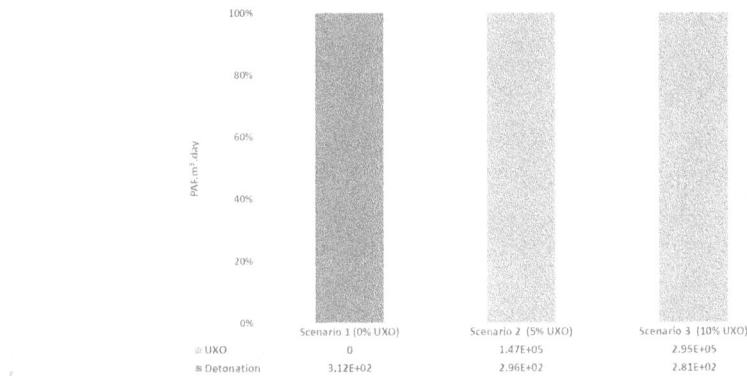

Figure 5.6. Impact comparison of the ecotoxicological impacts for an emission into natural soil, considering three scenarios (scenario 1: 0% UXO; scenario 2: 5% UXO; scenario 3: 10% UXO).

contributor to the total impact of the UXO is associated with the emission of nickel (87%) and RDX (20%). The quantities emitted by these two substances and their behaviour on the environment have a higher contribution to the total toxicological impact than the detonation emissions, even considering only 5% or 10% of the total impact of the UXO burdens. For the non-cancer effects, the potential toxicological impact increases only one order of magnitude. In this case, the main contributors to the total toxicological impact of UXO are associated with the emissions of TNT (81%) and 2,4-DNT (13%).

Figure 5.6 shows the ecotoxicological impacts for an emission into soil for the same three scenarios considered for human health impacts. The inclusion of UXO scenarios also shows an increase of approximately three orders of magnitude for the total ecotoxicological impacts. The emissions of copper (90%) and iron (9%) are the main contributors to the UXO impacts. Copper also contributes significantly to the total impact associated with the detonation emissions, but the quantity emitted in the UXO scenario is largely higher than the quantity of copper that is emitted from the ammunition detonation.

This analysis helps to show that the UXO burdens are significant to the contamination of military ranges. Only considering 5% and 10% of the total impact of UXO has a higher contribution to the toxicological impacts on human health and ecosystems than the toxicological impact associated with the emissions from detonation. Therefore, this analysis demonstrates that the malfunction of ammunition in training can pose a major impact on military ranges. However, it is important to indicate that the impacts resulting from UXO have an effect throughout a very long period of time for some substances, while the emissions from detonation pose an 'immediate' effect. The toxicological effect of UXO for a long period of time can influence the toxicological impacts shown in this study. Furthermore, the degradation products considered in this study can be different from what happens in reality as, for instance, RDX, PETN can degrade in other substances, and even the metals can bond with other organic substances and form metal–organic substances. These complex environmental mechanisms influence the behaviour of substances on the environment and, consequently, the potential toxicological impact.

5.4 Discussion of the application of life-cycle assessment methodology to manage military ranges

A traditional life-cycle assessment can be applied to the full life-cycle of military activities with the objective to quantify the environmental and toxicological impacts associated with those activities. This research focused on the military activities in shooting ranges or military training areas. Therefore, the toxicological impacts associated with the use of generic 155 mm ammunition were assessed, allowing one to ascertain what is contributing to the soil and air contamination, understand the influence of different environment characteristics to the toxicological impact and how that affects the intake fraction of different substances, and assess the impacts referent to unexploded ordnances. This information can help decision makers and range managers, with the inclusion of other complementary tools, to manage military ranges in a sustainable way.

Some methodological limitations that make the assessment of the toxicological impacts associated with the use of munitions difficult need to be discussed. Information regarding the contamination of military ranges is of great importance to develop LCA studies; however, to quantify and link the environmental impacts to the contamination of a specific type of ammunition is usually challenging, as most of the time training facilities have been used for long periods of time, and the majority of the items used were not recorded. The lack of data can be overcome with direct contact with range managers to ascertain the products and residues resulting from the use of specific ammunition, or by the continuous research of the emissions associated with the use of specific ammunition.

Another constraint is the quantification of the soil and groundwater contamination associated with UXO, due to the degradation of inert materials and energetic materials throughout time. These substances are not immutable due to biotic or abiotic degradation, creating different substances that can pose different potential toxicological impacts from the original energetic materials emitted. The toxicological burdens of these contaminants are also dispersed for a long period of time, which contributes to a different rate of toxicological impacts for different periods of the usage of the military range. The difficulty in accounting for this environmental mechanism is in identifying the degradation products and the rate of that degradation, which are dependent on time and environment conditions that vary for different training sites. The LCA methodology, as mentioned before, usually neither addresses site-specific impacts nor impacts dynamic with time.

This methodological limitation can be overcome with the inclusion of different degradation scenarios dependent on time and location in order to create eco-regions that simulate the site-specific conditions of different training sites for several countries. The eco-regions are implemented in the USEtox model to predict the toxicological impacts for that specific characteristic. Furthermore, a database with the environment conditions for each eco-region can be created and a framework explaining how to apply this information in USEtox can help range managers to control and take decisions for their training sites. The inclusion of the degradation products and their variation through time will allow one to ascertain what is contributing to the soil and water contamination; understand which types of ammunition are causing the contamination; estimate the amount of rounds that can be fired per year relative to a reference that estimates the maximum of impact permitted by the legislation; determine the potential human health impacts for training in indoor or outdoor conditions; and compare and select the appropriate technologies to remediate training sites that are contaminated. Ultimately, it is possible to develop tools that will allow shooting range managers to become more aware of the main environmental and toxicological impacts of the use of ammunition in order to define strategies to manage the ammunition burdens. This tool should also include the possibility to incorporate other aspects (e.g. economic, performance, safety, social factors) that can influence the selection of an appropriate sustainable decision.

References

[1] Francis R A 2011 The impacts of modern warfare on freshwater ecosystems *Environ. Manag.* **48** 985–99

[2] Tsuji L J S, Wainman B C, Martina I D, Sutherland C, Weberd J, Dumasd P and Nieboerb E 2008 The identification of lead ammunition as a source of lead exposure in first nations: the use of lead isotope ratios *Sci. Total Environ.* **393** 291–98

[3] Certini G, Scalenghe R and Woods W I 2013 The impact of warfare on the soil environment *Earth Sci. Rev.* **127** 1–15

[4] Thiboutot S, Brousseau P and Ampleman G 2015 Deposition of PETN following the detonation of seismoplast plastic explosive *Propell. Explos. Pyrotech.* **40** 329–32

[5] Hewitt A D, Jenkins T F, Walsh M E, Walsh M R and Taylor S 2005) RDX and TNT residues from live-fire and blow-in-place detonations *Chemosphere* **61** 888–94

[6] Walsh M R, Walsh M E and Hewitt A D 2010 Energetic residues from field disposal of gun propellants *J. Hazard. Mater.* **173** 115–22

[7] Walsh M R, Walsh M E, Ampleman G, Thiboutot S, Brochu S and Jenkins T F 2014 Munitions propellants residue deposition rates on military training ranges *Propell. Explos. Pyrotech.* **37** 393–406

[8] Walsh M R, Walsh M E and Voie Ø A 2014 Presence and persistence of white phosphorus on military training ranges *Propell. Explos. Pyrotech.* **39** 922–31

[9] Walsh M R, Walsh M E, Ramsey C A, Thiboutot S, Ampleman G, Diaz E and Zufelt J E 2014 Energetic residues from the detonation of IMX-104 insensitive munitions *Propell. Explos. Pyrotech.* **39** 243–50

[10] Walsh M R, Walsh M E, Ramsey C A, Brochu S, Thiboutot S and Ampleman G 2013 Perchlorate contamination from the detonation of insensitive high-explosive rounds *J. Hazard. Mater.* **262** 228–33

[11] Walsh M R, Walsh M E, Taylor S, Ramsey C A, Ringelberg D B, Zufelt J E, Thiboutot S, Ampleman G and Diaz E 2013 Characterization of PAX-21 insensitive munition detonation residues *Propell. Explos. Pyrotech.* **38** 399–409

[12] Clausen J, Robb J, Curry D and Korte N 2004 A case study of contaminants on military ranges: Camp Edwards, Massachusetts, USA *Environ. Pollut.* **129** 13–21

[13] ISO 2006 *ISO 14040: Environmental Management—Life Cycle Assessment - Principles and Framework* (Geneva: International Organization for Standardization)

[14] Ferreira C, Ribeiro J, Almada S, Rotariu T and Freire F 2016 Reducing impacts from ammunitions: a comparative life-cycle assessment of four types of 9 mm ammunitions *Sci. Total Environ.* **34–40** 566–67

[15] Ferreira C, Ribeiro J, Almada S and Freire F 2017 Environmental assessment of ammunition: the importance of a life-cycle approach *Propell. Explos. Pyrotech.* **42** 44–53

[16] Ferreira C, Ribeiro J, Mendes R and Freire F 2013 Life-cycle assessment of ammunition demilitarization in a static kiln *Propell. Explos. Pyrotech.* **38** 296–302

[17] Ferreira C, Ribeiro J, Roland C and Freire F 2019 A circular economy approach to military munitions: valorization of energetic material from ammunition disposal through incorporation in civil explosives *Sustainability* **11** 255

[18] Ferreira C, Freire F and Ribeiro J 2015 Life-cycle assessment of a civil explosive *J. Clean. Prod.* **89** 159–64

[19] European Commission - Joint Research Centre - Institute for Environment and Sustainability 2010 *International Reference Life Cycle Data System (ILCD) Handbook - General Guide for Life Cycle Assessment - Detailed Guidance* 1st edn EUR 24708 EN (Luxembourg: Publications Office of the European Union)

[20] Sala S, Marinov D and Pennington D 2011 Spatial differentiation of chemical removal rates from air in life cycle impact assessment *Int. J. Life Cycle Assess.* **16** 748–60

[21] Hollander A, Pistocchi A, Huijbregts M, Ragas A and Meent D 2009 Substance or space? The relative importance of substance properties and environmental characteristics in modeling the fate of chemicals in Europe *Environ. Toxicol. Chem.* **28** 44–51

[22] Pennington D W, Margni M, Ammann C and Jolliet O 2005 Multimedia fate and human intake modeling: spatial versus nonspatial insights for chemical emissions in Western Europe *Environ. Sci. Technol.* **39** 1119–28

[23] Manneh R, Margni M and Deschenes L 2010 Spatial variability of intake fractions for Canadian emission scenarios: A comparison between three resolution scales *Environ. Sci. Technol.* **44** 4217–24

[24] Manneh R, Margni M and Deschenes L 2012 Evaluating the relevance of seasonal differentiation of human health intake fractions in Life Cycle Assessment *Integr. Environ. Assess. Manag.* **8** 749–59

[25] Gandhi N, Huijbregts M A J, Meent D, Peijnenburg W J G M, Guinée J and Diamond M L 2011 Implications of geographic variability on comparative toxicity potentials of Cu, Ni and Zn in freshwaters of Canadian ecoregions *Chemosphere* **82** 268–77

[26] Rosenbaum K *et al* 2008 USEtox—the UNEP-SETAC toxicity model: recommended characterisation factors for human toxicity and freshwater ecotoxicity in life cycle impact assessment *Int. J. Life Cycle Assess.* **13** 532–46

[27] Onasch T B, Wood E C, Timko M T, Owens K C, Beardsley H M, Kolb C E, Fortner E C and Knighton W B 2008 *In situ* characterization of point of discharge fine particulate emissions project WP-0420 ESTCP Final Report (July)

[28] EPA 2009 Ordnance detonation emission factors developed based on exploding ordnance Emission Study Phase II Series 6, 5th edn ch 15 AP-42 *Report on Compilation of Air Pollutant Emission Factors vol 1* Stationary point and area sources US Army Corps, Aberdeen Proving Ground, Maryland

Chapter 6

Hazard assessment of exposure to ammunition-related constituents and combustion products

Monique van Hulst, Jan P Langenberg and Wim P C de Klerk

The awareness that ammunition-related compounds and the combustion products thereof may pose a health hazard to military personnel in the short or longer term is growing. Assessing the health hazard resulting from such exposures poses a challenge. Approaches used in industrial occupational hygiene are not as applicable to this type of exposure, due to the fact that military personnel are mostly exposed to complex mixtures and combinations of chemicals, in contrast to workers in the industry. A hazard index approach seems to be most suitable, albeit that for many of the components in the mixtures, exposure limits have not been established, often because toxicity data are absent. Since current *in silico* models are unable to adequately predict the composition of combustion products, experiments need to be performed to identify and quantify the components in the complex mixtures. Standard procedures for these experiments would make comparison of the results obtained by various laboratories easier. Ideally, measurements are done on site during the experiment, if that is not possible care should be taken that the composition of the sample does not alter in the time span between sampling and analysis. For hazard assessment of exposure to ammunition-related constituents and combustion products a stepwise approach is recommended: *in silico* prediction of the components of concern, followed by measurement of the actual composition, calculation of the hazard index, *in vitro* assessment of the toxicity of the emitted product. A limited *in vivo* study would be useful to confirm the predictions derived from the hazard index approach and *in vitro* toxicity tests.

doi:10.1088/978-0-7503-1605-7ch6

6.1 Introduction

For decennia, military personnel were only concerned about the physical harm that ammunition could cause to the human body. The fact that ammunition is composed of chemical substances that either as such, or after combustion, may exert toxic effects in the short or long term upon repeated exposure was long overlooked.

The awareness of the toxicity of ammunition-related compounds was perhaps first triggered by military smokes. Exposure to HC-smoke in confined spaces has caused several fatalities [1], whereas some of the original colored smokes appeared to contain rather toxic dyes, some of which appeared to be mutagenic in the Ames test [2].

In the aftermath of Operation Desert Shield/Desert Storm it was suggested that exposure to depleted uranium (DU) from kinetic energy weapons may have contributed to the so-called 'Gulf War illness' [3]. Although a causal relationship was not shown, the awareness that exposure to ammunition-related compounds may cause adverse health effects in the short or long term was created. Tungsten-based alloys were advocated as 'safe' alternatives for DU, with comparable performance as a weapon, but some of these alloys were shown be toxic and/or carcinogenic in animal models [4], casting doubts over their safety.

Exposure to lead from small caliber ammunition is known to cause adverse health effects [5, 6]. Therefore, lead-free ammunitions are being developed, often referred to as 'green' ammunitions. That such ammunitions are not necessarily less toxic has been shown by Voie *et al* [7]. Volunteers firing lead-free ammunition with an HK416 developed symptoms resembling 'metal fume fever', perhaps caused by high exposure to copper liberated by firing the lead-free ammunition.

These examples show that there is every reason to evaluate the toxicity of ammunition-related compounds and the combustion products thereof. This chapter addresses the state-of-the-art methodologies to assess the hazard posed by such compounds, as well as the current gaps and possibilities to fill these gaps.

6.2 Approaches to performing experiments, analyses and evaluations

6.2.1 *In silico* prediction of emitted products

The most convenient way to determine the emitted products during firing of ammunition would be the use of computer models and prediction tools. However, the complexity of the mixtures, the variation of combustion conditions, the effects of after-burning when the produced products leave the barrel, the influence of the friction of the projectile inside the barrel and the conditions outside of the barrel all influence the final composition (chemically, concentration and particle size) of the emissions. Koch and coworkers [8] wrote a review on thermodynamic codes used for the prediction of compounds formed when burning propellants. They compared the CERV, CHEETAH, EKVI, ICT, NASA, REAL and TANAKA codes by performing test cases on black powder, magnesium/teflon/viton (MTV), spectral flare, a propellant and some other compositions. Most of the models predicted similar emissions for the materials used in the case studies. However, only small numbers of volatile and condensed chemical species were predicted, due to limitations of the databases with components that are incorporated in the models. Jensen and coworkers [9] studied a model looking at atom interactions instead of the thermodynamic equilibrium that most of the models use. That particular software is called ReaxFF and it is based on 'molecular dynamics' (MD). They compared the predicted emission with the experimental results and noticed that the predicted and measured concentrations were more in agreement than when using the thermodynamic steady state models. This software has the advantage of taking into account that products are formed by the interaction between all components under the chosen combustion conditions. However, the measurements take a long time and the software is limited with respect to the number of atoms that can be entered into the calculation window. This means that compositions with a large number of different chemical components or compositions with a small percentage of a particular component cannot be calculated in the same ratios due to this software limitation.

For pyrotechnic smoke compositions Shaw and coworkers [10] used the program FactSage to predict combustion products. This software is able to model the combustion products without the limitations of adiabatic conditions and contains a large database of inorganic components.

6.2.2 Laboratory and field testing

Ammunition

For the determination of substances emitted during firing of ammunition it is important to perform tests under controlled conditions, to ensure that the wind direction and velocity do not influence the concentrations to be measured. Also, the humidity and temperature may influence which combustion products are formed as well as follow-on reactions when the combustion products leave the barrel. At the time of writing this chapter there are no internationally agreed guidelines for determining the emitted products under controlled circumstances. In the literature, experiments have been described with different setups and different amounts of ammunition being fired. Moxnes and coworkers [11], for example, used a plastic cylinder-shaped setup with a volume of around 1 m^3, where they fired 10 rounds of ammunition. Wingfors *et al* [12] used a setup with only two plastic boards on the side to make sure the emitted products would stay within the measuring range, firing only one round to make sure the emitted products would stay within the measuring range. If the signal was within the range, two additional rounds were fired and measured/sampled for 10 min.

Diaz and coworkers [13] performed a comparative study of the main combustion gases and volatile organic compounds (VOCs) between an indoor and outdoor setup. They fired six bags of M67 (single-based propellant) propelling charges. The concentrations measured for the indoor tests were higher than those observed during the outdoor test, which is not really surprising. During the outdoor test, for example, the carbon monoxide (CO) concentration measured was a factor of 10 lower than the carbon dioxide (CO_2) concentration, whereas for the indoor experiments the concentrations of both compounds were much higher and the CO concentration was almost a factor of 3 higher than that of CO_2. The authors also mention that, when firing a lower number of charges (so less propellant), the concentrations of VOCs and semi-volatile organic compounds (SVOCs) increase, whereas the concentrations of particles and gasses decrease. They suggest that when firing with less propellant the combustion temperature is lower, resulting in the formation of larger molecules such as VOCs and SVOCs. These observations demonstrate that the experimental conditions have a profound impact on the composition of the combustion products.

An example of a setup for performing experiments to measure the emission products for small caliber munitions is shown in figure 6.1, using a 'party tent' as the chamber from which air is analyzed, with a volume of ca. 27 m^3. The whole experiment was performed in a bunker. The front- and backside of the tent were closed during firing and measuring the emissions in order to eliminate the influences of air flow through the tent. Some of the measurement and sampling equipment was situated inside the tent, such as the particle size distribution measurement system, and the tubes withdrawing air from the tent into the infrared spectrometer, the gas chromatograph with mass spectrometer and the *in vitro* air–liquid interface systems.

TNO [14] used the setup shown in figure 6.1 to measure emitted products when firing different numbers of rounds of 9 mm ammunition. A comparison was made between firing 15 rounds and 150 rounds in one sequence. For example, for the measured nitric oxide (NO) and nitric dioxide (NO_2) concentrations, shown in figure 6.2, the ratio between the compounds NO and NO_2 was not the same, and did not have a constant value.

When firing 150 rounds of ammunition the NO concentration exceeds that of NO_2, which suggests that there is insufficient oxygen available to form NO_2. During the course of time both the concentrations of NO and NO_2 at the 15-rounds experiment decrease; however, at the 150-rounds experiments the NO concentration decreases

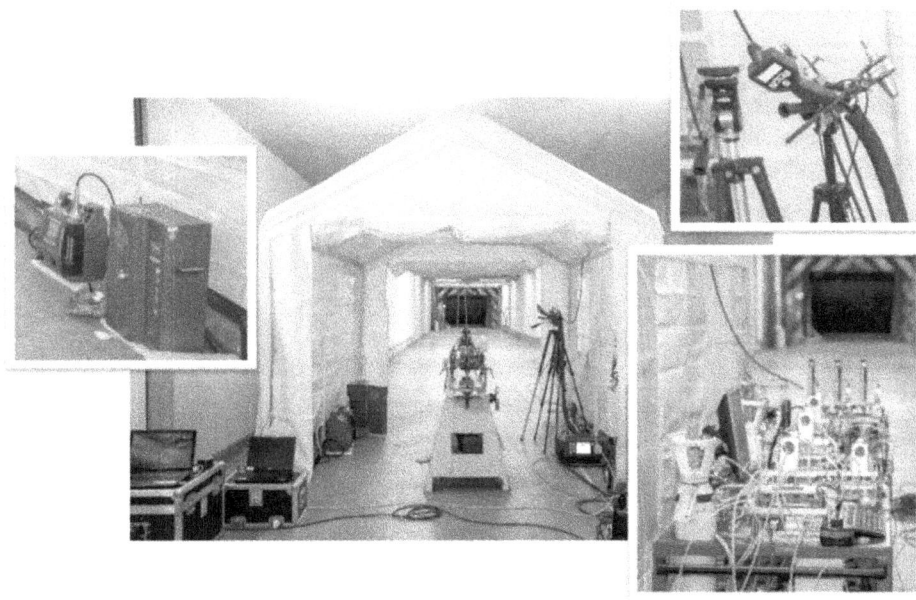

Figure 6.1. Example of a setup used to perform emission measurements during firing of ammunition (Source: TNO, The Netherlands).

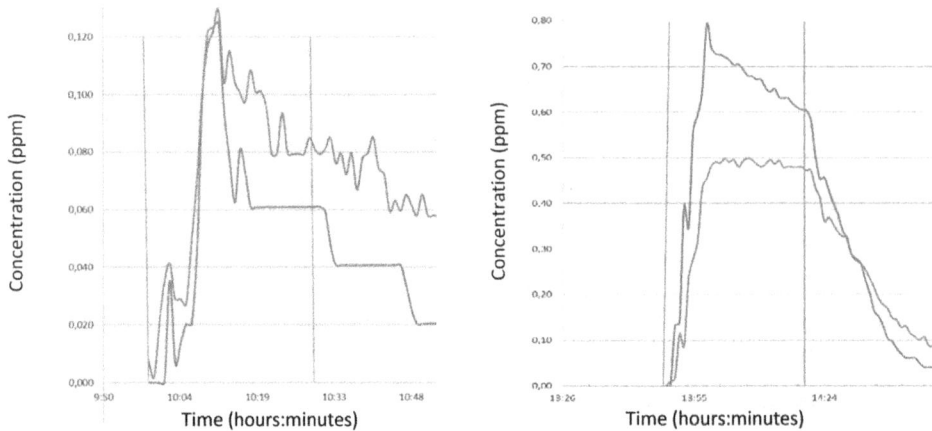

Figure 6.2. NO (blue) and NO_2 (red) concentrations after firing (a) 15 rounds and (b) 150 rounds. The blue vertical lines show the period of analyses [14] .

more rapidly than that of NO_2. Interestingly, the CO and CO_2 concentrations did not show this effect. Only the CO concentration increased significantly (results not shown).

These findings indicate that the volume of the test chamber in combination with the number of shots fired may influence the concentration of the various combustion products, the availability of oxygen being one of the critical parameters.

6.2.3 Military smokes

Experiments performed to determine exposure to military smokes show less variability than that observed for ammunition articles, due to the fact that smoke canisters are a more complete product, where the smoke formation is not dependent on the launch procedure. Hand grenades are thrown and form smoke from the canister. Climatic conditions may have an impact on the composition of the smoke that is formed, but also on the physical characteristics of the smoke particles. For instance, the humidity of the air will determine whether a sufficiently dense smoke screen will be produced from red phosphorous.

The key issue is to perform a representative experiment that is sufficiently reproducible to compare the results obtained for other smoke formulations. For this purpose, the experimental conditions need to be controlled as much as possible without influencing the smoke production.

At TNO, a variety of studies were performed to determine the smoke compositions including their potential health effects [15]. Of these studies, the most simple was with red phosphorous (RP) smoke from a launcher. In real life, the RP pellets are ejected from the canister to burn on the ground to form smoke. In our experiments this was simulated by igniting the RP pellets with an electrically heated wire in a closed chamber with subsequent analysis of the emitted products. When colored smoke hand grenades were studied, the complete canisters were used and ignited inside a bunker. It was ensured that the analytical equipment was situated at the same distance from the smoke generating source during each experiment.

Figure 6.3. Various smoke composition configurations for smoke formation: compressed pellet (left) and full-scale grenade (right).

If there is not yet a tangible final product to test, but only a smoke generating composition, it is important to test the composition as it would be in the final product. There is a large difference between the emitted products obtained from ignition of a powder or a compressed pellet. This is mostly due to the burning surface area and the amount of oxygen present in the sample due to packing. Also, confinement of the smoke forming composition is of influence on the formation of the emission products. See figure 6.3 for a representation of various smoke composition configurations.

6.2.4 Sampling and analysis

Preferably, analysis of the combustion products is done immediately and *in situ*. Unfortunately, not all compounds of interest can be measured on site, requiring sampling and transport to a laboratory, obviously, the sample should be stable in order to make the analysis representative of the composition at the moment the sample was drawn.

One of the conclusions drawn from the above-mentioned studies was that samples need to be stored properly immediately after being drawn in order to avoid artifacts. For example, red phosphorous smoke particles collected on scanning electron microscope (SEM) filters [14] appeared to attract water from the air resulting in the formation of little droplets (figure 6.4). In this case, the samples had to be stored in a desiccator directly after being drawn, in order to avoid droplet formation.

Another example is during a comparative study [14] on colored smokes, where particles were collected on the filters of an electrical low pressure impactor (ELPI) to study the morphology of the particles on the various stages of the ELPI. Surprisingly, for one of the smokes, needles were observed under the SEM for ELPI stages 2 and 6. These needles were considerably larger than the expected particle size range for these stages (figure 6.5). It was concluded therefore that these needles were formed on the filters upon storage.

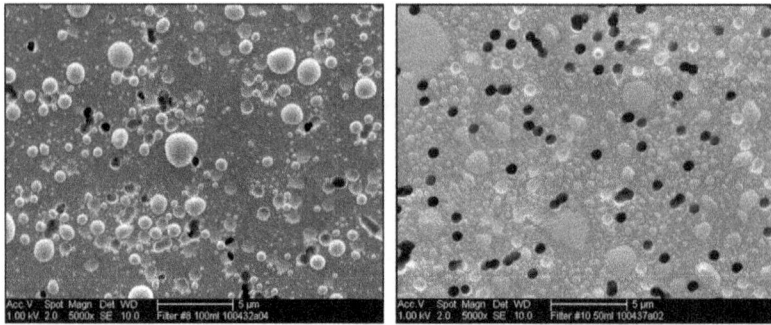

Figure 6.4. SEM filter placed in a desiccator immediately after sampling (left) and SEM filter after 10–15 min outside of the desiccator (right). For an explanation see the text. The black spots are the holes in the filter [14].

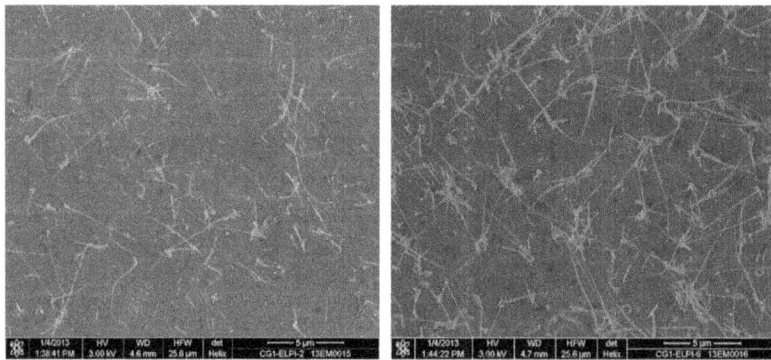

Figure 6.5. Collected smoke particles on an ELPI-plate (left) stage 2 and (right) stage 6 showing needle formation in the ELPI (plates collecting 0.07 and 0.32 μm aerodynamic size particles, respectively) [14].

Also, the sampling of volatile components and gasses (H_2S, CS_2 and SO_2) was investigated by TNO [16]. The effect of storage conditions and -duration were investigated for compounds adsorbed onto Tenax and for Cali-5-bond (2 l) and Tedlar (50 l) gas sampling bags. Storage at 0 °C seems to be best; however, 20 °C seems to be acceptable. At 40 °C and higher, after a few days only 20% of the original concentrations were left. When using a steel gas cylinder as a sampler, the decrease in concentration was 50% already after a few hours. The best way to analyze these gases is online/on site or at the latest within 24 h after sampling, while storing the samples under the proper conditions prior to analysis.

6.3 Hazard assessment and evaluation

6.3.1 General background/desktop approaches

In industrial occupational hygiene it is common practice to predict possible health effects from measured concentrations of a single chemical compound in the workplace, and relating these concentrations to threshold limit values (TLVs) derived for exposure scenarios, such as 8 h/day, 5 days/week, a working life, long or short term exposure limits (STEL-values) for 15 or 30 min exposures. These limit values are defined in such a way that no health effects in exposed workers should occur in the short or long term.

For military personnel such civilian industrial threshold values are often not applicable. Under operational conditions exposure may be 24 h/day for several months. The US Army Public Health Center (USAPHC) has derived 'military exposure guidelines' (MEGs), which are health-based chemical concentrations in air, water and soil for various military exposure scenarios during deployments, designed as decision aids for health risk assessors to evaluate the significance of field exposures to chemical hazards during deployments. The USAPHC technical guide containing the MEGs and the rationale behind them is freely accessible on the Internet [17].

Whereas TLVs and STEL-values are aimed at avoiding health effects, MEGs are defined for various effect levels, from 'negligible', via 'marginal' and 'severe' up to 'catastrophic'. The importance of the military mission will determine which effect level is considered to be acceptable.

MEGs are defined for one time exposure to a certain chemical. However, mostly, military personnel are not exposed to a single component, but to mixtures or combinations of chemical substances. Ammunitions are a complex mixture of chemicals and the emission products even more complex. The same applies to military smokes. The prediction of the toxicity of a mixture or combination of chemical substances is a largely unexplored area within toxicology.

For non-carcinogenic effects, the 'hazard index' (HI) approach is a basic approach for assessing the joint toxic action of a mixture of chemicals, assuming dose additivity. Exposures or doses for the various components of the mixture are scaled by a defined level of exposure generally regarded as 'acceptable' or 'safe' by the agency performing the assessment. An HI is defined as the sum of the quotients of the estimated dose of a chemical divided by its limit concentration or comparable value. When the hazard index for a mixture exceeds unity, concern for the potential hazard of the mixture increases. This straightforward approach is highly valuable, but it ignores potential interactions of toxic substances in mixtures at relevant exposure levels/doses.

The US Agency for Toxic Substances and Disease Registry (ATSDR) [18] has defined interactions as being deviations from the results expected on the basis of additivity of the doses of the individual components. In their terminology, interactions can be categorized as either less than additive (in the case of antagonism, inhibition, masking) or greater than additive (in the case of synergism, potentiation). In the noninteraction category dose additivity is applicable (no apparent influence on toxicity of each of the compounds in the mixture).

ATSDR proposes a tiered hazard assessment. If the hazard quotient of a single non-carcinogenic component in the mixture/combination exceeds a value of 0.1, then these compounds are addressed in the more refined Tier 2 and Tier 3 analyses. For non-carcinogens, dose additivity is assumed. For carcinogens the individual cancer risk estimates are determined, and response additivity is assumed. Interestingly, the ATSDR also takes other stressors into consideration such as noise and radiation, which may be relevant for military personnel.

The approach proposed by ATSDR is very thorough and is built on a solid scientific fundament. Yet, the approach is not simple, and requires availability of threshold limit values or at least some quantitative toxicological data. Unfortunately, such data are often lacking for ammunition-related compounds and combustion products thereof.

Prediction of the toxicity via computer models is not trivial, but *in silico* models for this purpose are under development. The more promising models not only incorporate molecular structure information, but also data on toxicokinetics and metabolism, as well as on toxicological pathways. Currently, the predictions of such models are not highly detailed, but compounds with a known similar toxicity will be identified, and the TLVs, STEL-values and/or MEGs of those compounds can be used as an approximation of the toxicity of the compounds of concern for which toxicological information is lacking.

Obviously, ATSDR is not the only organization that has defined approaches for desktop assessment of health hazards resulting from exposure to mixtures of chemicals. Comparable approaches are being proposed by the American Conference of Industrial Hygienists (ACIH) [19], the Occupational Safety and Health Administration (OSHA) [20], the National Institute for Occupational Safety and Health (NIOSH) [21] and the Environmental Protection Agency (EPA) [22].

Extrapolation to humans

The described desktop approaches are intended to make predictions of the health hazards resulting from exposure of humans to mixtures of chemical substances. The approaches based on dose additivity are relatively simple in use. Provided that threshold values are available, application of the dose additivity method does not require in-depth toxicological knowledge and can easily be used for any mixture of chemicals, if the concentrations of the components in a mixture are known. Because it is based on the assumption of dose additivity, the hazard index method is most appropriately applied to components that cause the same effect by the same mechanism of action. In practice, it may also be applied to components with different target organs (sometimes as a screening measure). The hazard index method does not take into account chemical interactions among the components of the mixture.

Many applications of the hazard index approach are intended for occupational hygiene purposes, i.e. for chronic exposure to the mixtures (e.g. 8 h a day, 5 days a week, a working life long), whereas for firing ammunition only acute toxicity resulting from short-term exposures have to be considered.

The applicability of the dose additivity hazard index method depends on the availability of threshold values that are applicable to these exposure conditions. If necessary, the hazard index method can be expanded with more refined procedures for chemicals with different toxicological targets and or pathways and to assess the impact of interactions between the chemicals in the mixture. This, however, requires in-depth insight into the toxicology of the various compounds in the mixture and data that will often not be readily available.

ATSDR performed the assessment of joint toxicological profiles of a number of certain priority mixtures of compounds that are of special concern for ATSDR. These mixtures are often encountered together at hazardous waste sites and may thus also be relevant for military personnel. The toxicological profiles that have been derived for these mixtures are available at the ATSDR website [23]. These are highly informative with respect to how the approach can be applied.

6.3.2 *In vitro* approaches

Exposure to ammunition-related compounds may occur via inhalation, via the skin and via ingestion. Of these, the respiratory route is by far the most prominent.

One of the established *in vitro* approaches for inhalation toxicity is by exposing cultured human lung cells in a controlled way to emission products resulting from ammunition articles. For this purpose a so-called 'air–liquid interface' (ALI) system is considered to be the most realistic procedure for exposing the lung cells to gases, vapors and aerosols, mimicking the real-life situation within the respiratory tract. Several ALI-systems are commercially available, among others CULTEX® and Vitrocell®.

An important feature of ALI systems is that the cells are exposed in a controlled way to the air sample under study, among others via the use of mass-flow controllers. The use of human lung cells in culture in combination with the ALI-system for the assessment of biological effects is an interesting *in vitro* approach. The sensitivity of the system has been shown to be high enough for studying relevant concentrations of single components such as NO_2, benzene and complex mixtures, such as diesel exhaust and cigarette smoke [24, 25].

In our experience the sensitivity for certain single military chemicals is not very impressive (for instance ammonia, where the ALI-system responded only at an exposure level that causes severe health effects in humans), but the system appeared to respond highly sensitively to mixtures of chemicals. This approach has been successfully applied to a variety of military and civilian toxicity issues, such as emissions from 'biodiesel' in comparison with regular diesel fuel, acute toxicity of colored and red phosphorus smokes [15], evaluating 'green' training ammunition [26] and so on.

For most purposes, collecting samples of emission products in gas sampling bags, which are subsequently connected to the ALI-modules, works very well. There is, however, a small risk that components in the mixture will (continue to) react whilst in the gas sampling bag. Also, deposition of components on the inside walls of the gas sampling bag may occur. In addition, contaminants from the material of the gas sampling bags may be released in the mixture to be analyzed. In such cases, the mixture that is led over the cultured lung cells in the ALI-system is not representative of the composition of the emission at the time of sampling. As an alternative we have developed a technique in which emission products are sampled from the experimental chamber (e.g. a fume hood, bunker or shelter tent) via a tube and led directly to the ALI-module, which is located just outside the experimental chamber. We did note some deposition of material inside the tube, but not to an alarming extent.

Cell types

The lung cell most widely used in ALI systems is the A549 cell [25, 27, 28]: human alveolar basal epithelial cells, isolated in 1972 from cancerous lung tissue of an 58-year-old Caucasian male. Alveolar basal epithelial cells are squamous in nature and responsible for the diffusion of substances, such as water and electrolytes, across the alveoli of lungs. They grow adherently, as a monolayer, *in vivo* as well as *in vitro*.

The fact that this is a cancer cell is considered to be a drawback. An implication of the A549 cell is that a more refined read-out of cells on the basis of toxicogenomics is not advised due to chromosome differences between these cancer cells and normal cells. On the other hand, the A549 cell is easily cultured and very robust when handled in exposure systems. In our experience A549 cells can be exposed in an air–liquid interface setting for at least 1 h before the cells become less viable. A lot of data have been generated on the effects of a variety of airborne toxicants on this type of cell, allowing comparison of the toxicity of (new) compounds or mixtures thereof with known compounds/mixtures.

As an alternative, Calu-3 cells can be used, which are derived from human bronchial submucosal glands. In the human lung, the submucosal glands are a major source of airway surface liquid, mucins, and other immunologically active substances and Calu-3 cells reflect these properties. Calu-3 cells can be grown at an air–liquid interface and demonstrate many characteristics of the bronchiolar epithelium, which, *in vivo*, serves as the barrier layer between inspired gas and underlying tissues.

Another option would be the use of primary bronchial epithelial cells (PBECs), which can be obtained from healthy bronchial tissue removed during surgery of lung tumors. Like Calu-3 cells, PBECs reflect toxicity at the bronchial level instead of the alveolar level, which may be interesting for various types of compounds. PBECs are more difficult to culture than A549 cells and somewhat less robust. A direct connection to a hospital is needed to guarantee the availability of bronchial tissue, which is a drawback in comparison with the readily commercially available A549s.

Read-out parameters

A series of read-out parameters has been developed. These parameters address different aspects of toxicity, thus giving some insight into the toxicity of the components and toxicological pathways. The most commonly used read-out parameters are:

- Alamar Blue conversion by the cells as a measure for mitochondrial activity and thus for cell viability;
- Lactate dehydrogenase (LDH) release from the cells into the medium as a measure of cell membrane integrity and thus also for cell viability;
- Release of interleukin-8 (IL-8) from the cells into the medium as an indication of inflammatory processes;
- The glutathione status of the cells, indicating whether the cells have encountered, and are able to deal with more, oxidative stress;
- The protein content of the cells as a general indicator of the cell condition.

These read-out parameters are robust and relatively easy to determine.

The range of read-out parameters is being expanded, incorporating among others more sophisticated approaches such as toxicogenomics and other '-omics' technologies (transcriptomics, metabonomics, proteomics,..). These provide a vast number of data but are rather laborious and (very) expensive.

Extrapolation to humans
It will be obvious that it is difficult to extrapolate a toxic effect observed in a single type of cell to an effect on a living human being. It should be realized that performing a toxicity test with, for instance, lung cells in an air–liquid interface system will only provide information on acute toxic effects on the lung cells themselves. In real life a variety of compounds (including some components present in combustion products and military smokes, such as CO, CO_2, HCN, etc) will not have a direct effect on the lung cells, but will pass into the blood via the alveoli and may cause considerable damage somewhere else in the body, sometimes after bioactivation in the liver. Such toxic effects after systemic absorption of chemicals can only be studied *in vitro* using additional models.

In vitro toxicity tests are highly suitable to obtain an impression of the toxicity of (mixtures of) compounds and to rank the toxicity of various (mixtures of) compounds, but not to derive a threshold value for humans. For compounds that exert their toxic effect via corrosion of the skin or epithelia it can be imagined that a reasonable prediction for humans can be made from the *in vitro* tests, for other mechanisms of action such a prediction will be highly unreliable. Attempts are being made to translate *in vitro* tests to humans. Wijte *et al* (2011) [29] have compared the toxicity of phosgene as measured in an ALI system on A549 cells with that measured in laboratory animals, and with the toxicity data reported for humans in the open literature, found a reasonable prediction of human lung toxicity of phosgene in humans on the basis of the ALI results.

6.3.3 *In vivo* approaches

Animal experiments are generally accepted as being indicative for the situation in humans. The advantages are obvious: animals are living organisms, all relevant processes such as absorption, distribution, metabolism and excretion are in place, and it is possible to perform experiments in a controlled way on sufficient numbers of animals to obtain statistically significant results. There are, however, also some drawbacks to using animals as a model for humans. Animal experiments are expensive, time-consuming and the societal acceptance is decreasing, in particular for defense-related research. Furthermore, it has to be realized that animals are not small humans. There are anatomical, physiological and biochemical differences between species that may have a profound impact on how various species respond to chemical substances. The choice of the animal model for the particular problem to be addressed is therefore essential.

For determination of the inhalation toxicity of chemicals the rat is a generally accepted animal model.

Still, it has to be realized that a rat is not a small human being. Differences between the two species occur with respect to absorption, distribution, metabolism and excretion of chemicals as well as in the pharmacological/toxicological interaction between chemicals and biological systems. The anatomical differences of the respiratory tract between rats and humans cannot be ignored: the nasal system of the rat is considerably more complex than that of humans. As a result more chemical compounds may be absorbed and/or detoxified in the nose of a rat than in that of a human, which may have an impact on the overall toxicity in the two species. However, since the inhalation toxicity of chemicals is mostly studied in rats, a huge database is available for this species, and the toxicity of ammunition emission products can be compared to those of other chemicals.

Protocolled animal studies
The Organization for Economic Co-operation and Development (OECD) has defined a series of protocols for determining inhalation toxicity of chemicals. The results from such OECD studies are accepted by regulatory organizations, for instance, within the context of REACH (Registration, Evaluation, Authorisation and Restriction of Chemicals), provided the tests are performed in compliance with the principles of Good Laboratory Practice (GLP). The OECD protocols are robust but require relatively large numbers of animals. In view of the costs of such studies and the societal pressure to reduce or even abandon the use of animals it is questionable whether such protocols are the preferred way to go when estimating the toxicity of military-related compounds and combustion products thereof.

Extrapolation to humans
The predictive value of animal studies for humans depends mainly on the validity of the model that was used as well as on the exposure conditions. When using animal data with the intention to establish toxicity predictions for humans, the animal data should be carefully evaluated with respect to their suitability for translating to humans. Even if the data appear to be suitable, safety factors may have to be used for translating animal data to humans, in order to remain on the safe side with the estimates.

6.4 Recommendations and way forward

Due to REACH legislation implemented in Europe to eliminate all chemicals forming hazards to the environment and human health from the market, most of the munition manufacturers and munition-related industries like propellant manufacturers are looking for alternative REACH-compliant chemicals. In the USA one of the materials addressed in the past is ammonium perchlorate, which is used in rocket propellant, because of perchlorates leaching from firing ranges into the drinking water streams.

The criteria used to look for alternative chemical components are often only based on the ingredients themselves instead of on the products, which are formed during the functioning of the munitions.

The difficulty lies in the fact that users get exposed to the combustion and emission products from the complete munition article including the particles formed by friction between the barrel and the projectile, and not to single intact ingredients. The inter-molecular reactions formed in a confined space can to a certain limit be addressed by models; however, the after-burning when products leave the barrel and interact with the environment (moisture, temperature difference, particles, etc) is most of the time not addressed in these models.

The composition of combustion products is strongly influenced by the conditions under which the combustion occurs. Relatively harmless chemical components may be transformed into highly toxic compounds, depending on the reaction conditions. However, these conditions are highly variable in real-life situations.

To provide some recommendation to address this challenge, in the following paragraphs an overview is provided with the current status of potentially promising new developments.

6.4.1 Prediction of emitted products

The prediction of emission products is difficult due to a large number of parameters that influence the combustion process in the cartridge, in the barrel and in the end when the composition has left the barrel. In order to predict the final exposure of military personnel the three reaction environments need to be evaluated sequentially. The output of step 1 is the input for step 2, and so on. Each of these steps requires a different calculation. The ignition process of the primer and the temperature of the gases and particles that ignite the propellant affect the combustion of the propellant in a qualitative and quantitative sense. But also the shape, size of the propellant and the density with which the propellant is packed in the cartridge will be of influence on the combustion products formed. When the projectile is pressed out of the cartridge the combustion products will be blown through the barrel of the weapon, where they are less compressed, resulting in cooling down of the emitted mixture, and possible contact with the particles scraped off from the projectile and from the barrel due to friction when passing through the barrel. The residence time of the combustion products inside the barrel depends on the barrel length and the velocity of the projectile. In the end, when the combustion products have left the barrel they come into contact with the outside air. Depending on temperature, pressure and humidity, the combustion products can react further. The final exposure of personnel will depend on the distance and position from the barrel, and obviously the number of rounds fired. For this purpose a dispersion model could be used.

The modeling of each of these subsequent steps is complex, especially because there are no models available that can predict all of these steps. Also, most of the existing models are limited in terms of the number of chemical components that are present in their databases and lack of capability to simulate non-ideal reaction conditions. The model ReaxFF does not have these limitations, but it is currently limited with respect to the number of atoms that the model can handle. As computational capacities are continuously improving, this limitation of ReaxFF may be remedied in the near future.

The best way to arrive at better predictions of the combustion products in a qualitative and quantitative sense seems to be via evaluation of the experimental data obtained by firing ammunition under various conditions. In this way, the influence of the various parameters can be identified and (hopefully) understood, after which it may be possible to develop a model suitable to predict emission products to which military personnel may be exposed. This requires protocols on how to perform the analyses of emission products in order to be able to compare data obtained from experiments performed by different laboratories. Such protocols should address realistic conditions for testing ammunition and military smokes, given the fact that the composition of the combustion products is markedly influenced by the experimental conditions, both in a qualitative and quantitative sense. Within the NATO-Applied Vehicle Technology panel, a group working on 'Hazard assessment of exposure to ammunition-related constituents and combustion products' was established. This group is tasked to provide guidance on the design of an approach for prediction of health hazards resulting from exposure of military personnel to ammunition-related constituents and combustion products thereof, in order to establish a basis for risk management of such exposures. The final report is expected in the first half of 2020.

6.4.2 Recommendations for experimental setup and analysis

The suggested approach is to start with a prediction of the potential products that may be formed during the functioning of the ammunition/smoke article. It should be kept in mind that most predictive models are based on a thermodynamic code, assuming ideal combustion conditions, which are highly unlikely to occur in real life. Such models do not include after-burning when the emitted products are coming into contact with the outside air (humidity, lower temperature) and oxidation reactions with the oxygen in the air may occur. So, it is important to be aware of the limitations of these predictive models and to keep an eye out for models that are able to make predictions for more realistic conditions.

The next step has to be analysis of the emitted products in a manner that is as realistic as possible. Ideally, the actual product as used by the military is tested. If that is not possible for safety or other reasons, a test design is needed that mimics the real-life situation as closely as possible but on a smaller scale.

Measurement of the particle size distribution in terms of numbers and size is essential, with special attention for ultrafine- and nanoparticles, which may penetrate deep into human airways. Also, a study of the composition of the particles is recommended. Several gaseous compounds that are most likely present, such as CO, CO_2, NO_x, HCN, SO_2 are preferably measured on site. In addition, it should be attempted to analyze which components are present in the combustion product or smoke, which requires sampling. These samples are then analyzed in a laboratory with state-of-the-art analytical equipment, among others such as comprehensive gas chromatography with mass spectrometric detection. Such a technique will provide data on the identity of the components and an approximate value of their concentrations. Via a HI approach an indication is obtained of the potential toxicity of the combustion product or smoke.

The analysis of the emitted products can be performed together with an *in vitro* study with, for example, cultivated human lung cells, exposed in an ALI system. This will provide information on the acute toxic effects on the exposed lung cells, but not on other organs, possible delayed effects or effects of repetitive exposures. Nevertheless, an experimental indication on the toxicity of the products is obtained, which can be compared with the outcome of the HI approach.

Finally, it might be necessary to perform a limited *in vivo* study. An example of this stepwise approach (*in silico, in vitro, in vivo*) was published in the proceedings of the 2015 International Pyrotechnics Seminar [30].

6.4.3 Developments in toxicity assessment

In an attempt to reduce or replace animal experiments, a variety of *in vitro* methods are being developed for all kinds of toxicity, for instance using stem cells, which can be differentiated to different kinds of cell types. Of these developments the 'human-on-a-chip' or 'body-on-a-chip' seems to be most promising. In these models the 'organs' on the chip are interconnected via microfluidics, thus mimicking the transport of compounds and metabolites thereof between various organs in the human body. For the military domain, a 'body-on-a-chip' project funded by the US Defense Threat Reduction Agency (DTRA) is ongoing [31]. The outcome of this project will be applicable to study the toxicity of ammunition-related compounds and combustion products thereof, as well as that of military smokes.

Meanwhile, the improvement of *in silico* models for the prediction of toxicological properties is underway, among others in the EU-TOXRISK project [32].

References

[1] Evans E H 1945 Casualties following exposure to zinc chloride smoke *Lancet* **23** 368–70
[2] National Research Council (US) Subcommittee on Military Smokes and Obscurants 1999 *Toxicity of Military Smokes and Obscurants* vol 3 (Washington DC: National Academies Press)
[3] Research Advisory Committee on Gulf War Veterans' Illnesses 2008 *Gulf War Illness and the Health of Gulf War VeteransScientific Findings and Recommendations* (Washington, DC: U.S. Government Printing Office)
[4] Kalinich J F, Emond C A and Dalton T K *et al* 2005 Embedded weapons-grade tungsten alloy shrapnel rapidly induces metastatic high-grade rhabdomyosarcomas in F344 rats *Environ. Health Perspect.* **113** 729–34
[5] Fischbein A, Rice C, Sarkozi L, Kon S H, Petrocci M and Selikoff I J 1979 Exposure to lead in firing ranges *JAMA* **241** 1141–4
[6] Kang K W and Park W-J 2017 Lead poisoning at an indoor firing range *J. Korean Med. Sci.* **32** 1713–6
[7] Voie Ø, Borander A K and Sikkeland L I B *et al* 2014 Health effects after firing small arms comparing leaded and unleaded ammunition *Inhal. Toxicol.* **26** 873–9
[8] Koch E C, Webb R and Weiser R 2011 *Review on Thermochemical Codes* O-138 MSIAC Report

[9] Jensen T L, Moxnes J F, Unneberg E and Dullum O 2014 Calculation of decomposition products from components of gunpowder by using reaxFF reactive force field molecular dynamics and thermodynamic calculations of equilibrium composition *Propell. Explos. Pyrotech.* **39** 830–7

[10] Shaw A P, Brusnahan J S, Poret J C and Morris L A 2016 Thermodynamic modeling of pyrotechnic smoke compositions *ACS Sustainable Chem. Eng.* **4** 2309–15

[11] Moxnes J F, Jensen T L, Smestad E, Unneberg E and Dullum O 2010 Lead free ammunition without toxic propellant gases *Propell. Explos. Pyrotech.* **38** 255–60

[12] Wingfors H, Svensson K, Hägglund L, Hedenstierna S and Magnusson R 2014 Emission factors for gases and particle-bound substances produced by firing lead-free small-caliber ammunition *J. Occup. Environ. Hyg.* **11** 282–91

[13] Diaz E, Avard S and Poulin I 2012 Air residues from live firings during military training *WIT Trans. Ecol. Environ.* **157** 399–408

[14] van Hulst M 2019 Determining the exposure of personnel when using ammunition *Specialists meeting AVT-322 Combustion products, exposure and related risks (20-22 May 2019, Slovakia)*

[15] van Hulst M, Langenberg J P, de Klerk W P C and Alblas M J 2017 Acute toxicity resulting from human exposures to military smokes *Propell. Explos. Pyrotech.* **42** 17–23

[16] van Deursen M, Groeneveld F R, van Zuijlen G A and Trap H C 2005 Monstername en analyses van toxische stoffen (een evaluatie van off-site analyses van omgevingslucht) Report TNO-DV2 2005 A079

[17] US Army Public Health Center Technical Guide 230 Environmental Health Risk Assessment and Chemical Exposure Guidelines for Deployed Military Personnel 2013 https://phc.amedd.army.mil/PHC%20Resource%20Library/TG230-DeploymentEHRA-and-MEGs-2013-Revision.pdf (Last accessed August 2018)

[18] ATSDR Guidance Manual for the Assessment of Joint Toxic Action of Chemical Mixtures 2004 http://atsdr.cdc.gov/interactionprofiles/ipga.html (Last accessed August 2018)

[19] American Conference of Governmental Industrial Hygienists 2019 *TLVs and BEIs, Threshold Limit Values for Chemical Substances and Physical Agents and Biological Exposure Indices, American Conference of Governmental Industrial Hygienists* (Cincinnati, OH: American Conference of Governmental Industrial Hygienists)

[20] Occupational Safety and Health Administration. OSHA Regulations (Standards – 29 CFR): Air contaminants. 910.1000 https://osha.gov/pls/oshaweb/owadisp.show_document?p_table=Standards&p_id=9991 (Last accessed September 17, 2018)

[21] NIOSH 1976 Criteria for a Recommended Standard for Occupational Exposure to Methylene Chloride, Cincinnati (OH) https://www.cdc.gov/niosh/docs/76-138/default.html (last accessed November 14, 2019)

[22] U.S. Environmental Protection Agency 1989 Risk Assessment Guidance for Superfund. Volume I. Human Health Evaluation Manual (Part A). Report EPA/540/1-89/001, U.S. Environmental Protection Agency, Office of Emergency and Remedial Response, Washington (DC) https://www.epa.gov/sites/production/files/2015-09/documents/rags_a.pdf (last accessed November 14, 2019)

[23] ATSDR (http://atsdr.cdc.gov/interactionprofiles/index.asp) (Last accessed August 2018)

[24] Ritter D, Knebel J W and Aufderheide M 2001 *In vitro* exposure of isolated cells to native gaseous compounds-development and validation of an optimised system for human lung cells *Exp. Toxicol. Pathol.* **53** 373–86

[25] Pariselli F, Sacco M G and Rembges D 2009 An optimized method for *in vitro* exposure of human derived lung cells to volatile chemicals *Exp. Toxicol. Pathol.* **61** 33–9

[26] van Hulst M, Langenberg J P, Duvalois W and de Klerk W P C 2016 Training ammunitions: Characterization of the emission products and effects on cultured lung cells (Abstract) *12th Int. Symp. on Protection against Chemical and Biological Warfare Agents (8–10 June 2016) (Stockholm, Sweden)*

[27] Lieber M, Smith B, Szaka L A, Nelson Rees W and Todaro G 1976 A continuous tumor-cell line from a human lung carcinoma with properties of type II alveolar epithelial cells *Int. J. Cancer* **17** 62–70

[28] Mazzarella G, Ferraraccio F, Prati M V, Annunziata A and Bianco A *et al* 2007 Effects of diesel exhaust particles on human lung epithelial cells; an *in vitro* study *Resp. Med.* **101** 1155–62

[29] Wijte D, Alblas M J, Noort D, Langenberg J P and van Helden H P M 2011 Toxic effects following phosgene exposure of human epithelial lung cells *in vitro* using a CULTEX® system *Toxicol. in Vitro* **25** 2080–7

[30] Pradines E, Glacial F, Medus D, Stiee E, Fedou F, van Hulst M and de Klerk W P C 2015 Methodological approach for the assessment of acute inhalation toxicity of smoke ammunitions by in silico, *in vitro* and *in vivo* modelling *11th Int. GTPS Seminar EUROPYRO 2015 and 41st Int. Pyrotechnics Seminar (4–7 May 2015) (Toulouse, France)*

[31] https://army.mil/article/112149/army_academia_develop_human_on_a_chip_technology (Last accessed August 2018)

[32] http://eu-toxrisk.eu/page/en/about-eu-toxrisk.php (Last accessed August 2018)

IOP Publishing

Global Approaches to Environmental Management on Military Training Ranges

Tracey J Temple and Melissa K Ladyman

Chapter 7

Review of remediation technologies for energetics contamination in the US

Harry D Craig

The use of energetic materials in munitions manufacturing, load, assemble and pack (LAP), and demilitarization operations in the US and the historical disposal practices at these facilities has resulted in soil and groundwater contamination. The closure and consolidation of these facilities since WWII and the promulgation of environmental regulations in the US required the need to develop soil and groundwater remediation technologies for these contaminants. Site investigations began in the 1980s and remediation projects in the 1990s, and a range of thermal, biological and chemical/physical technologies have been developed and utilized for treatment of energetic contaminated soils/sediments and groundwater.

Based on a review of 24 remediation projects at 22 facilities, 1.21 million cubic yards of soils have been treated by incineration (32.1%), composting (46.5%), the DARAMEND process (1.2%), alkaline hydrolysis (12.6%) and solidification/stabilization (7.9%). Groundwater contamination has been primarily treated by groundwater extraction and *ex situ* granular activated carbon (GAC) treatment. Most of the groundwater treatment systems began in the 1990s and remain operational today. Based on the results of recent pilot-scale treatability studies, *in situ* bioremediation will likely play a larger role in energetics groundwater remediation in the future. The development and application of these soil and groundwater remediation technologies over the past 30 years is described in this chapter.

7.1 Introduction

The production and use of energetic materials for munitions in the US corresponds with major conflicts in the 20th and 21st century. To accomplish munitions production, a network of government-owned government-operated (GOGO), government-owned contractor-operated (GOCO), and contractor-owned contractor-

7-1

operated (COCO) ammunition industrial sites have evolved over time. The government owns key facilities to manufacture explosives and propellants and to load, assemble and pack (LAP) munitions with contractors operating most facilities and operations.

In World War I the US expanded from six GOGO and two COCO private firms in 1914 to about 20 arsenals three years later. The 16 GOGO facilities combined with commercial producers totaled 92 plants manufacturing powder and high explosives; 28 made ammonium nitrate, 15 picric acid, 13 smokeless powder, and 11 trinitrotoluene (TNT). An additional 93 other LAP plants loaded shells, bombs, grenades, booster, fuzes and propellant charges [1]. During WWI, the US produced 273 million pounds of smokeless powder and 375 million pounds of high explosives. Soon after the war, munitions production lines were abandoned, quickly becoming dilapidated and near unusable [2]. The manufacturing plants erected by the government were mostly sold to the previous owners of the land they had built upon [1].

In WWII, the munitions industrial base experienced the largest expansion and then a reduction downward since that time. Prior to WWII, 6 GOGO facilities existed, and from June 1940 to Dec 1942, 84 ammunition plants were built, including 23 lAP plants, 12 for the manufacture of ammonia, magnesium, oleum and ammonium picrate, 9 for TNT manufacture, 2 for royal demolition explosive (RDX) manufacture and 4 for smokeless powder production. By the end of 1941, there was at least one of every essential type of government-owned ammunition plant incorporated into the industrial base to include TNT, DNT, tetryl, toluene, anhydrous ammonia, smokeless powder, bag loading and shell loading plants.

Immediately after the defeat of Germany, the Ordnance Department began closing down ammunition plants. After the defeat of Japan, the entire system was swiftly shut down. Around 50 plants, known to be excess, were transferred to the operating contractor or sold on the open market. Fourteen plants remained in active status, primarily engaged in demilitarization, renovation, or the production of fertilizer. The remaining plants not excessed were placed in inactive status, decontaminated, padlocked and left without maintenance money. In 1945, the number of GOCO plants was reduced from 84 to 38 and continued to decrease. In 2010, 16 ammunition plants remained in the US inventory [1]. The locations of major munitions facilities in the US closed since WWII are shown in figure 7.1.

7.2 Background

7.2.1 Analysis of energetic materials in environmental media

Prior to the 1970s no substantive environmental requirements generally applied to discharges from ammunition production, LAP activities, or demilitarization. Disposal of ammunition included open burning (OB), open detonation (OD), burial and ocean dumping. Ammunition storage depots conducted demilitarization of TNT and TNT containing energetic fillers (e.g. Composition B) with steam out/melt out processes that often produced large quantities of contaminated wastewaters that were discharged untreated into wastewater lagoons, impoundments, ditches and

Map of Key Munitions Facilities

GOGO, GOCO and year closed

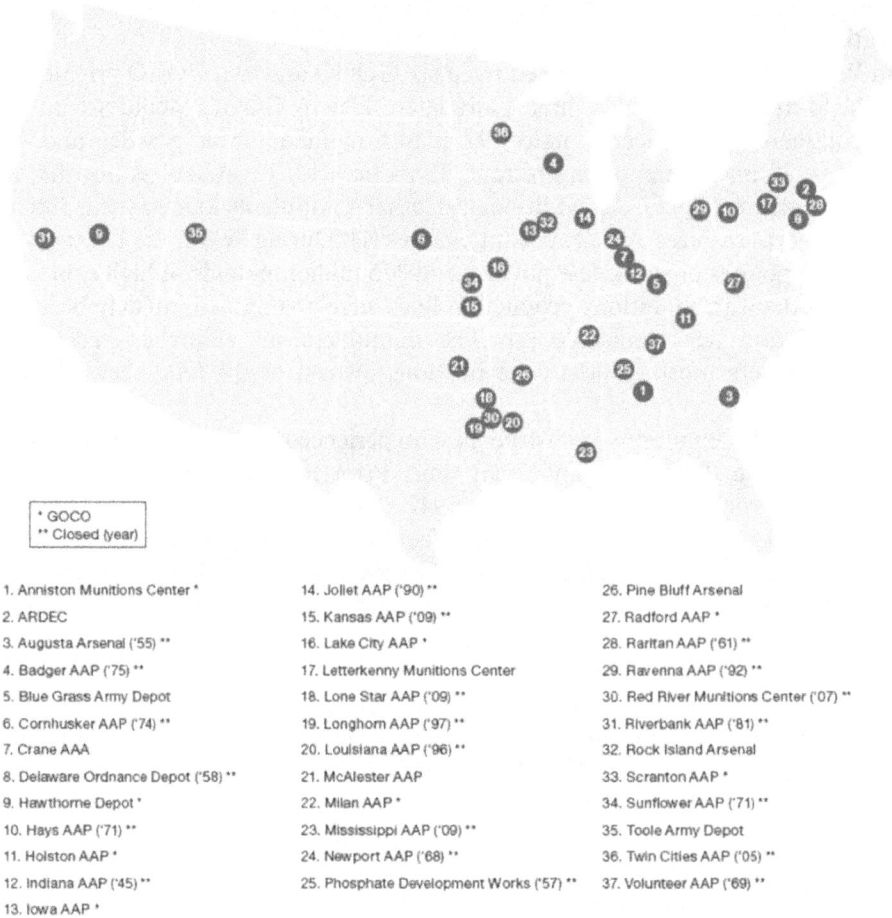

* GOCO
** Closed (year)

1. Anniston Munitions Center *	14. Joliet AAP ('90) **	26. Pine Bluff Arsenal
2. ARDEC	15. Kansas AAP ('09) **	27. Radford AAP *
3. Augusta Arsenal ('55) **	16. Lake City AAP *	28. Raritan AAP ('61) **
4. Badger AAP ('75) **	17. Letterkenny Munitions Center	29. Ravenna AAP ('92) **
5. Blue Grass Army Depot	18. Lone Star AAP ('09) **	30. Red River Munitions Center ('07) **
6. Cornhusker AAP ('74) **	19. Longhorn AAP ('97) **	31. Riverbank AAP ('81) **
7. Crane AAA	20. Louisiana AAP ('96) **	32. Rock Island Arsenal
8. Delaware Ordnance Depot ('58) **	21. McAlester AAP	33. Scranton AAP *
9. Hawthorne Depot *	22. Milan AAP *	34. Sunflower AAP ('71) **
10. Hays AAP ('71) **	23. Mississippi AAP ('09) **	35. Toole Army Depot
11. Holston AAP *	24. Newport AAP ('68) **	36. Twin Cities AAP ('05) **
12. Indiana AAP ('45) **	25. Phosphate Development Works ('57) **	37. Volunteer AAP ('69) **
13. Iowa AAP *		

Figure 7.1. Map of closed and operational munitions facilities in the US. Source: [2] figure 9, p 7.

playas [3, 4]. From the 1960s onward, washout and 'hogout' operations of ammonium perchlorate propellant containing rocket motors also resulted in the discharge of large amounts of perchlorate containing wastewaters into the environment, resulting in soil and groundwater contamination.

Initial application of analytical methods for explosives and propellant compounds in environmental media began in the 1970s to evaluate wastewater discharges at AAPs [5–7]. Further research was conducted on environmental contaminants in soil/sediments at AAPs and LAP facilities in the 1980s [8–10]. Extensive groundwater contamination downgradient of explosives wastewater lagoons was first reported in the 1980s [11] and perchlorate groundwater contamination in the late 1990s [12].

Nearly all explosives are manufactured by the nitration of organic compounds. TNT is manufactured by the stepwise nitration of toluene to 2,4,6-trinitrotoluene (TNT). Mono-nitrotoluenes (2-NT, 3-NT, 4-NT) and dinitrotoluenes (2,4-DNT, 2,6-DNT) are impurities in the TNT manufacturing process [13–15]. RDX is manufactured in the US by the Bachmann process by nitration of hexamine, and high melting explosive (HMX) is an impurity in the manufacture of RDX at a level of approximately 10% [14, 15]. HMX is also manufactured in the US by the Bachmann process under different conditions, with RDX as an impurity of the manufacture of HMX.

Environmental transformation processes for TNT include photodegradation and reductive biological and abiotic degradation. Common environmental transformation products include amino-dinitrotoluenes (2A-DNT, 4A-DNT) based on substitution of amino (NH_2) functional groups for nitro (NO_2) functional groups without ring cleavage. DNTs also biologically transform in the same manner with substitution of amino functional groups for nitro functional groups. 1,3,5-trinitrobenzene (1,3,5-TNB) is a major photodegradation product of TNT, and 3,5-dinitroanaline (3,5-DNA) is a biological degradation product of 1,3,5-TNB [16, 17]. RDX and HMX are resistant to aerobic biological degradation in the environment under aerobic conditions, but may biologically degrade under anaerobic conditions with the substitution of nitroso (NO) functional groups for nitro (NO_2) functional groups [18, 19] prior to ring cleavage. Environmental transformation products for picric acid (2,4,6-trinitrophenol) include picramic acid (2-amino-4,6-dintrophenol) and 2,4-dinitrophenol [20]. Tetryl also hydrolyzes into picric acid [21–23].

A systematic approach for analytical method development for environmental analysis of explosives and propellant compounds was developed in the mid-to-late 1980s and 1990s to address: (1) commonly occurring explosives contaminants, (2) manufacturing impurities, and (3) environmental transformation products of TNT, RDX, HMX, tetryl, and DNT isomers based on evaluation of HPLC-UV and GC-MS analytical methods [16, 24–26]. EPA Method 8330 was promulgated in 1994 based on a reverse phase HPLC-UV method for analysis of 14 target analytes in soils/sediments and water environmental media. Revisions to the method include the EPA Method 8330A in 1998 and EPA Method 8330B in 2006. EPA Method 8330B currently has 17 target analytes (3,5-DNA, nitroglycerin and PETN were added) and includes provisions for use of multi-increment sampling (MIS) design in appendix A of the method. EPA Method 8330 (all three versions) have a long history of use in the US on real-world sites, and has been used for the analysis of soils, sediments, groundwater, surface water, incinerator ash, compost residues, bioslurry reactor sludges, leachates from solidification/stabilization treatment and alkaline hydrolysis processes. Many of the site investigations for environmental contamination at AAPs, LAP plants, and depots in the US began in the last 1980s, with significant activities for both site characterization and remediation in the 1990s.

A GC-ECD analytical method for environmental analysis of explosives and propellant compounds (EPA Method 8095) has also been developed in 2007 with the same target analyte list as Method 8330B. A separate HPLC analytical method for

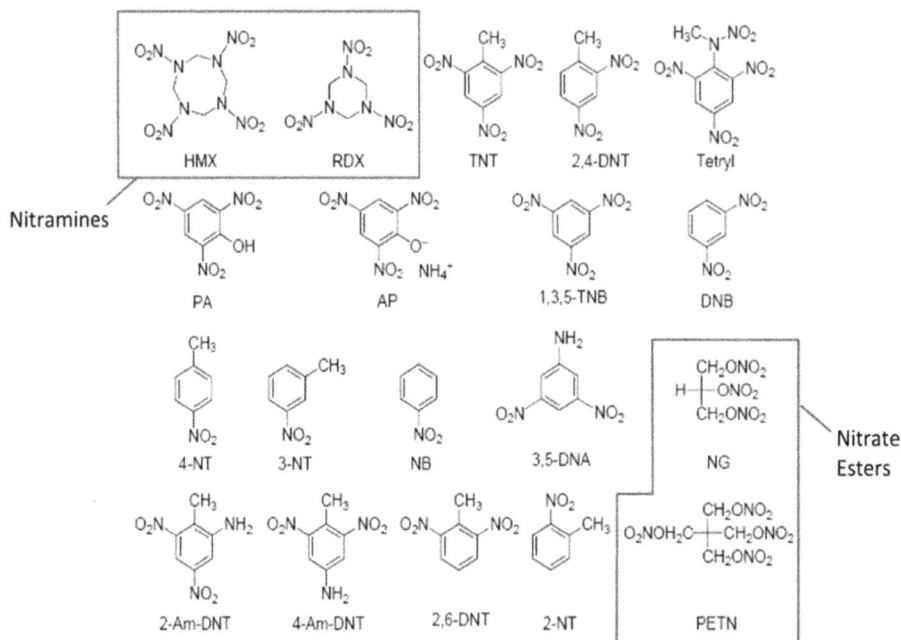

Figure 7.2. Conventional munitions contaminants. EPA (2012)/Reprinted with permission from [29]. Copyright (2001) American Chemical Society.

nitrophenols such as ammonium picrate (Explosive D) and picric acid has also been developed [27]. The chemical structures for common energetic materials uses in conventional munitions are shown in figure 7.2 [28].

EPA Methods 8330B and 8095 chromatographically separate 17 analytes in one analysis. Many explosive fillers have one to four compounds present in a mixture, which does not include the manufacturing impurities and environmental transformation products of the individual components. Different explosive compounds have different fate and transport properties, even if co-disposal occurs, such as TNT, RDX and HMX contaminated wastewaters. Determining which contaminants may be present together or separately at a specific location cannot be determined before a detailed site investigation is conducted.

The degree to which environmental degradation and transformation of explosives compounds occurs is often dependent upon environmental concentrations and the concentration of co-contaminants. Geochemical and microbial conditions also affect the degree to which degradation or transformation may occur. Environmental degradation cannot be adequately assessed unless parent compound and biological, abiotic, hydrolysis and photodegradation products are measured concurrently in the same sample(s). Explosives and propellant compounds exhibit extreme heterogeneity in soils and sediments due their solid state at ambient temperatures, and due to disposal practices such as OB/OD that result in significant soil disturbance.

7.2.2 Toxicology summary for energetic materials

Although energetic materials are well known for their explosive effects, there have been significant documented health effects from exposure to these compounds since the early 1900s. The first fatality in the US reported from chemical exposure to TNT occurred in 1917, shortly after significant TNT manufacturing operations began in WWI [30]. During WWI, 475 fatalities in US manufacturing operations were reported due to TNT exposure, with major routes of exposure being particulate inhalation and dermal absorption. In addition, 17 000 cases of TNT poisoning due to toxic liver jaundice and aplastic anemia occurred during WWI. By the beginning of WWII, industrial hygiene practices during manufacturing and loading operations had improved somewhat, but 23 fatalities occurred in the US from TNT exposure during manufacturing operations during WWII [31]. During the Vietnam era, there were more than 30 cases of RDX poisoning due to ingestion of C-4 explosive (91% RDX), including epileptic seizures and convulsions. More recent human occupational biomonitoring and epidemiological studies for energetic materials show increased risks for several health effects, including liver and bladder cancers, leukemia, heart attacks, cataracts and male reproductive toxicity.

Animal toxicity studies on the major classes of munitions compounds (e.g. nitroaromatics, nitramines and nitrate esters) were initiated in the 1970s and 1980s. Studies conducted by the US Army Biomedical Research and Development Laboratory (BRDL) in the mid-to-late 1980s demonstrated acute, subchronic and chronic mammalian toxicity effects for munitions compounds [32]. The US Army and US EPA Memorandum of Understanding (MOU) in the late 1980s developed regulatory Cancer Slope Factors (CSF) and non-cancer Reference Doses (RfD), and developed Lifetime Health Advisories (HA) for drinking water exposures based primarily on BRDL sponsored mammalian chronic toxicity studies [33]. These CSF and RfDs serve as the toxicological basis for developing environmental media (e.g. soil, groundwater, surface water, sediments, etc.) risk screening levels currently used as quantitative criteria for energetic materials site remediation.

7.2.3 Development of risk screening levels (RSLs) for energetic materials in soils and groundwater

Development of preliminary Remediation Goals (RG) for soil and groundwater in the US is generally based on risk assessment procedures for chronic (long term) carcinogenic and non-carcinogenic toxicity effects of individual compounds and cumulative effects for all detected compounds. Risk based screening levels (RSLs) provide default screening tables and a calculator to assist in decision making concerning contaminated sites and determine whether levels of contamination found at the site may warrant further investigation or site cleanup, or whether no further investigation or remedial action may be required. RSLs are often used as the performance criteria for bench and pilot-scale treatability studies to assess whether remediation technologies are sufficiently effective to meet RGs based on current and reasonable anticipated future land use(s). RSLs are derived from equations combining exposure assumptions with chemical specific toxicity values, as shown in figure 7.3.

A cancer slope factor (CSF) and the accompanying weight of evidence determination are the toxicity data most commonly used to evaluate potential human carcinogenic risks. Generally the CSF is a plausible upper-bound estimate of the probability of a response per unit intake of chemical over a lifetime. The CSF is used in risk assessments to estimate an upper-bound lifetime probability of an individual developing cancer as a result of exposure to a particular level of a potential carcinogen. Oral ingestion CSFs are toxicity values for evaluating the probability of an individual developing cancer from oral exposure to contaminant levels over a lifetime and are expressed in units of $(mg/kg\text{-}day)^{-1}$. A one in a million (1×10^{-6}) lifetime excess cancer risk is a common starting point for assessing carcinogenic risks.

A non-cancer reference dose (RfD) is an estimate (with uncertainty spanning perhaps an order of magnitude) of a daily oral exposure in the human population (including sensitive subgroups) that is likely to be without an appreciable risk of deleterious effects during a lifetime. RfDs can be derived from no observed adverse effect level (NOAEL), lowest observed adverse effect level (LOAEL), benchmark dose or categorical regression, with uncertainty factors generally applied to reflect limitations of the data used. Oral RfDs are expressed in units of mg/kg-day. Chronic

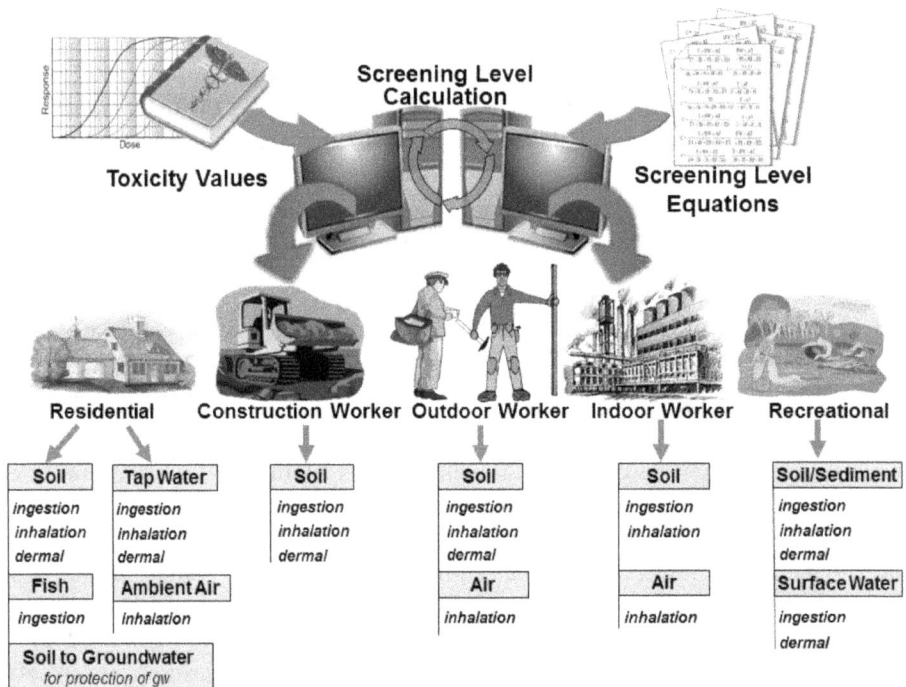

Figure 7.3. Methodology—risk screening levels. Source: EPA (2018), https://www.epa.gov/risk/regiona-screening-levels-rsls

oral RfDs are specifically developed to be protective for long term exposure to a compound. RfDs are generally the toxicity values used most often in evaluating non-cancer health effects at contaminated sites. Cancer and non-cancer toxicity values for common munitions compounds are shown in table 7.1.

RSLs present in table 7.2 are chemical specific concentrations for individual contaminants in soil and groundwater that may warrant further investigation or site cleanup based on current for future land use(s). RSLs generated from the RSL calculator may be site-specific concentrations for individual chemicals in soil, air, water and fish tissues. Given the very low vapor pressures for most energetic materials, volatile air emissions are not of concern for most of these individual compounds.

7.3 Remedial technologies for energetic materials and co-contaminants

7.3.1 Development and scale-up of remediation technologies

The legacy disposal practices at AAPs and Army/Navy depots, and demilitarization practices such as wastewater discharges and OB and OD disposal of excess, unserviceable, and obsolete munitions have resulted in soil and groundwater contamination by energetic compounds at a number of operational and closed defense facilities in the US. Based on environmental legislation promulgated in the US since the 1980s and the toxicity of primary contaminants of concern such as TNT, RDX, HMX and perchlorate, development of remediation technologies were necessary to treat these contaminants to acceptable risk based levels in soil and groundwater. A range of thermal, biological, chemical/physical and sorption technologies have been developed for this purpose over the last 30 years. The five primary technologies that have been used full scale to treat energetic contaminated

Table 7.1. Munitions compound toxicity values.

Compound	Cas no.	Cancer slope factor (CSF) $(mg/kg\text{-}day)^{-1}$	Reference dose (RfD) $(mg/kg\text{-}day)$
TNT	118–96–7	3×10^{-22}	5×10^{-4}
RDX (Hexogen)	121–82–4	1.1×10^{-1}	3×10^{-3}
HMX (Octogen)	2691–41–0		5×10^{-2}
Tetryl (CE)	479–45–8		2×10^{-3}
PETN	78–11–5		2×10^{-3}
Picric acid (2,4,6-TNP)	88–89–1		9×10^{-4}
2,4-DNT	121–14–2	$3.1 \times 10{-}1$	2.0×10^{-3}
2,6-DNT	606–20–2	1.5	3.0×10^{-4}
Nitroglycerine (NG)	55–63–0	$1.7 \times 10{-}2$	1.0×10^{-4}
Nitroguanidine (NQ)	556–88–7		1.1×10^{-1}
Perchlorate	14 797–73–0		7×10^{-4}
White phosphorus (WP)	7723–14–0		2×10^{-5}

Source: EPA (2018) Integrated Risk Information System (IRIS), www.epa.gov/iris

Table 7.2. Munitions compounds risk screening levels (RSLs) at 10^{-6} cancer risk and non-cancer risk of HI = 1.0.

Compound	Cas no.	Residential soil (mg kg^{-1})	Industrial soil (mg kg^{-1})	Tap water SL (µg l^{-1})
TNT	118–96–7	21	96	2.5
RDX (Hexogen)	121–82–4	26.1	28	0.7
HMX (Octogen)	2691–41–0	3900	57 000	1000
Tetryl (CE)	479–45–8	160	2300	39
PETN	78–11–5	130	570	19
Picric Acid (2,4,6-TNP)	88–89–1	57	740	18
2,4-DNT	121–14–2	1.7	7.4	0.24
2,6-DNT	606–20–2	0.36	1.5	0.049
Nitroglycerine (NG)	55–63–0	6.3	82	2.0
Nitroguanidine (NQ)	556–88–7	6300	82 000	2000
Perchlorate	14 797–73–0	55	820	14
White phosphorus	7723–14–0	1.6	23	0.4
2A-4,6-DNT	35 572–78–2	150	2300	39
4A-2,6-DNT	19 406–51–0	150	2300	39
1,3,5-TNB	99–35–4	2200	32 000	590
1,3-DNB	99–65–0	6.3	82	2.0
NB	98–95–3	5.1	22	0.14
2-NT	88–72–2	3.2	15	0.31
3-NT	99–08–1	6.3	82	1.7
4-NT	99–99–0	34	140	4.3
Picramic acid (2A-4,6-DNP)	96–91–3	6.3	82	2.0
2,4-Dinitrophenol	51–28–5	130	1600	39

Source: EPA (2018), https://www.epa.gov/risk/regional-screening-levels-rsls-generic-tables

soils/sludges are: (1) incineration, (2) composting, (3) the DARAMEND biological/chemical process, (4) alkaline hydrolysis, and (5) solidification/stabilization (S/S). The applications of these five technologies at 22 sites in the US are shown in table 7.3.

Some technologies such as incineration and S/S, had been previously developed for treatment of organics and inorganics in soils, respectively, and were adapted for use on energetic contaminated environmental media. The other technologies such as composting, the DARAMEND process and alkaline hydrolysis had more extensive laboratory, pilot and full-scale testing to demonstrate the effectiveness and costs for these technologies specifically applied to energetic materials contamination. For groundwater remediation, groundwater extraction and *ex situ* granular activated carbon (GAC) sorption (e.g. 'pump and treat') has been the primary method of treatment for the past 20 years. A more detailed discussion and example applications of these six technologies are described in the following sections.

Table 7.3. Quantities of energetics treated by soil remediation technologies in the US

Site	Incineration[1,2]	Composting[1,2]	DARAMEND[3]	Alkaline hydrolysis[4]	Solidification/ stabilization[5]
Cornhusker AAP, NE	46 000 tons				
Louisiana AAP, LA	119 000 tons				
Weldon Springs OW, MO	58 000 cy				
NOP Mead, NE	16 000 cy				
Alabama AAP, AL	32 000 cy				
Savanna AD, IL	42 000 tons				
Mass. Mil. Res., MA	65 000 cy				
Umatilla AD, OR		15 000 cy			25 000 cy
SUBASE Bangor, WA		4580 cy			
Camp Navaho, AZ		4000 cy			
NSWC Crane, IN		110 000 cy			
Hawthorne AAP, NV		64 000 cy			
Joliet AAP, IL		275 000 cy			
Milan AAP, TN		58 000 cy			
Newport AAP, IN		9000 cy			
Pueblo AD, CO		21 000 cy			
Sierra AD, CA		2000 cy			
NWS Yorktown, VA			4800 cy		
Tooele AD, UT			10 000 cy		
Iowa AAP, IA	10 000 cy			3375 tons	
Volunteer AAP, TN				150 000 tons	
Sunflower AAP, KS					70 000 tons
TOTALS	**388 000**	**562 500**	**14 800**	**153 375**	**95 000**

Notes: AAP—Army Ammunition Plant, AD—Army Depot, NOP—Naval Ordnance Plant, NSWC—Naval Surface Warfare Center, NWS—Naval Weapons Station, OW - Ordnance Works.
Sources: Adapted from [42][1], [37][2], DARAMEND Case Studies[3], [59][4] and [71][5].

7.3.2 Incineration

Incineration is an engineered process that employs thermal decomposition via thermal oxidation at high temperatures (usually 900 °C or greater) to destroy the organic fraction of the wastes and reduce volume. Incineration was one of the initial technologies identified for treatment of organic contaminants in soils and sludges in lieu of (1) landfilling without treatment, (2) storage in surface impoundments or (3) ocean disposal [34]. In the US, laws include the Resource Conservation and Recovery Act (RCRA) of 1976 and its subsequent amendments which have 'cradle to grave' provisions for controlling storage, transport, treatment and disposal of hazardous wastes [35]. Provisions of the Clean Air Act (CAA) of 1970 and Comprehensive Environmental Response Compensation and Liability Act (CERCLA) of 1980, commonly known as Superfund, also apply to the treatment of wastes via incineration. A typical process flow diagram for incineration includes a primary combustion chamber, secondary combustion chamber and air pollution controls, as shown in figure 7.4.

Given the thermal instability of energetics at elevated temperatures, incineration was one of the initial methods identified for the treatment of organic energetic contaminants in soil and sludges [36, 37]. Energetic contaminants in soil, most of which contain carbon, hydrogen, nitrogen and oxygen (CHNO explosives) degrade into combustion gasses, which are subjected to elevated temperatures with sufficient residence times. Energetic contaminants containing chlorine, such as ammonium, potassium and sodium perchlorates, may also have acid gas removal components due to the presence of hydrogen chloride (HCl) combustion gases. Transportable rotary kiln incinerators have been the most widely used system for energetics contamination in soils/sludges, a typical operating configuration is shown in figure 7.5.

Although incineration is an established technolgy to treat organic contaminants in soils, trial burns are required to establish site-specific performance for the specific contaminants in soils at a site [38, 38]. The required Destruction and Removal Efficiency (DRE) or precent removal is normally 99.99% for Principal Organic Hazarous Constituents (POHC). A representative example of *in situ* energetic contamination levels in soils at the former Naval Ordnance Plant (NOP) Mead, NE plant and excavated concentrations of energetics in soils in the trial burn are shown in tables 7.4 and 7.5, respectively.

7.3.3 Composting

Composting is an *ex situ* technology designed to treat excavated soils contaminated with a range of recalcitrant contaminants, including nitroaromatic and nitramine explosive compounds. The process involves mixing the contaminated soil with bulking agents such as wood chips, straw, hay or alfalfa, and organic amendments such as cattle and/or chicken manure, or other vegetative wastes. The selection of the specific compost ingredients depends on the contaminants to be treated, the physical/

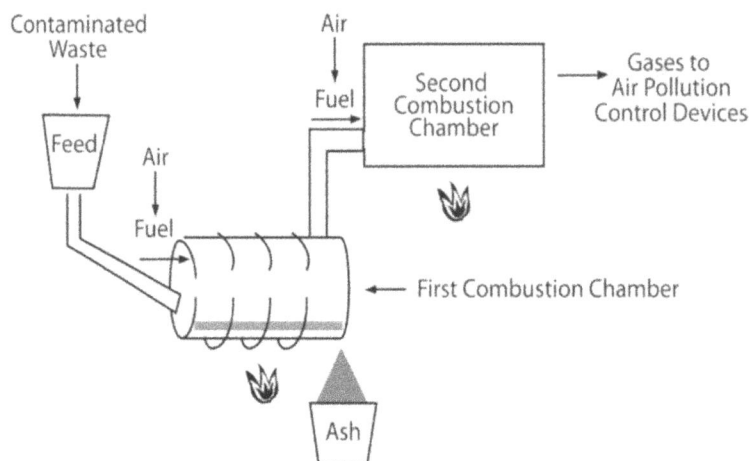

Figure 7.4. General incineration subsystems and typical process components. Source: EPA (2012) https://clu-in.org/download/Citizens/a_citizens_guide_to_incineration.pdf. Courtesy of H Shah, EPA: MACT EEE Combustion Basics Training.

Figure 7.5. Typical rotary kiln/afterburner combustion chamber. Source: EPA (2008), https://archive.epa.gov/region6/6pd/rcra_c/pd-o/web/pdf/a3-combustionequipment.pdf. Courtesy of H Shah, EPA: MACT EEE Combustion Basics Training.

Table 7.4. NOP mead, NE site—characteristics of untreated soil.

Sample location	Maximum TNT concentration found (mg kg^{-1})	Maximum RDX concentration found (mg kg^{-1})	Maximum TNB concentration found (mg kg^{-1})	Maximum DNT concentration found (mg kg^{-1})
Load line 1	133 000	39.6	338	28.9
Load line 2	176 000	23 270	430	119.3
Load line 3	29 700	40.4	95.3	14.8
Load line 4	131	22.7	6.0	17.6
Booster assembly area	7.0	ND	3.6	ND
Burning/proving grounds	313	1700	35.3	1.25
Administration area	0.314	ND	ND	ND
Primary area	0.45	ND	ND	ND

EPA (1998) Final Record of Decision, Operable Unit 1, Former Nebraska Ordnance Plant, Mead Nebraska, USEPA Region VII and USACE Kansas City District, August 1995.

chemical characteristics of the soil, and the availability of low-cost organic amendments. The goal is to achieve the desired bulk density, porosity, and organic amendments that can provide the proper balance of carbon and nitrogen (C/N) to promote biological activity in the compost. For most composting applications, it is necessary to provide amendments that will support mesophilic and thermophilic microbial activity [40]. Bioremediation, particularly composting, has been intensively investigated as an alternative treatment technology to incineration for explosives contaminated soils in the US [41, 42].

Table 7.5. NOP mead, NE—trial burn of contaminated soil feed analysis.

Test	Run 1	1D[b]	2	3	4	Average
Moisture, %	25	15.8	15.8	15.5	15.6	16.82
Heat value, BTU/lb	1200	2100	1.300	1600	330	1220
TNB, μg kg^{-1}	4500	5800	1.200	2700	7800	4213
DNB, μg kg^{-1}	<510[c]	<540	<490	<510	<440	<491
TNT, μg kg^{-1}	240 000	310 000	150 000	160 000	76 000	165 000
DNT, μg kg^{-1}	<510	<540	<490	<510	<440	<491
HMX. μg kg^{-1}	<510	<540	<490	<510	<440	<491
NT, μg kg^{-1}	<510	<540	<490	<510	<440	<491
RDX, μg ka^{-1}	<510	<540	<490	<510	<440	<491
Tetryl, μg kg^{-1}	<510	<540	13 000	1500	3000	4506
Arsenic, mg kg^{-1}	<13	5	5	5	NA	7.6
Barium, mg kg^{-1}	160	160	170	150	NA	160
Cadmium. mg kg^{-1}	0.7	0.8	1	0.8	NA	0.85
Chromium, mg kg^{-1}	6	6	8	7	NA	7
Lead, mg kg^{-1}	15	14	15	14	NA	14.5
Mercury, mg kg^{-1}	0.03	0.03	0.04	0.03	NA	0.033
Selenium, mg kg^{-1}	<13	<12	<12	<11	NA	−11.8
Silver, mg kg^{-1}	6	2	2	2	NA	2.7
Naphthalene, μg kg^{-1}	190	110	470	82	<390	273

Source: EPA (1998). Incineration of Explosives Contaminated Soil at the Former Nebraska Ordnance Plant, William J. Crawford, P.E. (USACE) and Kevin W. Birkett (USACE), January 8, 1998.

Principle of operation

Composting is a biological remediation technology that exploits the activity of a diverse range of microorganisms to degrade target contaminants. The biological activity is enhanced through the addition of readily degradable amendments to provide a sufficient supply of nutrients. As biological activity increases, compost temperatures reach described mesophilic (<45 °C) and thermophilic (45 °C to up 80 °C) conditions, the latter of which experiences the greatest degradation rates [43].

Composting is considered an aerobic technology because the compost is usually turned, mixed, or aerated to provide oxygen that enables extracellular hydrolysis and catabolic reactions. However, the high level of microbial activity can deplete oxygen between turning or aeration sequences, creating anaerobic conditions in portions of the compost material, and micro-anaerobic zones can exist in organically rich portions of the compost. These anaerobic processes can be beneficial for promoting degradation of contaminants such as nitramine explosives that are recalcitrant or only partially degrade under strict aerobic or anaerobic conditions. For contaminants that are readily mineralized under aerobic conditions, the development of anaerobic conditions is not desired, and turning, mixing and/or aeration frequencies are adjusted to minimize this potential.

Technology development

Initial proof of concept and scale-up from laboratory to pilot-scale tests for composting explosives contaminated soils and lagoon sediments began in the 1980s. Early data had promising degradation results (92% to 97%) for a range of explosives including TNT, RDX, HMX and tetryl, but soil loading rates were low, generally 10% or less in the compost mixtures [44, 45].

Follow-on pilot-scale treatability studies were conducted to evaluate mesophilic versus thermophilic conditions [46, 47]. A range of material handling approaches including static piles, mechanically agitated in-vessel reactors (MAIV), and mixed aerated and nonaerated windrows [48–50] were tested at Louisiana AAP (LAAP) and Umatilla Army Depot (UMDA) as shown in figure 7.6 [43]. TNT degradation was similar in aerated and nonaerated windrows, but RDX and HMX degradation were greater in mixed unaerated windrows at 30% soil loading rates [41, 50].

Windrow composting

Based on the results of these treatability studies, windrow composting was first selected as the preferred technology for the treatment of explosives contaminated washout lagoon soils at Umatilla Army Depot in lieu of incineration for 15 000 tons of contaminated soils as shown in figures 7.7 and 7.8 [50, 51]. Composting has been used to treat explosives contaminated soils at 10 Army Ammunition Plants (AAPs), Army Depots and Naval Ammunition Depots (NADs) in the US for soil quantities ranging from 1000 cubic yards to greater than 200 000 cubic yards [42]. Fifteen additional sites have soil explosives contamination quantities ranging from 1000 cubic yards to 55 000 cubic yards [42].

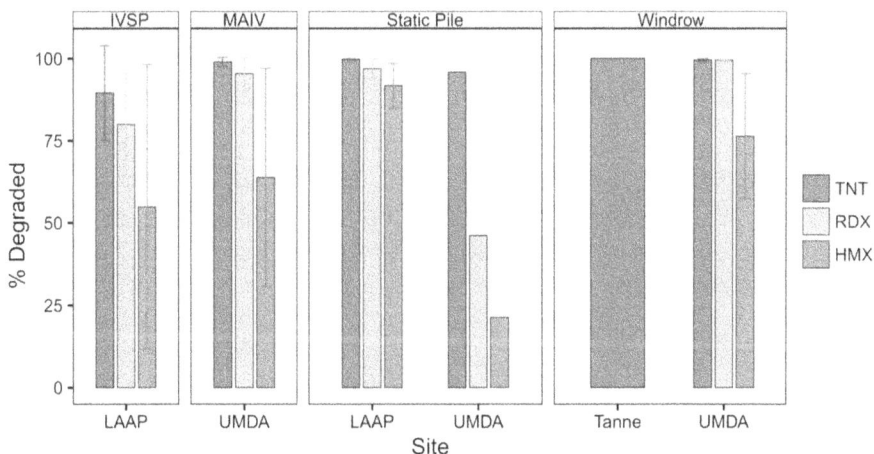

Figure 7.6. Compound reductions for different composting techniques using soils from contamination ammunition production sites. Error bars are ± 1 standard deviation. IVSP = in-vessel static pile; MAIV = mechanically agitated in-vessel. Data compiled by Bruns-Nagel *et al* [42] reproduced with permission.

Figure 7.7. Key steps of the windrow composting process at UMDA including: excavation of contaminated soil to 15 feet below ground surface (left), loading windrow machine with soil and amendments (middle), and periodically turning the windrows (right) [51].

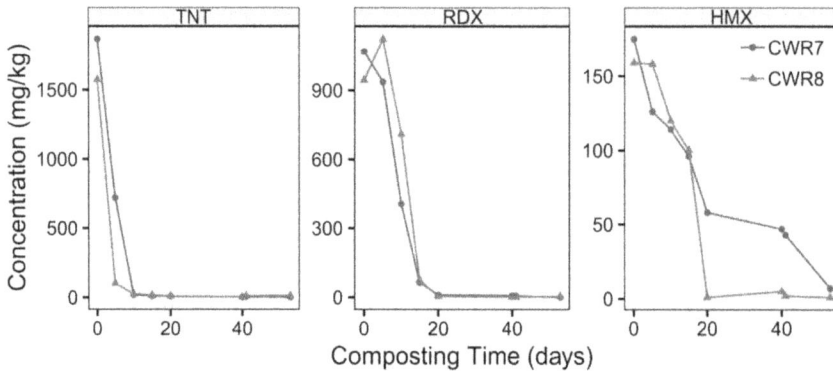

Figure 7.8. Degradation kinetics of three munitions contaminants in two UMDA windrows. Modified with permission from [50].

Implementation issues include amendment selection to promote thermophilic conditions, moisture control in the windrows to maintain optimal moisture levels, and frequency of windrow turning to optimize solid phase mixing, oxygen levels, temperature control and explosives degradation kinetics [50–52]. Composting consumes approximately one gallon of water per cubic yard per day during the active treatment phase, and must be added throughout the treatment process to maintain optimal moisture levels. The median degradation for explosives composting at 10 sites was 99.7% and the unit cost for composting treatment is $281 per cubic yard [40], which is approximately 50% lower than the estimated costs for on-site incineration treatment.

Advantages and limitations
One of the primary advantages associated with windrow composting is that degradation of explosive compounds is more rapid and efficient than that achieved with static piles or slurry phase biotreatment. Typical degradation kinetics indicate 90% to 99% degradation in 10 to 20 days as shown in figure 7.9 [53]. The end product can be humus-rich compost appropriate for reuse and re-vegetation. Plant uptake studies of treated compost residues show no uptake of explosives into plants grown in Umatilla composts [54]. A wide range of soil types can be treated through

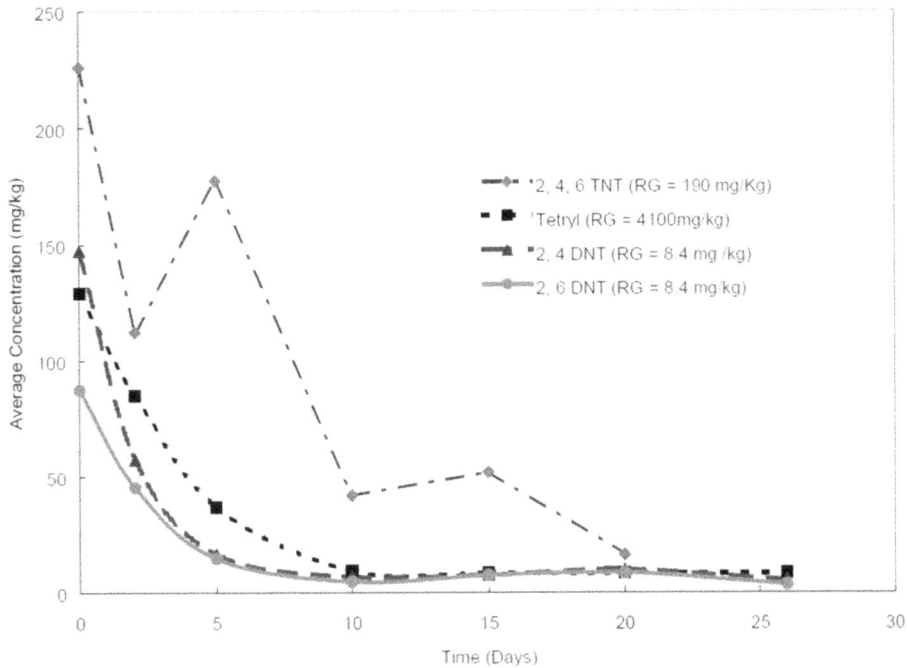

Figure 7.9. Energetics degradation kinetics for soil composting, Joliet AAP, IL. Reproduced with permission from [53].

addition and mixing of compost bulking agents and amendments. The treatment process is self-heating, so can operate year round without external heating.

The primary limitations associated with composting include space requirements, which can be moderate based on soil staging and treatment building footprints. The volume of final material can increase due to addition of amendments and bulking agents. The process may be susceptible to heavy metal concentrations. The contaminated soil in the compost mix is limited to approximately 30% by volume to achieve self-heating thermophilic conditions [40].

Technology cost drivers

Factors that drive the cost of composting include the following: composting is less land-intensive than land treatment, but still requires a moderate treatment footprint; batch treatment size will impact the number of batches; windrow widths and heights are limited by the size of the turning machinery, and lengths by the size of the buildings used; volatile emissions, primarily ammonia, may require venting; soils with low porosity may require bulking agents to increase the airflow through the compost pile; and soil screening may be required to remove large rocks, debris, or other oversize materials.

Composting may require nutrient amendments to optimize C/N ratios and water, in addition to bulking agents. O&M considerations include turnover frequency, maintaining moisture levels via water, maintaining nutrient levels via nutrient amendments; thermophilic aerobic conditions also must be maintained, which may require amendments with bulking agents, such as wood chips or sawdust. The need for containment structures also affects treatment costs. Windrow composting has been conducted using open windrows in arid climates, but inside structures in temperate and northern climates.

7.3.4 *Ex situ* and *in situ* biological/chemical reduction—DARAMEND process

The DARAMEND process is a reductive biological/chemical treatment process developed by Grace Bioremediation Technologies, Mississanga, Ontario, Canada, and is a patented technology (US Patents no. 5411664; 5480579; and 5618427). The DARAMEND process is a modification of conventional land farming methods with the addition of proprietary amendments, and includes operation of the soils through anaerobic, anoxic and aerobic phases of treatment [42]. The proprietary DARAMEND organic amendments are manufactured from plant materials (e.g. alfalfa) and may be supplemented with agricultural-grade or food-grade nutrients. The inorganic amendment most commonly applied is powdered metallic iron. Typical application rates per cycle for organic amendments are 0.5 to 10% (w/w) and 0.2 to 2.0% for inorganic amendments. The number and frequency of the cycles varies with contaminant type, concentrations and treatment criteria. Treatment cycles may be 10 to 20 days in length [42].

Mixing of the DARAMEND amendments and soil for the aeration phase is via the use of an agricultural tractor and rototiller attachment, with maximum mixing depths of 1 to 2 feet deep. The technology can be applied both *ex situ* and *in situ* to these mixing depths, as shown in figures 7.10 and 7.11 for *ex situ* application at Tooele Army Depot, UT and at NWS Yorktown, VA, respectively. For *ex situ* application the soil is placed in a treatment cell as shown in figure 7.10. In both *ex situ* and *in situ* applications, a temporary building is often constructed over the

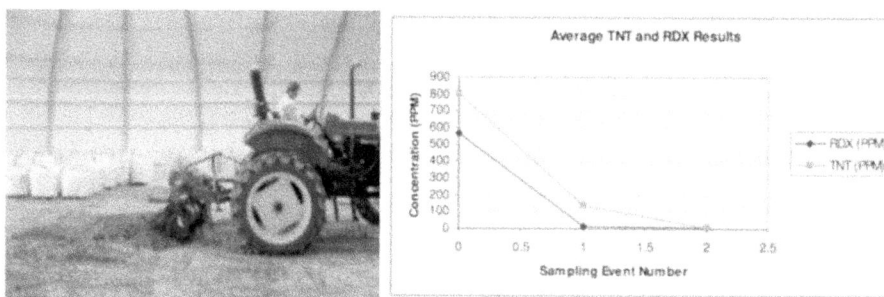

Figure 7.10. *Ex situ* treatment using the DARAMEND process, Tooele Army Depot, UT. Source: Peroxychem, Daramend Case Study, http://www.peroxychem.com/media/104116/daramend-case-study-tooele-army-depot-20–01-esd-14.pdf. Courtesy of Alan Seech, PeroxyChem LLC.

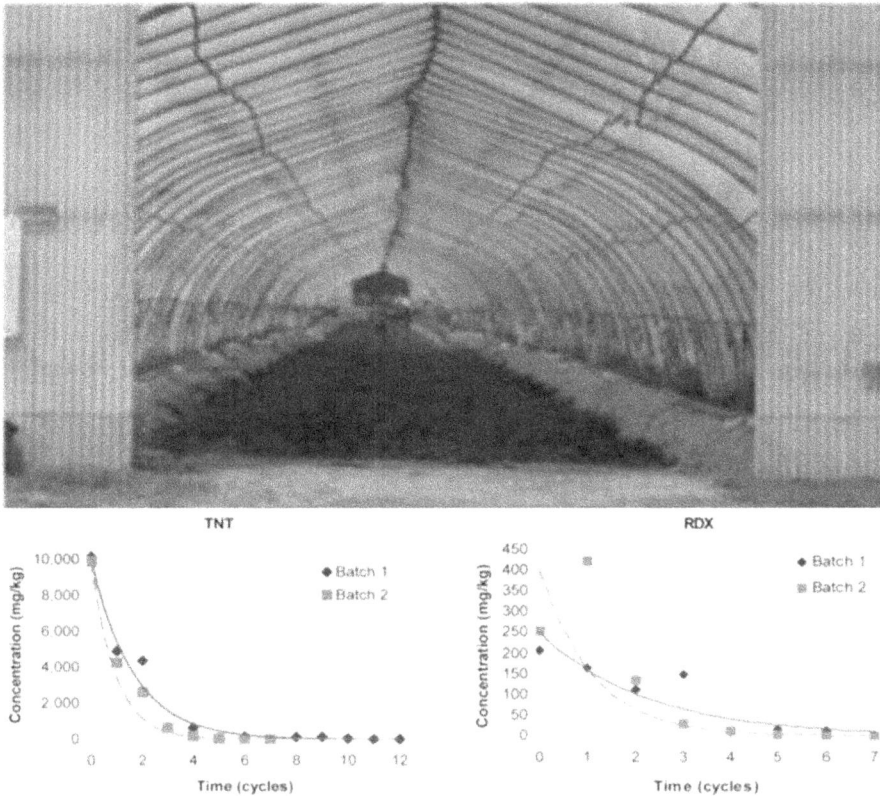

Figure 7.11. *Ex situ* treatment for the DARAMEND process, NWS Yorktown, VA. Source: Peroxychem, DARAMEND Case Study, http://www.peroxychem.com/media/174458/PeroxyChem-Daramend-Case-Study-Yorktown-TNT-RDX-Treatment-54-01-ESD-15.pdf. Courtesy of Alan Seech, PeroxyChem LLC.

treatment cells to allow for year round operation of the process. *In situ* applications typically have lower degradation rates as compared to *ex situ* applications. The DARAMEND treatment process is not self-heating.

The DARAMEND process offers two potential advantages over windrow composting relating to amendment addition: (1) a smaller volume increase in the treated soils, and (2) less local and seasonable variability in amendment supply [42]. Where composting amendments are less available or must be transported longer distances to a remediation site, the DARAMEND process may be more cost effective based on these factors. In addition, at sites where explosives have metals co-contaminants in soils, it may provide higher tolerance to metal concentrations for biological degradation [55], and less organic interferences if soils must be treated by a second treatment process such as solidification/stabilization for metals co-contamination. Organics and humic materials may interfere with the setting of cement materials for solidification/stabilization treatment [56, 57].

7.3.5 Alkaline hydrolysis (AH)

Alkaline hydrolysis is a chemical treatment process that occurs at high pH where energetic materials are decomposed into smaller molecules. Efficient treatment, particularly for RDX, occurs primarily at pH $\geqslant$ 12.0 [58]. The hydroxide sources used to elevate pH to these levels for bulk explosives and explosives contaminated soils and waters are typically calcium hydroxide and sodium hydroxide. Alkaline hydrolysis of energetic materials has been known since the late 1800s, but has not been significantly evaluated as a remediation technology for treatment of explosives in soil and water since the late 1990s [59]. Further research has demonstrated that the process occurs primarily in aqueous phase, and that soil properties such as CEC and organic carbon content have significant influence on the ability to achieve pH 12.0 or higher conditions in soils and porewaters. Soils have significant buffering capacity and depending on soil characteristics, require application of tons of hydroxide for each acre-ft (1613 cu yds) of soils to achieve pH > 12.0 conditions [60].

Initial field testing of alkaline hydrolysis treatment in soils was conducted at hand grenade ranges, where repeated detonations may result in the potential buildup of explosives residues in soils, such as TNT and RDX, and potential migration to shallow groundwater, particularly for nitramines such as RDX and HMX. Due to the significant heterogeneity of energetic residues in soils from these practices, it is difficult to establish statistically significant degradation rates as compared to control sites [61, 62]. The same soil heterogeneity issues exist at OB/OD disposal units where establishing baseline chemical concentrations for alkaline hydrolysis treatment evaluation remains a challenge [63]. Additional analytical method concerns have been raised regarding whether alkaline hydrolysis treated soils samples have been properly neutralized prior to shipment to the laboratory for analysis [64].

Application of alkaline hydrolysis treatment for legacy contamination at AAPs began in 2006 with the use of sodium hydroxide (NaOH) treatment of TNT and DNT contaminated soils at Volunteer AAP, TN. The basic *ex situ* treatment process for excavation of 150 000 tons of contaminated soils are placed in a lined basin on site. NaOH in pellet form is added at 1% to 2% rate based on starting contaminant concentrations, starting pH and soil buffering capacity, and treated in 300 cubic yard batches. Reagents are mixed into the soil with conventional equipment, and water is added to increase moisture content near saturation. The target treatment criteria is a pH value of 13.0 [59]. Soils were treated *in situ* as well as *ex situ*, with an estimated cost saving of 25% for *in situ* treatment. The estimated unit cost for *ex situ* treatment is $133/ton. A second full-scale alkaline hydrolysis treatment system was utilized at Iowa AAP to treat 3375 tons of energetic contaminated soils using application of 81 tons of NaOH to achieve a target pH of >12.5 [65].

7.3.6 Solidification/stabilization (S/S)

Solidification/stabilization (S/S) is an established technology that has been used primarily for the treatment of metals in soils and has also been used for the treatment of semi-volatile and non-volatile organic contaminants in soils [56]. Solidification refers to processes that encapsulate a waste to form a solid material and to restrict contaminant migration by decreasing the surface area exposed to leaching and/or coating the waste with low-permeability materials. Solidification can be accomplished by a chemical reaction between waste and binding (solidifying) reagents or by mechanical processes. Stabilization refers to processes that involve chemical reactions that reduce the leachability of a waste. Stabilization chemically immobilizes hazardous materials or reduces their solubility through a chemical reaction. The physical nature of the waste may or may not be changed by this process [66].

For heavy metals, cement-based S/S has been shown to be effective in immobilizing the contaminants even without any additional additives. In applying cement-based S/S for treating organic contaminants, the use of adsorbents such as organophilic clay and granular activated carbon (GAC), either as a pretreatment or as additives in the cement mix can improve organic contaminant immobilization in the solidified-stabilized wastes [67].

Metals can often occur as co-contaminants in soils at energetic contaminated sites, particularly at OB/OD disposal sites. Historical treatment practices would likely require a two-step soil treatment train, with incineration or possibly alkaline hydrolysis treatment for organic energetics contamination, followed by S/S treatment for metals contamination and final landfill disposal. Due to the cost and complexity of a multi-step treatment train, a single-step S/S treatment for both non-volatile organic energetics and metals is desirable [68].

Based on co-contamination of energetics (TNT, TNB, RDX, HMX) and metals (Pb, Cd) in soils above respective cleanup levels, bench and pilot-scale treatability studies were conducted on an OB/OD disposal site at Umatilla Army Depot, OR site [68, 69]. Leachability reductions were measured on untreated and treated soils based on the EPA Toxicity Characteristic Leaching Procedure (TCLP), EPA Method 1311 and HPLC-UV chemical analysis of leachate based on EPA Method 8330. Leachability reductions based on an optimized pilot-scale mix design of Portland cement and GAC are shown in table 7.6, and soils processing in figure 7.12. A second full-scale S/S project was conducted at Sunflower AAP, KS, for treatment of nitrocellulose (NC), nitroglycerin (NG) and lead (Pb) contaminated soils from the discharge of wastewaters from the manufacture of double base propellants [70, 71]. Leachability reductions for these contaminants from cement-based S/S treatment are shown in table 7.7.

Table 7.6. Reductions in leachable concentrations (% R) of explosives and metals in umatilla soils for S/S mixtures of cement and GAC based on TCLP extracts.

% Soil	% Cement	% GAC	Cd % R	Pb % R	TNT % R	RDX % R	TNB % R	UCS (psi)
79	8	0	99.6	99.8	NA	NA	NA	86.1
82	10	1.0	99.6	99.9	~90	<90	~97	60.9
78.5	10	1.5	98.6	99.9	~97	~92	~98	81.2
77.5	10	2.5	98.6	99.9	>99	>99	~98	117

Source: Channel et al (1996) [68, 69], UCS—unconfined compressive strength.

Figure 7.12. Solidification/stabilization of TNT/TNB/RDX/HMX/Pb/Cd contaminated soils, Umatilla Army Depot, OR – treatment plant (left) and landfill placement in cells (right).

Table 7.7. Solidification/stabilization soil treatability study results—Sunflower AAP, KS.

Parameter	Untreated soil conc. (mg kg^{-1})	Untreated soil TCLP leachate conc. (μg l^{-1})	Treated soil conc. (mg kg^{-1})	Treated soil TCLP leachate conc. (μg l^{-1})	Leachate conc. red. (%)
Pond sediments					
NC	3850	94	3470	< 1	> 98.9
NG	251	5	222	< 1	> 80.0
Pb	834	9520	874	< 200	> 97.9
Ditch soils					
NC	2410	55	1670	< 1	> 98.1
NG	65.2	2	192	< 1	> 50.0
Pb	1300	8600	1060	< 200	> 97.7

Source: [70].

7.3.7 Pump and treat w/granular activated carbon (GAC)

As a result of legacy disposal practices such as: (1) discharge of explosives contaminated wastewaters into unlined lagoons and impoundments, and (2) OB/OD munitions disposal, groundwater contamination has been documented at a number of defense-related facilities in the US. The US Army has 583 sites at 82 installations that are confirmed with explosives contaminated groundwater [72]. Groundwater contamination plumes range from tens of acres in size for OB/OD units from hundreds to thousands of acres for plumes from wastewater disposal sites. Groundwater plumes from surface disposal of wastewater have been reported up to 4 miles long, 2.5 square miles in size [11] and up to depths of 1500 feet below ground surface (bgs) [73]. An example of a large energetics groundwater plume is shown in figure 7.13 [74].

Extraction of contaminated groundwater and *ex situ* treatment has been one of the primary remediation methods for organic contaminants in groundwater in the

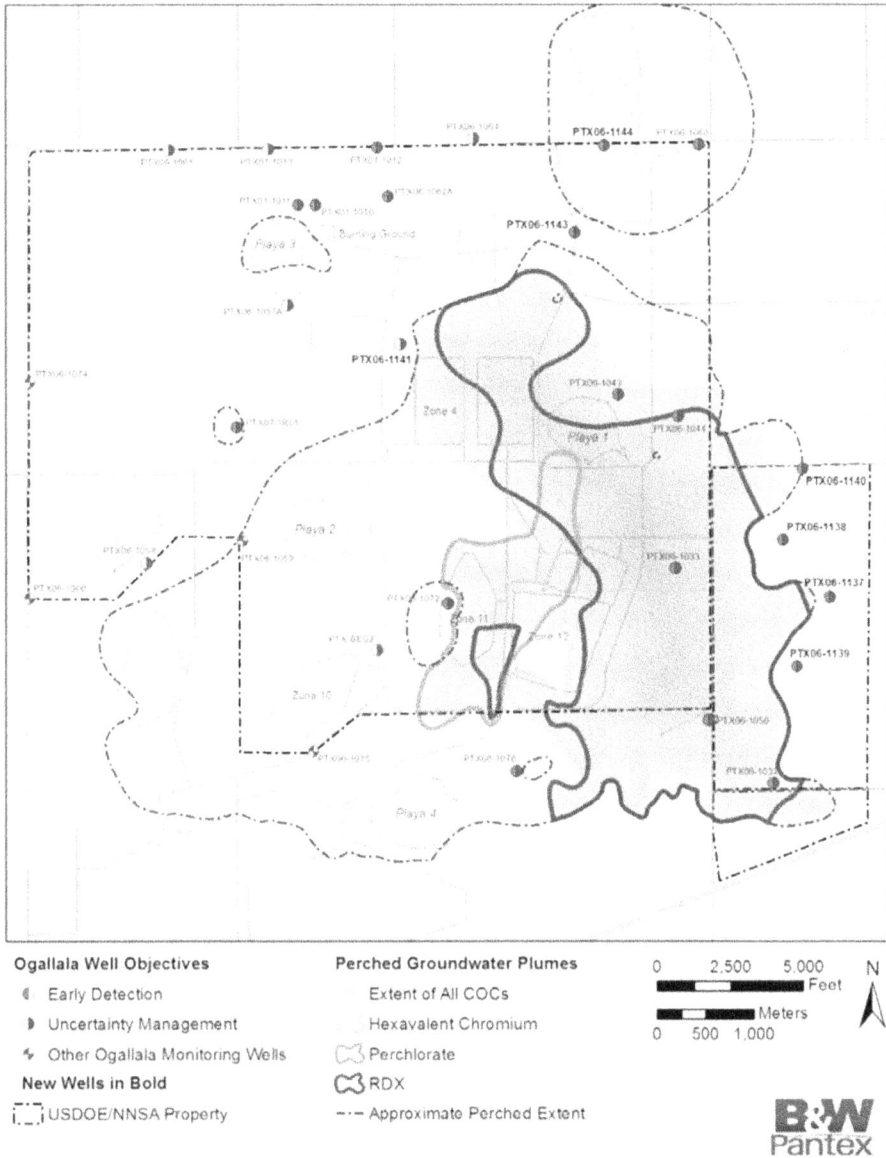

Figure 7.13. Extent of RDX and perchlorate groundwater contamination, DOE Pantex, Amarillo, TX (former WWII Pantex Ordnance Plant). Source: B&W Pantex (2009).

US for the past 30 years. GAC was initially utilized for sorption of explosives contaminated wastewaters at AAPs and depots in the 1980s, and was later extended to the treatment of explosives contaminated groundwaters in the 1990s. GAC loads based on a saturation front is shown in figure 7.14 and is affected by factors such as the specific GAC utilized, and empty bed contact time (EBCT) for the specific

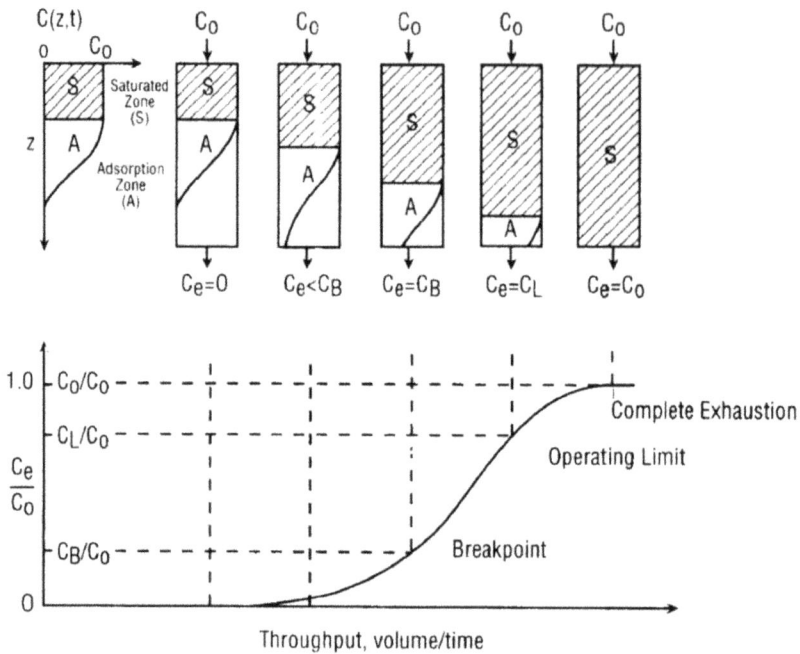

Figure 7.14. Breakthrough characteristics of fixed-bed GAC adsorber. Source: EPA (1991) [75].

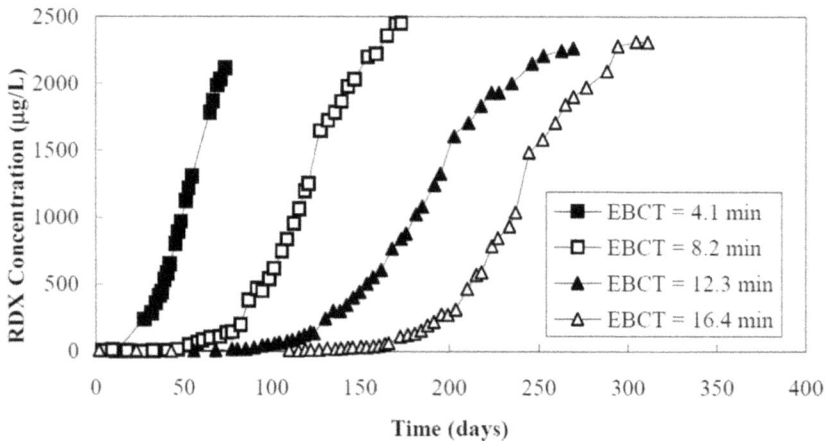

Figure 7.15. Comparison of full-scale RDX breakthrough curves for a range of empty bed contact times (EBCT) for Calgon F400 GAC. Source: [76].

organic contaminant(s) being sorbed [75]. figure 7.15 shows an example of the effects of EBCT on RDX breakthrough curves based on rapid small-scale column tests [76].

Granular activated carbon (GAC) has good sorption efficiency for most nitro-aromatic, nitramine and nitrate ester energetic compounds in water, but has limited sorption efficiency for ionic contaminants such a perchlorate. Spent GAC is typically thermally regenerated off-site at carbon regeneration facilities, and a combination of regenerated (~80%) and virgin (~20%) carbon returned to the site during carbon changeouts. For ionic contaminates such as perchlorate, ion exchange (IX) resins are used in conjunction with GAC in multi-step treatment trains for explosives and perchlorate removal in groundwater. Many of the pump and treat remediation systems for explosives groundwater contamination in the US started in the 1990s and remain operational as of this date. Nearly all *ex situ* treatment systems for explosives in groundwater in the US utilize GAC as the primary treatment method. Several advanced oxidation treatment methods, such as UV, ozone, peroxide and a combination of these technologies have been tested for groundwater treatment of energetics, but are generally not cost effective for either capital or operation and maintenance (O&M) costs as compared to the use of GAC. A representative mid-size groundwater pump and treat system for explosives contaminated groundwater is described below.

Umatilla Army Depot groundwater treatment system
Umatilla Army Depot was established as an army ordnance depot in 1941 at the beginning of WWII, and in 1988 the facility was closed under the Base Realignment and Closure (BRAC) legislation. Beginning in 1950 until 1965, Umatilla operated an Explosives Washout Plant on site, where munitions were opened and washed with hot water to removed and recover explosives. The plant was cleaned weekly, and wastewaters containing high concentrations of explosives were discharged via a trough and intermediate settling sump into two unlined lagoons as shown in figure 7.16. The lagoons received 85 million gallons of wastewater over a period of 15 years during plant operations [77]. Although lagoon sludges were removed regularly during operation to an OB ground, explosives contained in the wastewater percolated through the soils and into groundwater beneath the lagoons. The maximum concentration of TNT in the washout lagoons soils was 88 000 mg kg^{-1}, and 711 000 mg kg^{-1} TNT in intermediate settling sump sludges.

A site investigation of the Umatilla Explosives Washout Lagoons was initiated in 1988 to determine the extent of soil and groundwater contamination, which

Figure 7.16. Historical explosives washout lagoons—Umatilla Army Depot, OR.

documented a 330 acre groundwater plume in an unconfined sandy aquifer made up primarily of RDX with concentrations up to 6816 µg l^{-1}. TNT was also present in groundwater up to 3900 µg l^{-1}, but the TNT was generally confined to the area under and near the lagoons as shown in figure 7.17. In 1994, the cleanup decision for the groundwater unit selected groundwater extraction and GAC as the remedy. The contaminants of concern and groundwater cleanup levels are shown in table 7.8 [77].

The Umatilla groundwater treatment system was designed to pump 1300 gallons per minute (gpm) from 3 to 6 extraction wells to achieve hydraulic and chemical containment on the 330 acre plume, and started operation in 1997. The system is designed with two parallel 10 000 pound GAC units in series, with monitoring at the plant influent, between GAC and effluent sampling locations, as shown in figure 7.18. The system has operated for more than 20 years and has removed more than 12 000 pounds of explosives from groundwater. The system is estimated to run approximately 10 more years to achieve the groundwater remediation criteria.

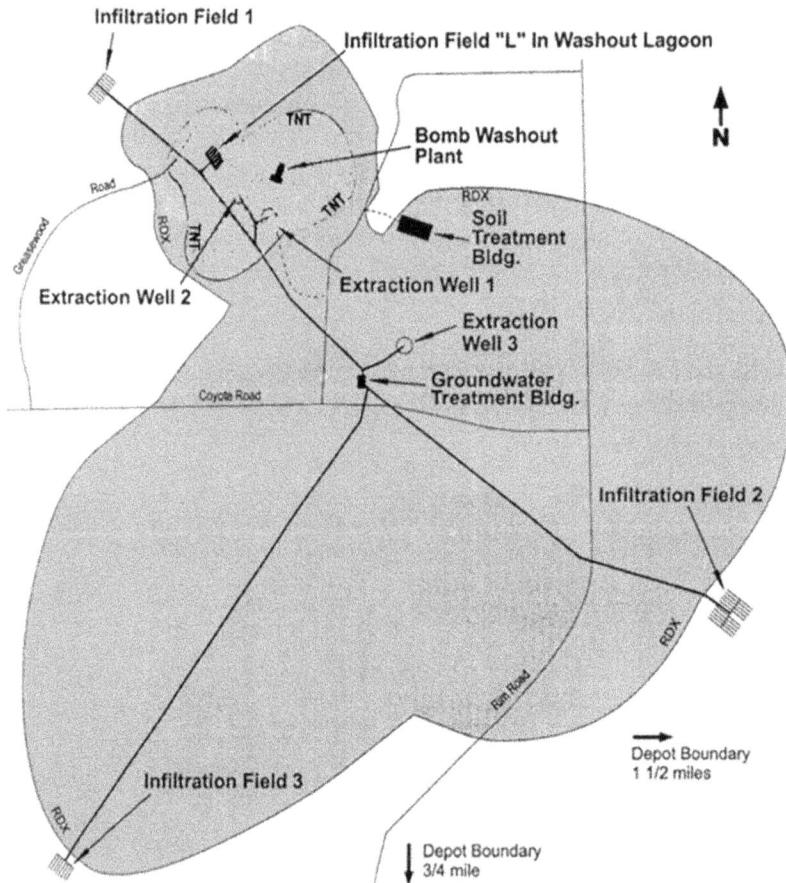

Figure 7.17. Explosives washout lagoons groundwater Plume, Umatilla Army Depot, OR.

Table 7.8. Groundwater remediation criteria—Umatilla Army Depot explosives washout lagoons groundwater.

Chemicals of concern	Cleanup criteria	Highest concentration
Hexahydro-1,3,5-trinitro-1,3,5-triazine (RDX)	2.1 µg l^{-1}	6816 µg l^{-1}
2,4,6-trinitrotoluene (TNT)	2.8 µg l^{-1}	3900 µg l^{-1}
1,3,5-trinitrobenzene (1,3,5-TNB)	1.8 µg l^{-1}	441 µg l^{-1}
1,3-dinitrobenzene (1,3,-DNB)	4.0 µg l^{-1}	24.4 µg l^{-1}
2,4-dinitrotoluene (2,4-DNT)	0.6 µg l^{-1}	497 µg l^{-1}
2,6-dinitrotoluene (2,6-DNT)	1.2 µg l^{-1}	5.3 µg l^{-1}
Octahydro-1,3,5,7-tetranitro-1,3,5,7-tetrazine (HMX)	350 µg l^{-1}	1448 µg l^{-1}

Source: EPA (2002).

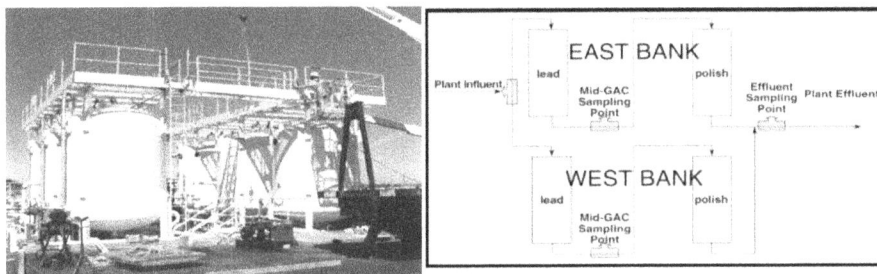

Figure 7.18. GAC groundwater treatment system—Umatilla Army Depot, OR.

7.3.8 *In situ* bioremediation (ISB)

Groundwater in situ *bioremediation*

Based on the success of *ex situ* anaerobic co-metabolism using carbon substrates to biologically degrade explosives and perchlorate in soils, laboratory and pilot-scale studies have been conducted to evaluate the feasibility of *in situ* biological degradation of energetic compounds in groundwater [72, 78]. The potential advantages of utilizing *in situ* biological approaches for explosives treatment are: (1) reduced cost and infrastructure compared to traditional pump and treat methods, (2) complete degradation of explosives on site rather than mass transfer to a second medium such as GAC, and (3) reduced overall remediation timeframes [79]. A range of carbon substrates have been tested including acetate, lactate, molasses, cheese whey, corn sirup, fructose, glycerol, ethanol and emulsified vegetable oil (EVO). The potential implementation approaches include: (1) active or semi-passive plume treatment, (2) semi-passive biobarrier, (3) passive injection biobarrier, (4) passive trench biowall, and (5) passive trench zero valent iron (ZVI) permeable reactive barrier (PBR) [78]. Trenching implementation approaches for ISB are not generally feasible at >50 ft. bgs. More recently, a pilot-scale *in situ* bioremediation test was conducted for a shallow (<50 ft. bgs) passive injection biobarrier for co-mingled RDX and perchlorate in groundwater using EVO carbon substrate, demonstrating the ability to degrade these energetic compounds concurrently under anaerobic reducing conditions [80].

For deeper plumes (e.g. >50 ft bgs) or those that are large or very thick, passive approaches may not be technically feasible and/or cost prohibitive (e.g. injecting passive carbon substrates at closely spaced intervals to >50 ft bgs). Active or semi-passive treatment systems may be technically and economically more attractive under these conditions and better suited for heterogeneous geologies, as ground-water recirculation improves mixing and distribution of injected carbon substrates within the subsurface as shown in figure 7.19 [79]. *In situ* bioremediation has also been tested as an additional technology used in conjunction with currently operating groundwater pump and treat systems with GAC such as the NOP Mead, NE [81] and the DOE Pantex Plant, TX [74] sites.

Soils in situ *bioremediation (ISB)*

In addition to groundwater remediation testing, pilot-scale *in situ* bioremediation (ISB) studies have been conducted for surface and vadose zone soils at OB/OD units and grenade ranges, where deposition and accumulation of energetic residues due to repeated detonations results in soil contamination and may serve as a source of leaching to shallow groundwater, particularly for nitramines (RDX, HMX) and perchlorate. Carbon substrates utilized for these tests include dilute molasses [82] and waste glycerol [83] at OB/OD units, and glycerol + lignosulfonate at grenade ranges [84]. These pilot tests demonstrate that ISB would more likely be able to treat energetics in vadose zone soils to greater depths as compared to other *in situ*

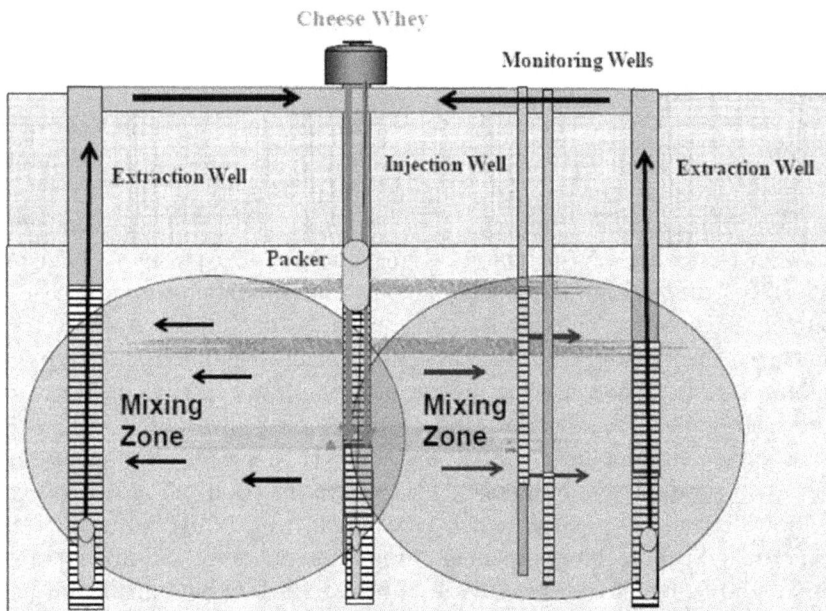

Figure 7.19. Schematic of ISB carbon substrate extraction-reinjection design. Source: [79].

technologies such as alkaline hydrolysis, where the depth of treatment, typically 6 to 12 inches bgs, is limited by the significant buffering capacity of soils. At sites where surface infrastructure or treatment systems may not be desirable or feasible, such as range contamination, *in situ* bioremediation for surface and shallow vadose zone soils may be a viable treatment option in the future.

7.4 Conclusions and further work

Based on the need for remediation technologies to treat energetic contaminants in soil and groundwater, a range of thermal, biological and chemical/physical technologies have been developed and implemented over the past 30 years. Incineration was the first technology utilized to treat highly contaminated soils/ sediments at AAP and depot wastewater lagoons beginning in the late 1980s and was the primary soil treatment method for a period of about 10 years. Biological treatment was intensively evaluated in the 1990s and composting had largely replaced incineration as the primary soil treatment from the mid-1990s onward based on costs, public acceptance and potential beneficial reuse of treated soils. The DARAMEND process was developed to provide both biological and chemical reduction of energetics in soils and has some advantages over composting with regard to amendment additions in some field applications. More recently, development and utilization of alkaline hydrolysis as a primary treatment method for energetics in soils has been implemented based on cost and relative simplicity of the treatment process. Solidification/stabilization (S/S) has been implemented as a primary treatment method where metals are co-contaminants with organic energetics, and a single-step treatment process for organics and inorganics is utilized prior to landfill disposal of treated wastes. A summary of the quantities treated by each of the respective treatment methods is shown in table 7.9.

Nearly all of the full-scale soil remediation technologies utilized to date in the US rely on solid phase materials handling processes in *ex situ* treatment systems, which provides for significant process control and flexibility in system implementation. Although slurry phase degradation such as soil bioslurry reactors have been

Table 7.9. Summary of applications of energetic soil remediation technologies in the US (1.21 million cubic yards).

	Incineration	Composting	DARAMEND	Alkaline hydrolysis	Solidification/ stabilization
No. sites	8	10	2	2	2
Total quan.(cu yds)	388 000	562 500	14 800	153 375	95 000
Mean site quan.(cu yds)	48 500	56 250	7400	76 688	47 500
% of total	32.1	46.5	1.2	12.6	7.9

intensively studied at laboratory scale and to limited extent at pilot scale, none have been implemented at full scale with soil quantities greater than 1000 tons.

Remediation of energetics contamination in groundwater in the US begin in the mid-1990s and most of these treatment systems remain in operation to this day. Nearly all of the groundwater treatment systems for energetics such as TNT, TNB, RDX and HMX in the US utilize pump and treat systems with GAC as the primary treatment method. Advanced oxidation processes (AOPs) have been evaluated but are generally not found to be cost effective relative to operation of GAC systems for parts per billion ($\mu g \, l^{-1}$) level contaminants in groundwater. Where perchlorate is a co-contaminant, ion exchange (IX) resins are often used in series with GAC in groundwater treatment systems. More recently, *in situ* bioremediation has been evaluated as an alternative to or as a complementary technology to GAC treatment systems. Existing monitoring and extraction well networks are often used to test the effectiveness of added carbon substrates for biostimulation, or additions of specific microbial populations for bioaugmentation. *In situ* bioremediation will likely play a larger role in energetics groundwater contamination remediation in the future, particularly at sites where use of *ex situ* technologies may not be desirable or feasible such as artillery/mortar range impact targets, grenade ranges, or OB/OD training and disposal units.

References

[1] JMC 2010 *History of the Ammunition Industrial Base* (Rock Island, IL: Joint Munition Command (JMC) History Office)

[2] Haraburda S S 2017 *Conventional Munitions Industrial Base* (Arlington, VA: The Institute of Land Warfare, Association of the U.S. Army) LWP No. 113

[3] EPA 2014 Technical Fact Sheet—2,4,6-Trinitrotoluene (TNT), https://epa.gov/sites/production/files/2014-03/documents/ffrrofactsheet_contaminant_tnt_january2014_final.pdf

[4] EPA 2014 Technical Fact Sheet—Hexahydro-1,3,5-trinitro-1,3,5-triazine (RDX), https://epa.gov/sites/production/files/2014-03/documents/ffrrofactsheet_contaminant_rdx_january2014_final.pdf

[5] Forsten I 1973 Pollution abatement in a munitions plant *Environ. Sci. Technol.* **7** 806–10

[6] Walsh J T, Chalk R C and Merritt C 1973 Application of liquid chromatography to pollution abatement studies of munitions wastes *Anal. Chem.* **45** 1215–20

[7] Patterson J, Brown J, Duckert W, Polson J and Shapira N I 1976 *State of the Art: Military Explosives and Propellants Production Industry* vol I, II, and III (Cincinnati, OH: U.S. EPA Industrial Environmental Research Laboratory) EPA-600/2-76-213, (NTIS PB-265 385)

[8] Spanggord R J, Gibson B W, Keck R G, Thomas D W and Barkley J J 1982 Effluent analysis of wastewater generated in the manufacture of 2,4,6-Trinitrotolune, 1. characterization study *Environ. Sci. Technol.* **16** 229–32

[9] Spanggord R J, Mabey W R, Mill T, Chou T-W, Smith J H, Lee S and Roberts D 1983 Environmental Fate Studies on Certain Munitions Wastewater Constituents, Phase IV Model Lagoon Studies, prepared by SRI International, Menlo Park, CA for U.S. Army Medical Research and Development Command, Ft. Detrick, MD

[10] Burrows E P, Rosenblatt D H, Michell W R and Parmer D L 1989 *Organic Explosives and Related Compounds: Environmental and Health Considerations* TR-8901 Technical ReportFt. Detrick, MD: U.S. Army Biomedical Research & Development Laboratory

[11] Spaulding R F and Fulton J W 1988 Groundwater munitions residues and nitrate near Grand Island, Nebraska U.S.A *J. Contam. Hydrol.* **2** 139–53

[12] Urbanski E T 2002 Perchlorate as an environmental contaminant *Environ. Sci. Pollut. Res.* **9** 187–92

[13] Urbanski T 1984 *Chemistry and Technology of Explosives* vol 1 (New York: Pergamon)

[14] Army 1984 *Military Explosives*, Headquarters, Department of the Army, TM 9-1300-214

[15] AEHA 1985 Water Pollution Aspects of Explosives Manufacturing, U.S. Army Environmental Hygiene Agency (AEHA), Aberdeen Proving Ground, MD, Technical Guide (TG)–140

[16] Walsh M E and Jenkins T F 1992 *Identification of TNT Transformation Products in Soil* 92-16 Special ReportU.S. Army Corps of Engineers, Cold Regions Research and Engineering Laboratory (Hanover, NH)

[17] Grant C L, Jenkins T F and Golden S M 1993 *Evaluation of Pre-extraction Analytical Holding Times for Nitroaromatic and Nitramine Explosives in Water* 93-24 Special Report U.S. Army Corps of Engineers, Cold Regions Research and Engineering Laboratory (Hanover, NH)

[18] McCormick N G, Cornell J H and Kaplan A M 1981 Biodegradation of hexahydro-1,3,5-trinitro-1,3,5-triazine *Appl. Environ. Microbiol.* **42** 817–23

[19] Harkins V R, Mollhagen T, Heintz C and Rainwater K 1999 Aerobic biodegradation of high explosives: Phase I—HMX *Biorem. J.* **3** 285–90

[20] Nipper M, Carr R S, Biedenbach J M, Hooten R L and Miller K 2005 Fate and effects of picric acid and 2,6-DNT in marine environments: toxicity of degradation products *Mar. Pollut. Bull.* **50** 1205–17

[21] Hoffsommer J C and Rosen J M 1973 Hydrolysis of explosives in sea water *Bull. Env. Contam. Toxicol.* **10** 78–9

[22] Kayser E G and Burlinson N E 1982 *Migration of Explosives in Soil* (Silver Spring, MD: Naval Surface Warfare Center (NSWC) White Oak) NSWC TR 82–56

[23] Kayser E G, Burlinson N and Rosenblatt D W 1984 *Kinetics of Hydrolysis and Products of Hydrolysis and Photolysis of Tetryl* (Silver Spring, MD: Naval Surface Warfare Center (NSWC) White Oak) NSWC TR 84–88

[24] Jenkins T F, Leggett D C, Grant C L and Bauer C F 1986 Reversed-phase high-performance liquid chromatographic determination of the nitroorganics in munitions wastewater *Anal. Chem.* **58** 170–75

[25] Jenkins T F, Walsh M E, Schumacher P W, Miyares P H, Bauer C F and Grant C L 1989 Liquid chromatographic method for the determination of extractable nitroaromatic and nitramine residue in soil *J. AOAC* **72** 890–99

[26] Walsh M E, Jenkins T F, Schnitker S P, Elwell J W and Stutz M H 1993 *Evaluation of SW846 Method 8330 for Characterization of Sites Contaminated with Residues of High Explosives* 93–5 Special Report U.S. Army Corps of Engineers, Cold Regions Research and Engineering Laboratory (Hanover, NH)

[27] Thorne P G and Jenkins T F 1995 *Development of a Field Method for Quantifying Ammonium Picrate and Picric Acid in Soil and Water* 95–2 Special Report U.S. Army Corps of Engineers, Cold Regions Research and Engineering Laboratory (Hanover, NH)

[28] EPA 2012 *Site Characterization for Munitions Constituents* (Washington, D.C.: Office of Solid Waste and Emergency Response) EPA-505-S-11-001

[29] Meyers R A (ed) 2001 Encyclopedia of analytical chemistry. Applications, theory and instrumentation *J. Am. Chem. Soc.* **123** 3621–837

[30] Martland H S 1917 Trinitrotoluene poisoning *J Am. Med. Assoc. (JAMA)* **Vl. LXVIII** 835–37

[31] Yinon J 1990 *Toxicity and Metabolism of Explosives* (Boca Raton, FL: Lewis Publishers)

[32] Dacre J C 1994 Hazard evaluation of army compounds in the environment *Drug Metab. Rev.* **26** 649–62

[33] Roberts W C and Hartley W R 1992 *Drinking Water Health Advisory: Munitions* (Boca Raton, FL: CRC Press)

[34] EPA 1981 *Engineering Handbook for Hazardous Waste Incineration* (Washington, D.C.: Office of Water and Waste Management) SW-881

[35] Dempsey C R and Oppelt E T 1993 Incineration of hazardous waste: a critical review update *Air Waste* **43** 25–73

[36] Major M A and Amos J C 1992 *Incineration of Explosive Contaminated Soil as a Means of Site Remediation* TR 9214 Technical Report U.S. Army Biomedical Research and Development Laboratory (Ft. Detrick, MD)

[37] EPA 1993 *Handbook: Approaches for the Remediation of Federal Facility Sites Contaminated with Explosive or Radioactive Wastes* (Washington, D.C.: Office of Research and Development) EPA/625/R-93/013

[38] EPA 1998 *Incineration at the Former Nebraska Ordnance Site, Mead, Nebraska* (Washington, D.C.: Office of Solid Waste and Energency Response, Technology Innovation Office)

[39] Zupko A J, Johnson N P, Leibbert J M, Empson W B, Mroz D L and Bales F E 1999 Incineration of Explosives-Contaminated Soil and Reactive RCRA Waste at the Former Weldon Spring Ordnance Works, Missouri, USA, Technical paper #9008, IT3 Conference Incineration and Thermal Treatment Technology, Orlando, Florida, May 10-14, http://citeseerx.ist.psu.edu/viewdoc/download;jsessionid=DD33908118D6994A9BE9D3E584CC0306?doi=10.1.1.26.492&rep=rep1&type=pdf

[40] EPA 2003 *Application, Performance, and Costs for Biotreatment Technologies for Contaminated Soils* (Washington, D.C.: U.S. Environmental Protection Agency) EPA/600/R-03/037 (NTIS PB2003-104482)

[41] Craig H D, Sisk W E, Nelson M D and Dana W H 1995 Bioremediation of explosives-contaminated soils: a status review *Proc. of the 10th Annual Conf. on Hazardous Waste Research* (Kansas State University)

[42] Jerger D E and Woodhull P 2000 Applications and costs for biological treatment of explosives-contaminated soils in the U.S *Biodegradation of Nitroaromatic Compounds and Explosives* ed Spain *et al* (Boca Raton, FL: Lewis Publishers)

[43] Bruns-Nagel, Steinbach K, Gemsa D and von Löw E 2000 Composting (humification) of nitroaromatic compounds *Biodegradation of Nitroaromatic Compounds and Explosives* ed Spain *et al* (Boca Raton, FL: Lewis Publishers)

[44] Isbister J D, Doyle R C and Kitchens J F 1982 *Engineering and Development Support of General Decon Technology for the U.S. Army's Installation Restoration Program. Task 2—Composting of Explosives* (Atlantic Research Corporation) http://dtic.mil/dtic/tr/fulltext/u2/a291281.pdf

[45] Doyle R C, Isbister J D, Anspach G L and Kitchens J F 1986 *Composting Explosives/Organics Contaminated Soils* (Atlantic Research Corporation) http://dtic.mil/dtic/tr/fulltext/u2/a169994.pdf

[46] Williams R T, Ziegenfuss P S and Sisk W E 1992 Composting explosives and propellant contaminated soils under mesophilic and thermophilic conditions *J. Ind. Microbiol.* **9** 137–44

[47] Garg R, Grasso D and Hoag G 1991 Treatment of explosives contaminated lagoons sludge *Hazard. Waste Hazard. Mater.* **8** 319–40

[48] Weston R F 1988 *Field Demonstration—Composting Explosives-Contaminated Sediments at the Louisiana Army Ammunition Plant (LAAP)* AMXTH-JR-TE-88242 ReportU.S. Army Toxic and Hazardous Materials Agency http://dtic.mil/dtic/tr/fulltext/u2/a202383.pdf

[49] Weston R F 1991 *Optimization of Composting for Explosives Contaminated Soils* CETHA-TS-CR-91053 Final Report U.S. Army Toxic and Hazardous Materials Agency http://dtic.mil/dtic/tr/fulltext/u2/a246345.pdf

[50] Weston R F 1993 *Windrow Composting Demonstration for Explosives-Contaminated Soils at the Umatilla Depot Activity* CETHA-TS-CR-93043 Report U.S. Army Environmental Center (Hermiston, Oregon)

[51] Emery D D and Faessler P C 1997 First Production Level Bioremediation of Explosives Contaminated Soils in the U.S. *Ann. NY Acad. Sci.* **829** 326–40

[52] ACOE 2011 Soil Composting for Explosives Remediation: Case Studies and Lessons Learned, U.S. Army Corps of Engineers, Public Works Technical Bulletin 200-1-95 https://wbdg.org/FFC/ARMYCOE/PWTB/pwtb_200_1_95.pdf

[53] Scalzo A M, Thompson M A, Holz A, Murray W, Mally D and Mondal K 2001 *Proc. of Int. Containment and Remediation Conf. and Exhibition (Orlando, FL)*

[54] AEC 1998 *Results of a Study Investigating the Plant Uptake of Explosives Residues from Compost of Explosives-Contaminated Soil Obtained from the Umatilla Army Depot Activity, Hermiston, OR* SFIM-AEC-ET-CR-98043 Report U.S. Army Environmental Center (AEC)

[55] Elgh-Dalgren, Waara K S, Düker A, von Kronhelm T and van Hees P A W 2009 Anaerobic bioremediation of a soil with mixed contaminants: explosives degradation and influence on heavy metal distribution, monitored as changes in concentration and toxicity *Water Air Soil Pollut.* **202** 301–13

[56] EPA 1993 Engineering Bulletin: Solidification/Stabilization of Organics and Inorganics, Office of Emergency and Remedial Response (Washington DC) EPA/540/S-92/015, https://clu-in.org/download/contaminantfocus/dnapl/Treatment_Technologies/SS-Eng-Bulletin.pdf

[57] Kogbara R B 2014 A review of the mechanical and leaching performance of stabilized/solidified contaminated soils *Environ. Rev.* **22** 66–86

[58] Hwang S, Felt D R, Bouwer E J, Brooks M C, Larson S L and Davis J L 2006 Remediation of RDX-contaminated water using alkaline hydrolysis *J. Environ. Eng.* **132** 256–62

[59] Johnson J L, Felt D R, Martin W A, Britto R, Nestler C C and Larson S L 2011 *Management of Munitions Constituents in Soil Using Alkaline Hydrolysis* (Vicksburg, MS: U.S. Army Corps of Engineers, Engineer Research and Development Center) ERDC/EL TR-11–16, https://apps.dtic.mil/dtic/tr/fulltext/u2/a552790.pdf

[60] Davis J L, Larson S L, Felt D R, Nestler C C, Martin A W, Riggs L, Valente E J and Bishop G R 2007 *Engineering Considerations for Hydroxide Treatment of Training Ranges* (Vicksburg, MS: U.S. Army Corps of Engineers, Engineer Research and Development Center) ERDC/EL TR-07-3 https://apps.dtic.mil/dtic/tr/fulltext/u2/a470974.pdf

[61] Thorne P G, Jenkins T F and Brown M K 2004 *Continuous Treatment of Low Levels of TNT and RDX in Range Soils Using Surface Liming* (Hanover, NH: US Army Corps of Engineers, Cold Regions Research and Engineering Laboratory) ERDC/CRREL TR-04-4, https://apps.dtic.mil/dtic/tr/fulltext/u2/a423744.pdf

[62] Larson S L, Davis J L, Martin W A, Felt D R, Nestler C C, Fabian G, O'Conner G, Zynda G and Johnson B 2008 *Grenade Range Management Using Lime for Dual Role of Metals Immobilization and Explosives Transformation* (Vicksburg, MS: U.S. Army Corps of Engineers, Engineer Research and Development Center) ERDC/EL TR-08-24, http://el.erdc.usace.army.mil/elpubs/pdf/trel08-24.pdf

[63] Martin W A, Felt D R, Larson S L, Fabian G L and Nestler C C 2012 *Open Burn/Open Detonation (OBOD) Area Management Using Lime for Explosives Transformation and Metals Immobilization* (Vicksburg, MS: U.S. Army Corps of Engineers, Engineer Research and Development Center) ERDC/EL TR-12-4, https://apps.dtic.mil/dtic/tr/fulltext/u2/a556181.pdf

[64] Larson S L, Felt D R, Waisner S, Nestler C C, Coyle C G and Medina V F 2012 *The Effect of Acid Neutralization on Analytical Results Produced from SW846 Method 8330 after the Alkaline Hydrolysis of Explosives in Soil* (Vicksburg, MS: U.S. Army Corps of Engineers, Engineer Research and Development Center) ERDC/EL TR-12-14, https://lrh.usace.army.mil/Portals/38/docs/fuds/PBOW/Effect%20of%20Acid%20Neutralization%20during%20AH%20process.pdf

[65] TetraTech 2009 Technical Memorandum: Soil Treatment Results for Alkaline Hydrolysis in Trench 6 Inert Disposal Area for Iowa Army Ammunition Plant, https://mvs.usace.army.mil/Portals/54/docs/fusrap/Iowa/Volume%205/6968%20Technical%20Memorandum%20Soil%20Treatment%20Results%20for%20Alkaline%20Hydrolysis%20in%20Trench%206%20Inert%20Disposal%20Area.pdf

[66] EPA 2000 *Solidification/Stabilization Use at Superfund Sites* (Washington, D.C.: Office of Solid Waste and Emergency Response) EPA-542-R-00-010, https://epa.gov/sites/production/files/2015-08/documents/solidifcation-stabilization-sf-sites.pdf

[67] Paria S and Yuet P K 2006 Solidification-stabilization of organic and inorganic contaminants using Portland cement: a literature review *Environ. Rev.* **14** 217–55 http://dspace.nitrkl.ac.in/dspace/bitstream/2080/346/3/environ%20rev-06-paria-published%20version.pdf

[68] Channel M 1996 *An Evaluation of Solidification/Stabilization for Treatment of Contaminated Soils From the Umatilla Army Depot Activity* (Vicksburg, MS: U.S. Army Corps of Engineers, Waterways Experiment Station) Technical Report EL-96-4, https://apps.dtic.mil/dtic/tr/fulltext/u2/a307251.pdf

[69] Channel M, Wakeman J and Craig H 1996 *Solidification/Stabilization of Metals and Explosives in Soils, Proc. of the 11th Annual Conf. on Hazardous Waste Research* (Kansas State University) https://engg.ksu.edu/HSRC/96Proceed/channel.pdf

[70] Lear P R, Walker K and Meier J 2003 *Stabilization of Metals and Propellants in Soils from the former Sunflower Army Ammunition Plant* (National Defense Industrial Association) https://ndiastorage.blob.core.usgovcloudapi.net/ndia/2003/environ/lear.pdf

[71] Bates E and Hills C 2015 *Solidification and Stabilization of Contaminated Soil and Waste: A Manual of Practice*, Hygge Media https://clu-in.org/download/techfocus/stabilization/S-S-Manual-of-Practice.pdf

[72] Wani A H, O'Neal B R, Davis J L and Hansen L D 2002 *Treatability Study for Biologically Active Zone Enhancement (BAZE) for the In Situ RDX Degradation in Groundwater* (Vicksburg, MS: U.S. Army Corps of Engineers, Engineer Research and Development Center) ERDC/EL TR-02-35, https://apps.dtic.mil/dtic/tr/fulltext/u2/a408128.pdf

[73] Heerspink B P, Pandey S, Boukhalfa H, Ware D S, Marina O, Perkins G, Vesselinov V V and WoldeGabriel G 2017 Fate and transport of hexahydro-1,3,5-trinitro-1,3,5-triazine

(RDX) and its degradation products in sedimentary and volcanic rocks, Los Alamos, New Mexico *Chemosphere* **182** 276–83

[74] B&W Pantex 2009 *Pantex Plant Ogallala Aquifer and Perched Groundwater Contingency Plan* (Babcock & Wilcox Technical Services Pantex) https://pantex.energy.gov/sites/default/files/185705.pdf

[75] EPA 1991 *Engineering Bulletin: Granular Activated Carbon Treatment* (Washington, D.C.: Office of Emergency and Remedial Response (OERR)) EPA/540/2-91/024, https://cfpub.epa.gov/si/si_public_record_Report.cfm?Lab=NRMRL&dirEntryId=49879

[76] Henke J L and Speitel G E 1998 *Performance Evaluation of Granular Activated Carbon System at Pantex: Rapid Small Scale Column Tests to Simulate Removal of High Explosives from Contaminated Groundwater* (Amarillo, TX: Amarillo National Resource Center for Plutonium) ANRCP-1998-10, https://osti.gov/biblio/290851

[77] EPA 2002 *Dynamic Field Activity Case Study: Treatment System Optimization: Umatilla Chemical Depot* (Washington, D.C.: Office of Solid Waste and Emergency Response (OSWER)) EPA/540/R-02/007

[78] Stroo H F and Ward C H (ed) 2009 *In Situ Bioremediation of Perchlorate in Groundwater* (New York: Springer)

[79] Hatzinger P and Lippincott D 2012 *In-Situ Bioremediation of Energetic Compounds in Groundwater* Project ER-200425 Final Report ESTCP https://apps.dtic.mil/dtic/tr/fulltext/u2/a581233.pdf

[80] Hatzinger P B and Fuller M E 2016 Passive Biobarrier for Treating Co-mingled Perchlorate and RDX in Groundwater at an Active Range, ESTCP Project ER-201028, Final Report https://apps.dtic.mil/dtic/tr/fulltext/u2/1022312.pdf

[81] Wade R, Davis J L, Wani A H and Felt D 2010 Biologically Active Zone Enhancement (BAZE) for *In Situ* RDX Degradation in Groundwater, ESTCP Project ER-0110, Final Report, https://apps.dtic.mil/dtic/tr/fulltext/u2/a604028.pdf

[82] Payne Z M, Lamichhane K M, Babcock R W and Turnbull S J 2013 Pilot-scale *in situ* bioremediation of HMX and RDX in soil pore water in Hawaii *Environ. Sc.: Proc. Imp.* **15** 2023–29

[83] Jugnia L-B, Beaumier D, Holdner J, Delisle S, Greer C W and Hendry M 2018 Enhancing the potential for *in situ* bioremediation of RDX contaminated soil from a former military demolition range *J. Soil Sedi. Contam.* **26** 722–35

[84] Borden R C, Won J and Yuncu B 2017 Natural and enhanced attenuation of explosives on a hand grenade range *J. Environ. Qual.* **46** 961–67

IOP Publishing

Global Approaches to Environmental Management on Military Training Ranges

Tracey J Temple and Melissa K Ladyman

Chapter 8

Characterization and monitoring of energetic compounds on training ranges: case studies in Alaska, United States

Samuel A Beal, Tom A Douglas, Gary W Larsen, Matthew F Bigl, Marianne E Walsh and Michael R Walsh

Energetic compounds (e.g. DNT, TNT and RDX) deposited during military training can have an impact on range sustainment, primarily through the potential migration of these compounds off the installation. Characterization and monitoring studies on two installations in the state of Alaska, United States, provide critical field-based data on energetic compound release pathways and subsequent fate and transport in the environment. Here we summarize significant findings from over 15 years of characterization and monitoring on Joint Base Elmendorf-Richardson and Fort Wainwright, Alaska. Energetic compound concentrations in soil varied by orders of magnitude between different range activities, with some of the highest concentrations observed in soils around low-order detonations, at a propellant burn site and on a demolition range. The infrequent occurrence of low-order detonations provided test cases for the spatial heterogeneity of compound deposition to ranges and also the degradation kinetics of RDX in an estuarine impact area. Monitoring of surface soil concentrations on an artillery firing point over a 15-year period demonstrated relatively stable propellant compound concentrations despite continuous use for training exercises. Subsurface soils from this firing point had an order of magnitude lower propellant compound concentrations than surface soils, which suggested minimal vertical migration. Groundwater and surface water monitoring between 2012 and 2016 on and around Fort Wainwright did not yield concentrations above reporting limits, indicating no significant migration of energetic compounds off of the installation.

8.1 Introduction

Live-fire training is a vital component of military readiness, but associated environmental concerns can challenge the sustainment of training ranges. One such concern is the potential migration of energetic compounds, originally deposited on ranges, off the installation through surface and groundwater. In a notable example, potential migration of high explosives, propellants and lead into a drinking water aquifer in Massachusetts, United States, prompted restrictions on training activities and environmental remediation costs of hundreds of millions of dollars [1, 2].

Energetic compounds can be released into the environment during training in a number of different ways. On impact areas, artillery rounds that infrequently fail to detonate (duds) eventually corrode or rupture from collision with fragments of nearby detonations, exposing their entire energetic mass for dissolution [3, 4] Infrequent low-order and partial detonations result in micron to centimeter-scale particles scattered around a few hundred square meters from the site of impact [5]. High-order detonations of conventional high explosives typically consume >99.99% of their energetic contents during detonation, but residues are still usually detectable and may accumulate [6–8]. Demolition exercises and blow-in-place operations for unexploded ordnance disposal also can leave behind unconsumed energetic material [9]. At firing points, propellant compounds are deposited as small particles, and the amount of propellant residues has been found to vary dramatically between different munition types, with anti-tank rockets and larger caliber mortars exhibiting some of the highest observed residues per round [10]. Unused artillery propellant is sometimes disposed of by burning on or near firing points, and high concentrations of propellant compounds have been found in soils near burn sites utilizing fixed pans [11].

Mobilization of energetic compounds into surface or groundwater is controlled by compound-specific solubility in water, compound accessibility to water within the particle matrix [12], and adsorption/transformation processes on the surface and within soil. The nitroaromatic compounds (i.e. TNT [2,4,6-trinitrotoluene], 2,4-DNT [2,4,-dinitrotoluene], 2,6-DNT [2,6-dinitrotoluene] and DNAN [2,4-dinitroanisole]) tend to have low solubility and high affinity for soil; the nitramine compounds (i.e. RDX [1,3,5-hexahydro-1,3,5-trinitrotriazine] and HMX [octahydro-1,3,5,7-tetranitro-1,3,5,7-tetrazocine]) have low solubility and relatively low affinity for soil; and the nitrate ester compound NG (nitroglycerin), nitroazole compound NTO (3-nitro-1,2,4-triazol-5-one), and nitrimine compound NQ (nitroguanidine) have high solubility and low affinity for soil [13, 14]. Numerous potential degradation products and pathways have been demonstrated from photo, biotic and abiotic transformations.

Of primary interest here are the environmental concentrations and the fate and transport of energetic compounds on training ranges covering different range usages (i.e. artillery firing points, impact areas, burn sites, small arms ranges, multi-purpose combat ranges and demolition ranges). Here we summarize major findings from a compilation of over 15 years of environmental characterization and monitoring data at two installations in Alaska, United States.

8.2 Studied ranges

The two installations are Joint Base Elmendorf-Richardson (JBER), located in Anchorage, Alaska, and Fort Wainwright (FWA), located within and surrounding Fairbanks and Delta Junction, Alaska (figure 8.1). Both installations have sub-arctic climates, but the maritime influence on JBER contributes to substantially warmer winters and wetter conditions than on FWA. Although FWA is one of the northernmost US installations, it is not continuously underlain by permafrost, and both JBER and FWA have subsurface soil and geology similar to more temperate installations.

Live-fire training on JBER involves artillery, mortars, small arms, anti-tank rocket, grenade and engineer demolition ranges. Multiple artillery and mortar firing points surround a central impact area on Eagle River Flats, an estuarine saltmarsh, which has been in use since the 1940s.

Training on FWA is more varied and expansive over two training areas. On Donnelly Training Area (DTA), training involves artillery, mortars, small arms, multi-purpose combat facilities, propellant burn sites and aerial bombing. The impact areas used for artillery and bombing are located on the opposite side of the braided channel of the Delta River from the firing points. Multi-purpose combat ranges on DTA include small arms and mobile Stryker vehicle-based mortar firing points. Training on Yukon Training Area (YTA) includes a multi-purpose combat range using Stryker vehicle-based mortars and artillery.

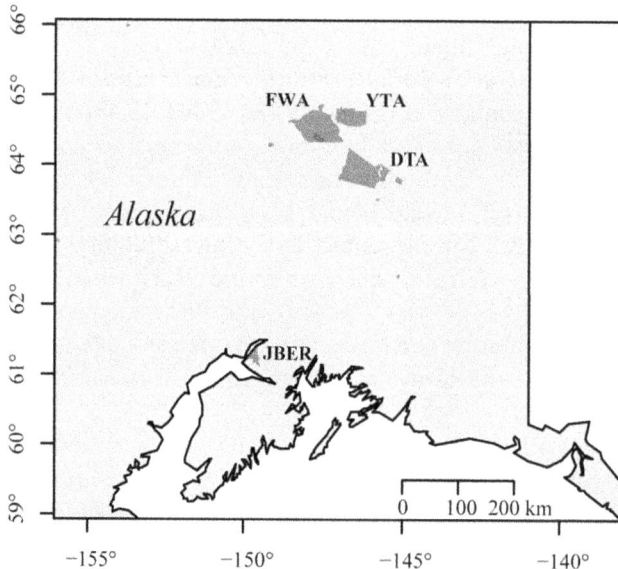

Figure 8.1. Installation boundaries of Joint Base Elmendorf-Richardson (JBER) and Fort Wainwright (FWA), including Yukon Training Area (YTA) and Donnelly Training Area (DTA), which are part of FWA.

8.3 Methods

The data presented here were compiled from samples collected between 1990 and 2016. With the exception of 24 samples collected in 1990 and 77 samples collected in 2001, all samples were collected and analyzed by researchers at the US Army Cold Regions Research and Engineering Laboratory (CRREL). Over this period of time soil sampling approaches advanced with increasing acknowledgment of the heterogeneity of energetic compounds deposited on ranges [15]. Early usage of discrete and composite sampling was replaced in 2003 by multi-increment sampling that provides a more accurate mean soil concentration for a given area [16]. All sampling was performed during the summer months. Soil sample processing generally followed that described by Cragin *et al* [17], Hewitt *et al* [18], Taylor *et al* [19], Walsh *et al* [20], and US Environmental Protection Agency (USEPA) Method 8330B appendix A. In brief, samples were air dried, dry sieved for the <2 mm soil fraction, pulverized with a puck mill, and representatively sub-sampled before extraction with acetonitrile. Water samples were generally filtered and solid-phase-extracted following USEPA Method 3535. Analysis typically followed USEPA Method 8330 using high performance liquid chromatography with ultraviolet detection. USEPA Method 8095 using gas chromatography with electron capture detection was typically used for confirmation.

8.4 Results and discussion

Over 15 years of characterization research on Alaskan ranges has produced a greater practical understanding of release pathways, areas most greatly impacted, and eventual fate and transport of energetic compounds.

8.4.1 Range activities

The soil concentrations of approximately 3000 samples are presented by range usage activity in figure 8.2. The focus of this analysis is on the compounds that are typically found on ranges, including NG, DNT, HMX, RDX and TNT. Other energetic compounds and degradation products were detected in less than approximately 10 percent of tested samples, including 4-amino-2,6-dinitrotoluene (10% of $n = 1655$), 2-amino-4,6-dinitrotoluene (10% of $n = 1643$), 1,3,5-trinitrobenzene (4% of $n = 848$), methyl-2,4,6-trinitrophenylnitramine (i.e., tetryl; 3% of $n = 776$), pentaerythritol tetranitrate (2% of $n = 189$), 3,5-dinitroaniline (1% of $n = 326$) and nitrobenzene (0.3% of $n = 324$). The following compounds were not detected in any samples: 1,3-dinitrobenzene (0% of $n = 764$), 2-nitrotoluene (0% of 324), 3-nitrotoluene (0% of $n = 324$) and 4-nitrotoluene (0% of 324). Even within areas where certain compounds are expected to be deposited (e.g. TNT on impact areas, NG on firing points), soil concentrations were frequently below detection, owing to the heterogeneous nature of these compounds on ranges.

The compound NG, commonly used in double and triple-based propellants, was most concentrated and frequently detected at artillery firing points, small arms ranges, multi-purpose combat ranges and a propellant burn point. 2,4-DNT, used in

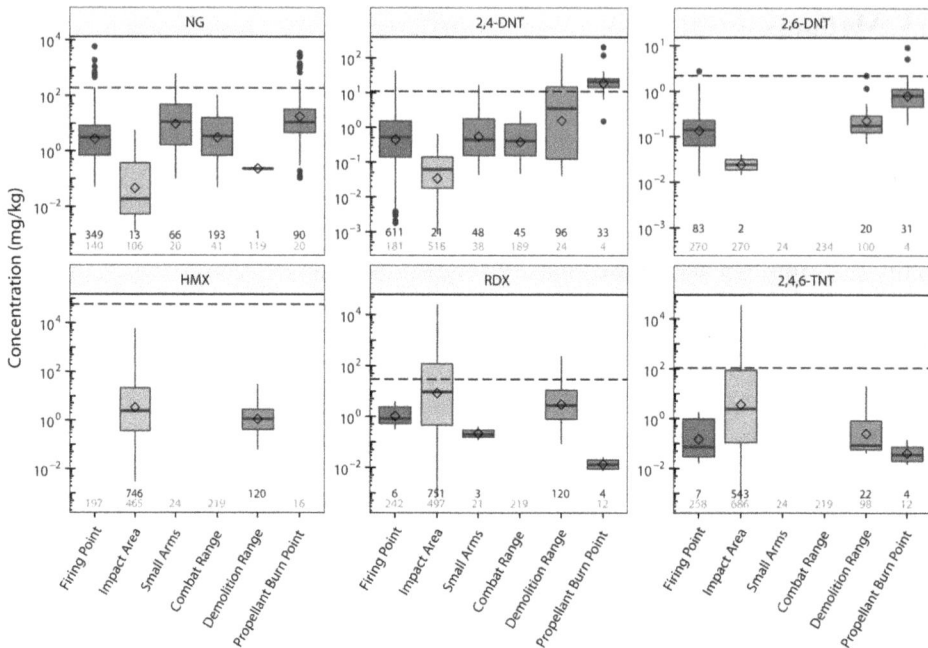

Figure 8.2. Boxplots of energetic compound soil concentrations by range activity. The number of samples with concentrations above the reporting limit are indicated in black text, and the number of samples below the reporting limit (not plotted) are indicated by the red text. Diamonds are group means. Dashed lines represent the USEPA regional soil screening levels for ingestion of industrial soil.

single and double-based propellants, was commonly detected at artillery firing points, small arms ranges and multi-purpose combat ranges, but concentrations on average 10 to 100-fold greater were found at a demolition range and propellant burn point. 2,6-DNT was infrequently detected, given the isomer's lower content in propellant, but had the greatest concentrations at artillery firing points, a demolition range and a propellant burn point.

The high explosive (HE) compounds HMX, RDX and TNT were typically detected at impact areas and at a demolition range, but positive detections elsewhere point to the multi-use nature of some ranges. The high variability and high concentrations of HE on impact areas is the result of sampling predominantly in and around impact craters.

8.4.2 Potential point sources

Efficiency testing of a wide range of munitions has identified that the main sources of HE compounds on ranges are derived from low-order and dud artillery rounds [21]. Propellant deposition rates depend highly on the weapons system fired, with, for example, orders of magnitude variability between small arms munitions and anti-tank rockets [10, 11] Research on these Alaskan ranges identified two situations where compound concentrations in soil can be highly elevated.

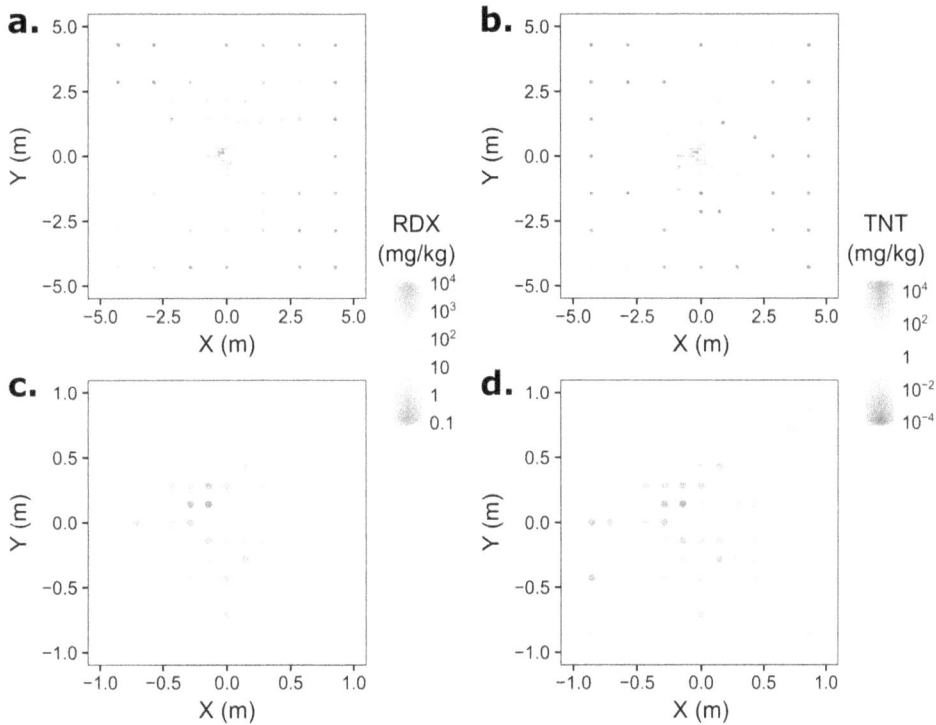

Figure 8.3. Map of surface soil concentrations of RDX (a. and c.) and TNT (b. and d.) in a 10 m × 10 m plot centered on the low-order detonation crater on JBER of a 120-mm mortar round containing Composition B (unpublished work by A D Hewitt).

Low-order detonations

Low-order detonations can produce micron to centimeter-sized particles deposited over an area of a few hundred square meters from the site of detonation [5, 21]. In March 2006, training with 120-mm mortars containing Composition B (TNT, RDX, and HMX) on JBER produced five low-order detonations on the Eagle River Flats impact area (figure 8.3) [22]. One of these low-order detonations was selected for a study on depositional heterogeneity and sampling optimization by Hewitt *et al* (unpublished). In August 2006, 196 discrete soil samples were collected in nested grids up to 10 m by 10 m surrounding the detonation site using a 3 cm diameter and 3 cm deep sampling corer. The concentrations of TNT and RDX were, as expected, greatest toward the center of the detonation site, but were detected throughout the grid as far as 7 m from the center. Concentrations within the grid spanned over five orders of magnitude for both TNT and RDX, and exhibited order of magnitude variability between samples separated by centimeters within the central 1 m by 1 m grid. The highly heterogeneous distribution of HE compounds surrounding this low-order detonation demonstrates the necessity of multi-increment sampling approaches in representatively sampling ranges.

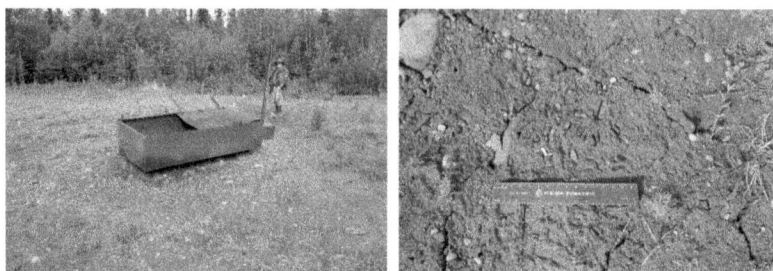

Figure 8.4. A fixed propellant burn pan located on DTA with intact centimeter-sized propellant grains at the soil surface surrounding the pan.

Propellant burn site

While training with mortars and howitzers, units do not always utilize the full number of propellant charges issued with each round. In the US, these charges are typically burned on site by igniting in piles on the ground or in large fixed burn pans. Environmental conditions have an effect on the consumption efficiency of burns and propellant grains can be ejected when uncontained [23]. From figure 8.2, it is apparent that high concentrations of the propellant compounds NG and DNT were found at the propellant burn site. Around the fixed burn pan at this site, unburned propellant grains were visible at the soil surface (figure 8.4). These grains may have been spilled prior to burn events or ejected during burns due to the pan's lack of a screened covering. To improve propellant burn operations, Walsh *et al* developed a screened burn pan that is portable between firing points and increases burn efficiency to >99.9% [11]. Replacement of the bagged propellant with newer modular charges for some artillery applications is reducing the need for propellant burn operations.

8.4.3 Fate and transport on ranges

While concentrations in range soil may be a concern to potential ecological and human receptors on ranges, a greater concern for range sustainment is migration of harmful compounds off the installation through surface and groundwater. As energetics are initially deposited as solid particles, dissolution is the first process governing fate and transport. Different energetic compounds have widely varying solubilities in water, but access of water to compounds within the matrix of energetic particles is also critical. NG, for example, is highly soluble, but in propellant particles it is bound to relatively insoluble nitrocellulose, thereby limiting its dissolution [12, 24]. Once dissolved, compound transport is affected by compound-specific adsorption onto organic matter and mineral surfaces. Both in solid and dissolved states, compound transformation by interaction with minerals, microbes and sunlight can also occur. Several findings reported below provide insight into the fate and transport of compounds on the studied Alaskan ranges.

Firing points

A number of different artillery firing points on JBER and FWA were sampled on an annual to semi-annual basis to examine potential accumulation of energetic compounds in surface soils. The most sampled site is Firing Point Sally located within DTA on FWA. Firing Point Sally has been used repeatedly for artillery training, and the site has a 2 to 5 cm thick organic soil horizon overlying loess and glacial till. The time series record of propellant compounds in soil at this site is shown in figure 8.5. Despite continual training on the site over the 15-year period, soil concentrations of NG and 2,4-DNT remained relatively stable and generally within the replicate sample variability of any given year. Given continual loading of propellants by training exercises, the stable concentrations indicate a balance with dissolution, advection and/or transformation.

Between 2007 and 2013 at Firing Point Sally, soil samples were collected at 3 to 6 cm depth along with the standard samples collected at 0 to 3 cm depth. Median concentrations of NG and 2,4-DNT within the upper 3 cm of soil were an order of magnitude greater than those at 3 to 6 cm depth (figure 8.6). This vertical segregation of concentrations suggests dissolution and downward advection occurred, but the majority of energetic mass remained within the surface soil layer or was otherwise transformed before migrating below 3 cm depth.

Low-order craters over time
While studying disturbance effects on white phosphorus in subsurface sediment at Eagle River Flats, JBER, two C4 blocks unintentionally detonated low-order, resulting in two craters with visible dispersed particles of C4 [5]. These large particles were taken off site for disposal, and the areas within a 1 m radius of the crater centers were sampled over the course of four years. In both craters, concentrations of RDX and HMX decreased by orders of magnitude over the four year period (figure 8.7). The more rapid initial decrease in concentrations for Crater 1 was attributed by Walsh *et al* [5] to potential differences in sediment moisture and associated redox conditions that might favor abiotic and biotic degradation processes. Assuming little to no advection of compounds outside the 1 m radius around the craters, recovery from low-order detonation on Eagle River Flats appears to be relatively rapid, potentially a result of the anaerobic conditions of the estuarine sediments.

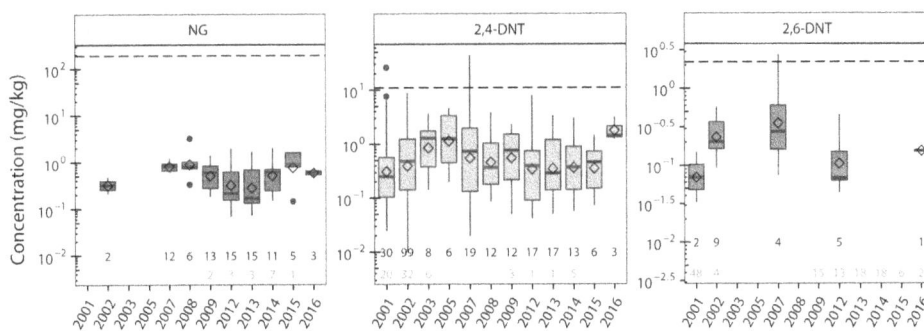

Figure 8.5. Time series of propellant compound concentrations in surface soil at Firing Point Sally, DTA, FWA. The number of samples with concentrations above the reporting limit are indicated in black text, and the number of samples below the reporting limit (not plotted) are indicated by the red text. Diamonds are group means. Dashed lines represent the USEPA regional soil screening levels for ingestion of industrial soil.

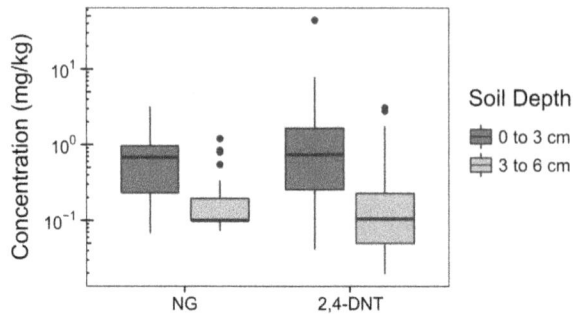

Figure 8.6. Comparison of propellant compound concentrations with soil depth at Firing Point Sally. For this comparison, non-detected samples are reported as the estimated method detection for NG and 2,4-DNT (0.1 and 0.05 mg kg^{-1}, respectively).

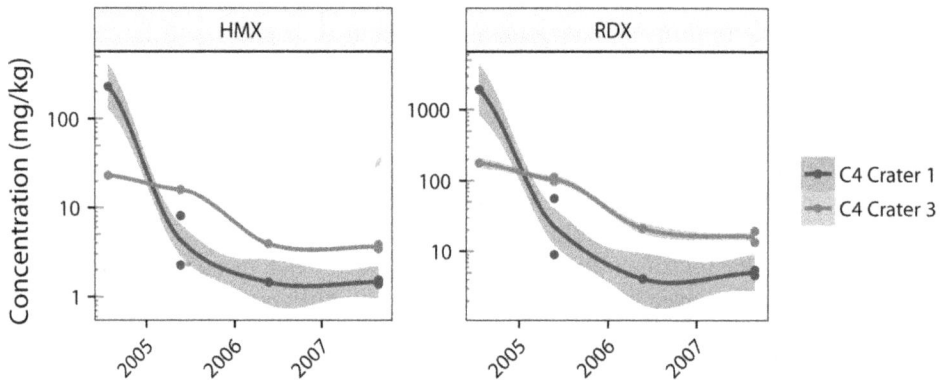

Figure 8.7. Soil concentrations of HMX and RDX in two low-order C4 detonation craters on Eagle River Flats, JBER (more detail on craters in Walsh *et al* [5]).

Groundwater and surface water monitoring

To examine potential compound migration, surface and groundwater monitoring was conducted within and surrounding FWA between 2012 and 2016. Surface water samples were collected from lakes and ponds on DTA and from rivers and streams downstream of DTA and YTA. Groundwater samples were collected from five wells within a realistic combat training facility on DTA. Depth to groundwater in these wells ranged from 2 to ~30 m during the sampling events. In 276 surface and groundwater samples over the four year monitoring period, no samples had reproducible concentrations of energetic compounds above laboratory reporting limits (table 8.1). This monitoring suggests that off-site migration has not been occurring on FWA.

8.5 Conclusion

The characterization and monitoring of energetic compounds on the two Alaskan installations has led to an improved understanding of range impacts, depositional

Table 8.1. Reporting limits for surface and groundwater monitoring on and around FWA.

Analyte	Reporting limit ($\mu g\ l^{-1}$)
Octahydro-1,3,5,7-tetranitro-1,3,5,7-tetrazocine (HMX)	0.2
1,3,5-Trinitrobenzene	0.05
1,3,5-Hexahydro-1,3,5-trinitrotriazine (RDX)	0.1
1,3-Dinitrobenzene	0.05
2,4,6-Trinitrotoluene (TNT)	0.05
2,4,6-Trinitro-phenylmethylnitramine (Tetryl)	0.2
Nitroglycerin (NG)	0.2
2,4-Dinitrotoluene (2,4-DNT)	0.05
2,6-Dinitrotoluene (2,6-DNT)	0.05
2-Amino-4,6-dinitrotoluene	0.2
4-Amino-2,6-dinitrotoluene	0.2

distributions, and fate and transport behavior. Relatively high concentrations of propellant compounds, particularly DNT, were found at a propellant burn site, demolition range and some firing points. Relatively high concentrations of HE compounds were found on impact areas largely focused around impact craters from low-order detonations, but high concentrations were also found on a demolition range where training is more spatially confined. The sampling of low-order detonations on Eagle River Flats, JBER, has produced important data on the spatial distribution of energetic compounds and the environmental recovery rate in this estuarine environment. Despite ongoing training on FWA, concentrations of energetic compounds in firing point soils have remained relatively stable, and monitoring of groundwater and surface waters has revealed no evidence for off-site migration. Although encouraging, the results from these ranges do not necessarily translate to other ranges that encounter different munitions usage and different environmental, hydrologic and geologic conditions. However, the results of this extended monitoring and characterization work can inform the managing of ranges so that they are sustained for training operations well into the future.

References

[1] DEP ARC 2010 Defense environmental programs annual report to congress, prepared by the office of the under secretary of defense for acquisition, technology, and logistics. Available online at: https://denix.osd.mil/arc/arcfy2010/

[2] EPA 1997 Administrative order on Consent (AOC) For Response Action, Docket No: SDWA I-97-1019. Available online at: https://semspub.epa.gov/src/document/01/448135

[3] Taylor S, Bigl S and Packer B 2015 Condition of *in situ* unexploded ordnance *Sci. Total Environ.* **505** 762–69

[4] Walsh M R, Walsh M E, Poulin I, Taylor S and Douglas T A 2011 Energetic residues from the detonation of common US ordnance *Inter. J. Energ. Mater. Chem. Propul.* **10** 169–86

[5] Walsh M E, Collins C M, Ramsey C A, Douglas T A, Bailey R N, Walsh M R, Hewitt A D and Clausen J L Energetic residues on Alaskan training ranges: studies for US army Garrison Alaska 2005 and 2006. ERDC/CRREL TR-07-9, Hanover, NH, 2007. Available online at: http://www.dtic.mil/dtic/tr/fulltext/u2/a471042.pdf

[6] Jenkins T F, Walsh M E, Miyares P H, Hewitt A D, Collins N H and Ranney T A 2002 Use of snow-covered ranges to estimate explosives residues from high-order detonations of army munitions *Thermochim. Acta* **384** 173–85

[7] Hewitt A D, Jenkins T F, Walsh M E, Walsh M R and Taylor S 2005 RDX and TNT residues from live-fire and blow-in-place detonations *Chemosphere* **61** 888–94

[8] Walsh M R, Walsh M E, Ramsey C A, Thiboutot S, Ampleman G, Diaz E and Zufelt J E 2014 Energetic residues from the detonation of IMX-104 insensitive munitions *Propellants Explos. Pyrotech.* **39** 243–50

[9] Hewitt A D, Jenkins T F, Walsh M E and Brochu S 2009 Validation of sampling protocol and the promulgation of method modifications for the characterization of energetic residues on military testing and training ranges. ERDC/CRREL TR-09-6, Hanover, NH Available online at: www.dtic.mil/dtic/tr/fulltext/u2/a607234.pdf

[10] Walsh M R, Walsh M E, Ampleman G, Thiboutot S, Brochu S and Jenkins T F 2012 Munitions propellants residue deposition rates on military training ranges *Propell. Explos. Pyrotech.* **37** 393–406

[11] Walsh M R, Thiboutot S, Walsh M E and Ampleman G 2012 Controlled expedient disposal of excess gun propellant *J. Hazard. Mater.* **219–220** 89–94

[12] Taylor S, Dontsova K, Bigl S, Richardson C, Lever J, Pitt J, Bradley J P, Walsh M E and Simunek J 2012 Dissolution rate of propellant energetics from nitrocellulose matrices. ERDC/CRREL TR-12-9. Hanover, NH Available online at: http://acwc.sdp.sirsi.net/client/search/asset/1011489

[13] Pennington J C 2006 *Distribution and Fate of Energetics on DoD Test and Training Ranges* TR-06-13 Final ReportERDC (Vicksburg, MS)

[14] Mark N, Arthur J, Dontsova K, Brusseau M and Taylor S 2016 Adsorption and attenuation behavior of 3-nitro-1,2,4,-triazol-5-one (NTO) in eleven soils *Chemosphere* **144** 1249–55

[15] Jenkins T F, Hewitt A D, Grant C L, Thiboutot S, Ampleman G, Walsh M E, Ranney T A, Ramsey C A, Palazzo A J and Pennington J 2006 Identity and distribution of residues of energetic compounds at army live-fire training ranges *Chemosphere* **63** 1280–90

[16] Hewitt A D, Jenkins T F, Walsh M E, Walsh M R, Bigl S R and Ramsey C A 2007 Protocols for collection of surface soil samples at military training and testing ranges for the characterization of energetic munitions constituents ERDC/CRREL TR-07-10, Hanover, NH Available online at: http://www.dtic.mil/dtic/tr/fulltext/u2/a471045.pdf

[17] Cragin J H, Leggett D C, Foley B T and Schumacher P W 1985 *TNT, RDX, and HMX Explosives in Soils and Sediments* 85-15 Report CCREL (Hanover, NH)

[18] Hewitt A D, Bigl S R, Walsh M E, Brochu S, Bjella K and Lambert D 2007 Processing of training range soils for the analysis of energetic compounds ERDC/CRREL TR-07-15, Hanover, NH

[19] Taylor S, Jenkins T F, Bigl S R, Hewitt A D, Walsh M E and Walsh M R 2011 Guidance for Soil Sampling for Energetics and Metals. ERDC/CRREL TR-11-15, Hanover, NH Available online at: http://www.dtic.mil/dtic/tr/fulltext/u2/a583020.pdf

[20] Walsh M R, Walsh M E, Gagnon K, Hewitt A D and Jenkins T F 2014 Subsampling of soils containing energetics residues *Soil Sedim. Contam.: Inter. J.* **23** 452–63

[21] Walsh M R, Temple T, Bigl M F, Tshabalala S F, Mai N and Ladyman M 2017 Investigation of energetic particle distribution from high-order detonations of munitions *Propell. Explos. Pyrotech.* **42** 932–41

[22] Walsh M E, Taylor S, Hewitt A D, Walsh M R, Ramsey C A and Collins C M 2012 Field observations of the persistence of comp B explosives residues in a salt marsh impact area *Chemosphere* **78** 467–73

[23] Walsh M R, Walsh M E and Hewitt A D 2010 Energetic residues from field disposal of gun propellants *J. Hazard. Mater.* **173** 115–22

[24] Jenkins T F 2008 *Characterization and Fate of Gun and Rocket Propellant Residues on Testing and Training Ranges* TR-08-1 ERDC (Vicksburg, MS)

IOP Publishing

Global Approaches to Environmental Management on Military Training Ranges

Tracey J Temple and Melissa K Ladyman

Chapter 9

Heavy metal contamination on small arms shooting ranges

R Keiser and O Hausheer

During the last century, live-fire military training has led to contamination by heavy metals such as lead, antimony and copper. Due to their potential toxicity, this has become a public health concern. Indeed, the spent ammunition and their weathering products have released heavy metals into the environment. This is particularly a problem on small arms shooting ranges, where heavy metal soil contamination has reached high levels. This chapter provides information about a step-by-step process for small arms ammunition contaminated areas from first assessment to remediation, especially regarding Swiss regulations. To address the problem, the Swiss government's guidelines recommend a process beginning with a historical investigation (mainly gathering data from existing papers), which gives a first indication of the potential pollution. This is followed by a technical investigation, which consists of a detailed field examination and includes sampling and analysis to determine the effective level of contamination as well as expansion of the contaminated area. There is subsequently a risk assessment, which evaluates the potential environmental and health consequences. As a result, a remediation concept will be established that aims for the best compromise between cleaning the contaminants, environment preservation and the lowest possible costs.

For small arms shooting ranges (military and civilian), the best way to prevent pollution is to improve construction properties of artificial backstops, avoid the use of areas with special values and adapt military training exercises. In conclusion, even though contamination through military live-fire training cannot be completely avoided, it is important that military personnel are aware that environmental protection is an issue of national importance. Therefore, we recommend environmental communication and education. Current investigations focus mainly on lead

doi:10.1088/978-0-7503-1605-7ch9

or copper. Further research should focus on other contaminants such as antimony, since knowledge of their environmental behavior is still insufficient.

9.1 Introduction

Contamination of shooting range soils by lead (Pb), antimony (Sb) and copper (Cu) released from corroding ammunition has become an issue of public concern over recent years (figure 9.1). Military exercises on small arms shooting ranges include weapons up to 0.50 caliber firearms, such as handguns, assault rifles and machine guns. Due to its low cost, availability and high density, Pb is the material of choice for the core of small arms ammunition. In addition, 2–5% of each bullet consists of metallic Sb (depending on the manufacturer) as a hardening agent and Cu as a coating agent. In recent years, copper has also been used for several types of special ammunition, such as frangible ammunition. The content of one bullet (in gram) of Swiss standard ammunition is shown in table 9.1. To demonstrate the pollution potential, a magazine for an assault rifle that contains 20 shots would release 60 g (GP 90) or 160 g (GP 11) of Pb into the environment within seconds.

A military shooting range is an outdoor area designed for real life training for soldiers. The elements of a range (i.e. firing point, target area, target location) and their definitions, which are used in this chapter, are illustrated in figure 9.2. Bullets should move through the target and accumulate in the backstops, (i.e. bullet collection structure). There are various designs of different kind of backstops, e.g. earth walls, steep hillsides (minimum of 30° incline) or (as of more recently) artificial

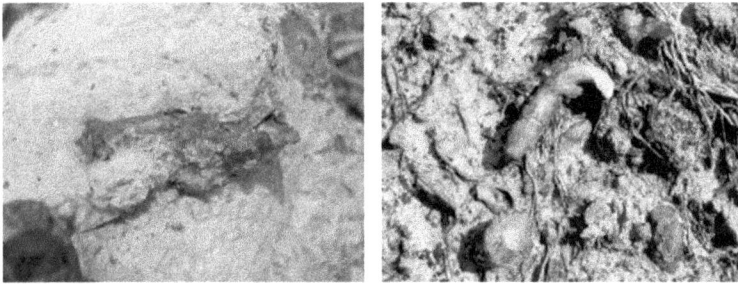

Figure 9.1. Munition released into the environment during military shooting exercises after years in the soil.

Table 9.1. Pollutant composition of Swiss military ammunition per bullet.

Bullet	Pb [g]	Cu [g]	Sb [g]
5.6 mm Gewehrpatrone 90	2.99	0.1	0.06
7.5 mm Gewehrpatrone 11	8.378	0.257	0.171
9 mm Pistolenpatrone 41	6.1	0.2	0.3
9 mm Pistolenpatrone 14	6.1	1.17	0.3

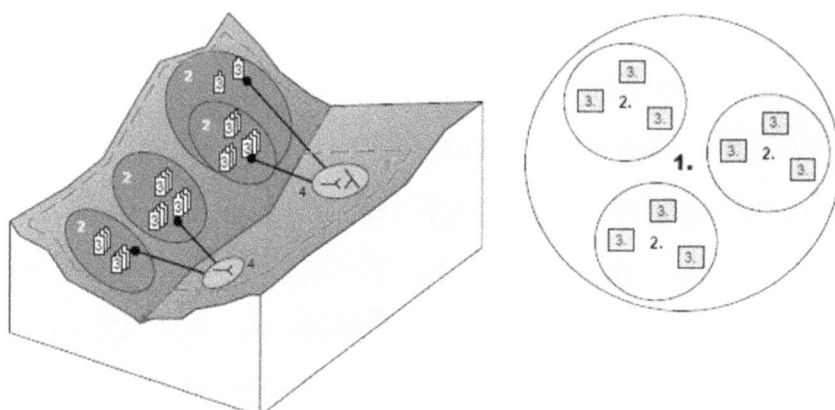

Figure 9.2. Elements of a shooting range: (1) shooting range; (2) target area; (3) target location; (4) firing point. Adapted from [1]. Copyright, Federal Authorities of the Swiss Confederation.

backstops. Because of missed shots or ricochets, bullets also end up outside of the target area. In theory, the heavy metal load decreases rapidly with distance to the target location while spreading to the topsoil of the range area. The targets may or may not be fixed.

The weathering of spent bullets can pose a threat to human health and the environment. At certain levels, these elements can be toxic to organisms (e.g. already low levels of Pb and Sb lead to nerve damage). Organisms can be exposed either by direct uptake or via plants, which in turn exposes the food chain. Another exposure pathway is via soil solution into groundwater or surface water, which can be used as drinking water. Heavy metals are chemical elements that cannot degrade in the environment, which can lead to high soil levels. Thus, the problem with shooting range contamination is not an acute or a short-term environmental risk. Rather, the environmental impact of the shooting activity builds up over a period of dozens or hundreds of years. Especially problematic for the environment are shooting ranges where the target area is in contact with water, such as wetlands or target areas where the groundwater table is close to the surface.

In conclusion, it is important to evaluate the environmental risk potential of current and past shooting practices, to put in effort to minimize the identified risk and to reduce the overall environmental exposure of heavy metals.

9.2 Methods for contaminated site management

Because of the amount of heavy metals per small arms bullet and the large amount of bullets fired, target areas of small arms shooting ranges are often considered to be contaminated sites. The investigation of the pollution status of these sites and possible remediation efforts require a systematic approach. Firstly, all target areas of shooting ranges are registered in a database. In a further step, the historical investigation includes the sites operating history and gives a first hypothesis of the pollution load and what effects it has on the environment. These findings are the

Figure 9.3. Management steps for the evaluation of contaminated sites on small arms shooting ranges.

basis of the classification if the site is considered safe or not. If it cannot be excluded that a target area poses an environmental risk, then up-to-date technical investigation procedures are required to evaluate the actual pollution load of the site.

All findings are to be documented in a report and described in a way that is uniform, transparent and reproducible. The results from the technical investigation are the basis of the risk assessment and the remediation concept for the site. Mistakes or uncertainties during the historical and/or technical investigation can result in environmental risks remaining undetected or be the cause for high costs during potential future remediation works. Thus, all uncertainties should always to be documented. Figure 9.3 shows the different procedures, with a selected summary of the main conclusions of every step. The principles are similar for all contaminated site management of small arms shooting ranges; however, the following sections will be from a Swiss point of view.

9.2.1 Historical investigation

For all contaminated sites, the historical investigation gives a first clue to the potential of pollution of the area thus determining the need for monitoring and/or clarifying if further investigations are required. A very central step for the historical investigation of shooting ranges is the establishment of contact with people in charge of the shooting range, who have knowledge on past and current shooting activities. With them, an on-site inspection is carried out and they are to be mentioned by name in the report. They are the contact for any further information if needed.

During the on-site visit, the area should be photographically documented. All ownership structures and contractual arrangements should also be noted.

The evaluation of the history of the site includes a documentation of all shooting target areas (former and current) of the shooting range with notation of the location. It is very important to gather information on the shooting activity, especially on which weapons and ammunition were used. This information combined with an estimation of the number of rounds fired over the years (i.e. shooting intensity) allows an estimation of the pollution load entering the environment through shooting activities.

The evaluation of the obtained information concerning the pollution situation of the site results in a pollution hypothesis. The areas of the shooting range can be divided into different categories:

- Heavily contaminated areas;
- Medium or variable contaminated areas;
- Slightly contaminated areas;
- Areas not contaminated (e.g. flyover areas).

Important factors for the classification include the extension of the target area and the total amount of pollution entering the environment, which should be defined in certain guidelines. If classified as at most slightly contaminated, then these sites do not pose an environmental risk and they can be considered safe for any further civilian usages.

The environment of the shooting range is best described by documenting the hydrogeology, geology and the vegetation of the area. For this task, also all available geographical information (e.g. web mapping), including historical photographs of the site can be helpful on evaluating the surrounding area. Any structural changes in the landscape assists in understanding the pollution situation of the site.

Apart from describing the environment, all natural resources that are exposed by the contaminated site are examined in detail. With shooting practices, the first point of entry for pollutants into the environment is via the soil. Depending on the pollution load and the usage of the land, the exposure of the soil compartment is evaluated. Through chemical weathering, pollutants can also be leached out of the soil and end up in the groundwater. If this is the case, then this compartment is also considered exposed. Surface water is considered an exposed compartment if streams pass through heavily contaminated areas of shooting ranges or are at a distance of < 10 m to these areas. Air, as the fourth natural compartment, is not exposed to this kind of heavy metal contamination and is thus excluded from the assessment. From the person of contact all information on past and current usages of the environment (i.e. pasture for grazing animals, spring tapping) need to be verified.

The findings of the historical investigation are summarized in a pollution suspicion matrix as shown in table 9.2 with an example of one target site from a Swiss military shooting range.

This matrix delivers a compacted overview of the findings. The potential of pollution must also be presented graphically, which is shown with a different example of a Swiss shooting range in figure 9.4.

Table 9.2. Suspicion matrix of the pollution potential of a shooting target site.

	Time period	Used ammunition	Pollutants	Location of pollutants	Potential of pollution	Distribution pathways
Lower Tannmatt: western sector	Before 1900 until approx. 1980	GP 11	Pb, Cu, Sb	Soil, underground, groundwater	Medium to high	Leachate, groundwater, grazing animals

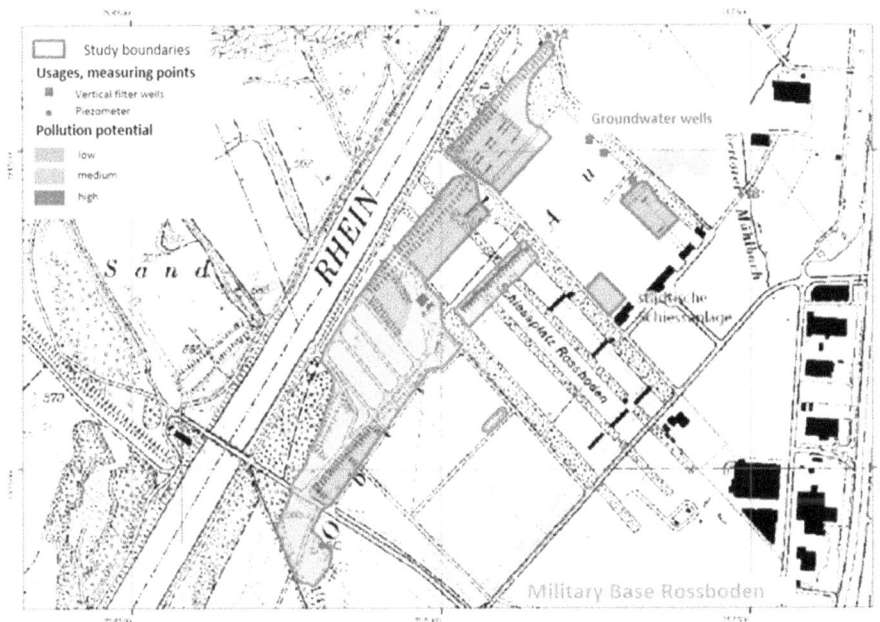

Figure 9.4. Example of a pollution map of a Swiss shooting range Rossboden. Adapted from [1]. Copyright, Federal Authorities of the Swiss Confederation.

For sites that require a further investigation, i.e. which pose a potential threat to the surrounding environment, an investigation strategy should be proposed during the historical investigation. To obtain a better picture of the heavy metal contamination of the target areas, the contaminations levels of Pb, Sb and Cu should be measured and the lateral and vertical extension of the polluted area mapped out. For this, fieldwork using a mobile X-ray fluorescence (m-XRF, figure 9.5) device is carried out, which allows an efficient, low-cost on-site evaluation of the contamination content of the soil. The sampling strategy also includes a sampling grid that may be distributed as targeted or systematic (figure 9.6), depending on the pollution potential, on the orography of the site and the quality of information obtained within the historical investigation. In general, target areas of shooting ranges are

Figure 9.5. Mobile XRF Niton XL3t. Adapted from [1]. Copyright, Federal Authorities of the Swiss Confederation.

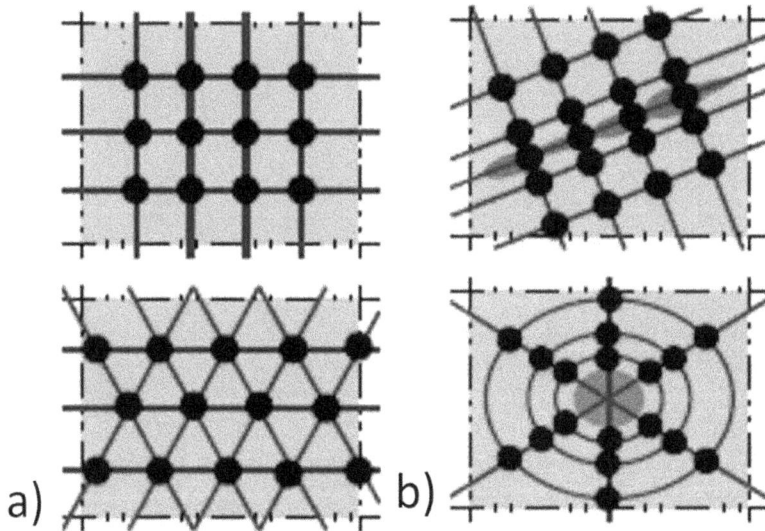

Figure 9.6. Sampling grid: (a) systematic, (b) targeted. Adapted from [1]. Copyright, Federal Authorities of the Swiss Confederation.

investigated using a systematic grid; however, the selection should always be described. Suitable guidelines for the grid size within a systematic sampling grid are given in table 9.3.

Apart from the soil samples taken to evaluate the extension of the pollution, also reference soil samples for laboratory analysis of the total content are taken from at least six locations within the pollution gradient ($100 \ \mathrm{mg \ kg^{-1}} < x < 2000 \ \mathrm{mg \ kg^{-1}}$ Pb according to XRF field measurements). The measurement with the m-XRF is suitable for fieldwork and enables a first impression of the pollution; however, only

Table 9.3. Guidelines for the grid size within a systematic sampling grid.

Pollution potential	Examples	Grid size
Heavy pollution	Areas around fixed targets. Backstops, target areas and local area	5 m to 10 m10 m to 20 m
Medium pollution	In-between area, area with material from former backstop	20 m to 40 m
Light pollution	peripheral areas, flyover areas	> 40 m, may be targeted distribution

Figure 9.7. The XRF beam measures surface concentration (as shown in red). In the laboratory, the total content is assessed using chemical digestion techniques and analysis with mass spectrometry (framed in blue). Adapted from [1]. Copyright, Federal Authorities of the Swiss Confederation.

the surface of the sample (figure 9.7) is measured. With the laboratory analysis of the reference samples, this error can be corrected for.

Cu is measured simultaneously to Pb (with m-XRF and in the laboratory); however, Sb cannot be measured directly in the field. As an approximation, 2%–5% of Pb is the total content of Sb in the soil. However, Sb has a different environmental behavior than Pb. Amongst other characteristics Sb has a much higher solubility, so the actual soil Pb/Sb ratio is dependent on local water conditions and permeability of the soil. Thus, only the measurement of the reference soils gives a proper indication of soil Sb levels.

If there is suspicion of groundwater pollution, then groundwater samples are to be taken downstream of the shooting range with the following priority:

- Piezometer tubes in the aquifer;
- Groundwater pumping stations, spring water catchments;
- Uncaptured spring water and other water outlets, drainage pipes.

All sampling procedures must be carried out correctly and documented using a sampling protocol.

All findings of the historical investigation, the suggestions for the investigation strategy of the technical investigation and technically clear plans describing the spatial context of the contaminated areas and the proposed sampling locations are to be documented in a report. The annex includes the inventory of used documents, the mentioned plans, all available data sheets on shooting activities, and a photo documentation of the site. The Swiss Federal Department of Defence, Civil Protection and Sport has released an example for a table of contents of a report on the historical investigation for all military shooting ranges [1]

9.2.2 Technical investigation

The technical investigation is a detailed *in situ* examination, which involves sampling the soil in order to determine the lateral and vertical expansion of the contaminated area. Generally, it is conducted after shooting ranges have closed or if there are construction plans, meaning that the pollution load does not increase after the investigation. It can also be conducted if there is a change of use of the shooting range (e.g. agricultural use).

The first step is to mark (figure 9.8) and measure the points as suggested with the investigation strategy developed within the historical investigation. The grid should be adapted to the current conditions. The location of all measuring points is assessed with a Global Positioning System (GPS), so that an exact position can be portrayed on maps. Field measurements with an m-XRF have a reliable measuring range from 100–2000 mg Pb/kg soil depending on the soil matrix. It is important that the measurement is performed with a 'homogeneous' soil sample (i.e. no particulate metal or gravel). In addition, the measurement is dependent on the type of m-XRF, so the brand name and exact type is to be noted.

Detailed information on the preparation and measurement of soil samples with a m-XRF requires special training. To summarize, there are three important steps:

1. Samples from 0–20 cm depth are dug out, and bigger grains and plant residues are sorted through. If there are visible bullet fragments, the sample has to be rejected.
2. The sampling material is packed into a plastic bag. For every measuring campaign, it is important to use the same sort of plastic bags (without color or inscription), as different plastics absorb X-rays differently.

Figure 9.8. Marking the measuring/sampling grid in the landscape.

3. The samples are mixed by kneading and then pressed flat to provide a homogeneous, dense surface for measurement. One sample is measured three times, each for at least 20 s. If the measured Pb and Cu contents vary less than ±20% from the average, then the measurements may be used, otherwise they are to be rejected.

Samples with visible bullet fragments deliver unreliable m-XRF results; however, these samples will always contain very high pollutant values, so the exact results do not have to be recorded. Even though the m-XRF saves the data digitally, an ongoing, careful documentation of field measurements is indispensable.

All sampling points in the sampling grid are measured. The Pb and Cu content is documented and delivers a first image of the contamination. The edge of the sampling grid must be adjusted until only low Pb values are documented (example in figure 9.9) or the measurements are below detection limit.

In a next step, the existing sampling grid is compacted and additional sampling points are measured in the middle of cells with big differences in the pollution content (examples for additional measuring points would be J3, L3, etc, as shown in figure 9.9). In areas with an unclear distribution, the sampling grid may be compacted once further. This continuous assessment of the pollutant distribution allows a reasonable and verifiable image of the contamination situation.

Per target area, the vertical distribution (for a soil profile in the field, see figure 9.10) of the contamination is assessed in heavily contaminated areas (>1000 mg Pb/kg according to XRF field measurements). Sample extraction should not be located directly behind a target. Soil samples are taken from 0–20 cm, 20–40 cm, 40–60 cm, 60–80 cm and 80–100 cm and measured with m-XRF and analyzed in the laboratory. From experience, values at 80 cm and lower are

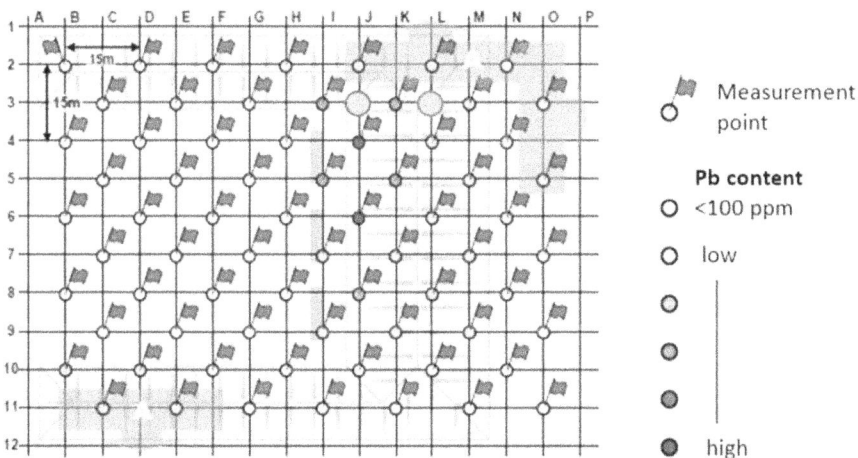

Figure 9.9. An example of the systematic sampling grid (grid size 15 m) with measured samples. The sampling grid must be adjusted near the edges of area A, until only low Pb values are documented. In addition, the sampling grid is compacted where the pollutant distribution is unclear, e.g. measuring points J3 and L3. Adapted from [1]. Copyright, Federal Authorities of the Swiss Confederation.

Figure 9.10. Soil profile.

insignificant. If this is not the case and it is possible, then samples must be taken from lower depths. All material must be described by its lithological and geotechnical characteristics.

The XRF field measurements and the total concentrations analyzed in the laboratory of the reference samples are plotted against each other. The generated regression coefficients are then used to correct the XRF field values for all soil samples. Pb and Sb must be measured in all reference soil samples. If there is suspicion of high Cu pollution, then Cu is also measured. The regression should have a R^2 value > 0.9, at worst >0.85. Only if properly justified, can outliners be excluded. In general, laboratory values are higher than field values because of various effects, e.g. water content of the field sample. For one shooting range, which may consist of various target areas, only one set of reference samples may be collected, if all areas are based on the same bedrock and soil type. Pollution maps should always illustrate the corrected XRF values.

Also, if requested within the scope of the investigation, strategy, groundwater and surface water samples are analyzed for their Pb, Sb and Cu content (figure 9.11). Whenever possible a reference sample upstream (i.e. uninfluenced by the site) should be sampled. Elution tests aim at estimating the maximal amount of contamination in the leachate (worst case scenario). There are various elution tests available with standardized procedures. The principle is that a sample of the contaminated site is flushed with an eluent (e.g. water or carbon dioxide saturated water) and the eluate is analyzed for its pollution content. From experience these tests can be unreliable, as they do not produce reproducible results, so that a need for this has to be justified (e.g. when no groundwater samples can be taken in an area downstream of a contaminated site).

Data errors (see figure 9.12) occur mainly during sampling. Therefore, it is particularly important to perform the sampling carefully in order to ensure

Figure 9.11. Groundwater sampling with groundwater pumping from a borehole next to the backstop.

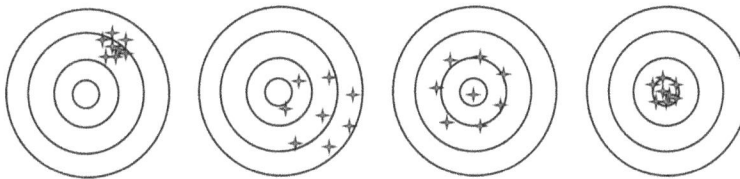

Figure 9.12. Measurement errors from left to right: high precision and false, low precision and false, low precision and correct, good precision and correct.

representative and reproducible results. Generally, standardized protocols are available for taking samples from governmental environmental agencies. Adaptations from these guidelines can be necessary depending on the site and the pollutant analyzed; these deviations should, however, always be justified.

In order to evaluate and interpret the measurement results, the contamination contents are compared to all legal guideline values for groundwater, surface water and soil. With this, a well-founded conclusion for the contamination status of the site can be delivered.

The technical investigation ends with a report that includes a summary of the historical investigation, a transparent description of all conducted measurements, documentation of all results and a well-founded conclusion of the pollution status with a first suggestion of measures. The main results are displayed graphically on maps. There are three important maps (for Pb and Cu) that should always be included: a map of measurement results (figure 9.13(a)), a pollution map (figure 9.13(b)), where the measurement results are compared to legal guidelines and an action plan (figure 9.13(c)), which gives a suggestion for further steps. For concrete examples, see Case Study Flumserberg, section 9.3. For all plans, always

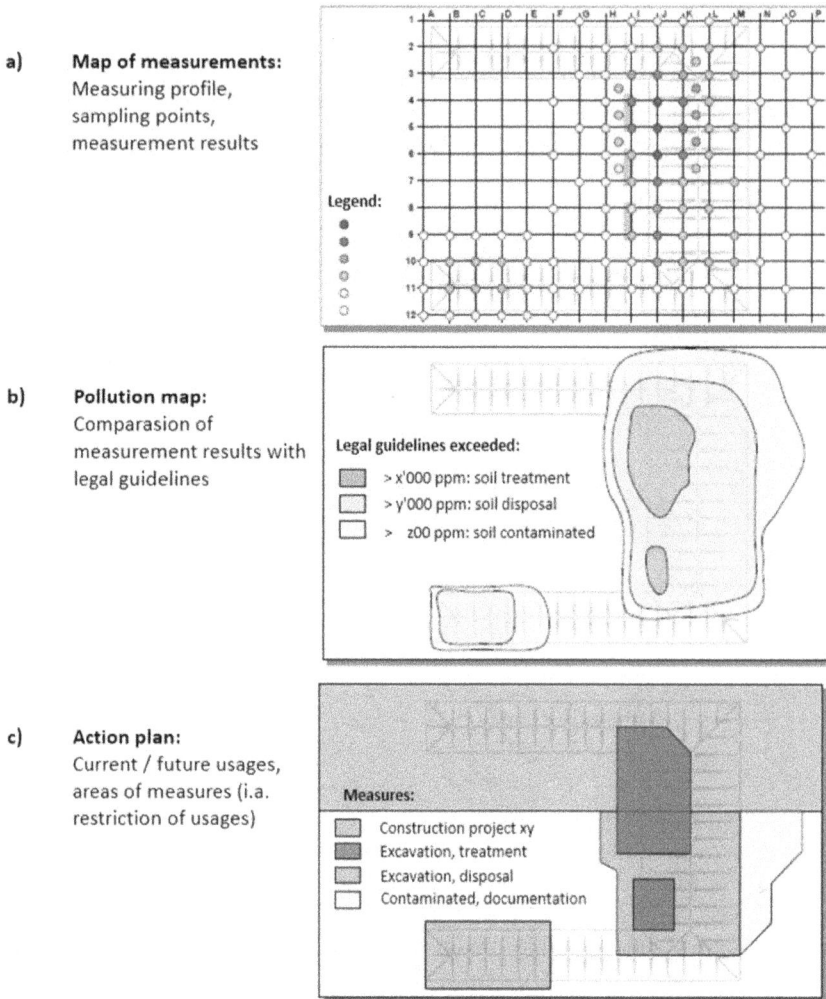

a) Map of measurements:
Measuring profile,
sampling points,
measurement results

b) Pollution map:
Comparasion of
measurement results with
legal guidelines

c) Action plan:
Current / future usages,
areas of measures (i.a.
restriction of usages)

Figure 9.13. Illustration of the measurement results in three maps: (a) plan of measurements, (b) pollution plan, and (c) action plan. Adapted from [1]. Copyright, Federal Authorities of the Swiss Confederation.

the corrected values are shown (i.e. not the field measurements). The overall assessment of the pollution should always be considered in the light of possible knowledge gaps and the uncertainty of measurements. Thus, the final chapter of the report should include a detailed discussion on the most important plausible causes of errors regarding the pollution potential of the contaminated site. The Swiss Federal Department of Defence, Civil Protection and Sport has released an example for a table of contents of a report on the technical investigation for all military shooting ranges [1].

9.2.3 Risk assessment

The risk assessment evaluates the potential environmental and health consequences of the contaminated area based on the results of the technical investigations. The aim of the assessment is to establish the need for remediation. The assessment is based on:

- The pollution potential according to the technical investigation;
- The potential for release of the pollutants to the environment;
- The exposure of each environmental compartment: groundwater, surface water and the soil.

General rules for a risk assessment of target areas of small arms shooting ranges do not exist, as most targets sites are unique in both the type of shooting activity and the environmental setting. Thus, the above-mentioned bullet points must be described for every individual target site. However, the environmental risks requiring remediation are usually stated in legal guidelines, as explained for water bodies and soils separately in the following sections.

In the case of water, a clear risk exists, when pollutant concentrations in the surface water or groundwater exceed legal values. For groundwater captures used for drinking water downstream of the target site, already the detection of the contaminants may be considered a potential health threat. However, the contaminants must be directly linked to a shooting range. Thus, for the evaluation of the results it is important to take into consideration a background sample taken upstream from the polluted site and to consider all contaminated sites and geogenic sources in the area. It is to be noted that in practice it is difficult to analyze water in the immediate discharge of the site or from nearby surface waters. For groundwater, there can be a concrete threat of contamination because of insufficient retention from the contaminated site. For example, the smaller the distance between the contaminated site and the groundwater table and the less clay and iron minerals content a soil has (i.e. insufficient retention), the higher the threat of leaching of contaminants into groundwater. In addition, there is a risk for contamination, if through erosion heavily contaminated soil may enter the surface water. This can be the case for fixed targets near streams or if streams pass through or near heavily contaminated areas of shooting ranges.

In Switzerland, soils are not generally considered exposed to pollutants if the target areas are only used for shooting practices. In this case the soil is considered as part of the construction (this point of view does not count for water bodies). This point changes, when the shooting ranges close. Also if there are other usages of the target area, such as pastureland (see figure 9.14), then soil concentrations are to comply to legal limits as not to pose a risk to the grazing animals. In general, target locations within areas also used for pastureland should be fenced off. For soils of target areas not used for any agricultural purpose (e.g. forest areas or high mountain regions), there may exist no legal guideline values. When heavily contaminated shooting ranges are closed, the soils are automatically considered an exposed natural resource.

Figure 9.14. Target areas have to be fenced off to limit exposure of heavy metal pollution to grazing sheep.

For the evaluation of the risk associated with shooting ranges used as grasslands for grazing animals, the Swiss Federal department of Defence, Civil Protection and Sport has released a guideline [3]. The risk associated to the exposure of the contaminants to grazing animals is based on two main aspects. First, the duration of the land use and secondly the relation of the area exposed to a pollution load in comparison to the unexposed area. This gives an approximation for the total amount of contamination that the animals can potentially take up (which depends further on the type of contaminant and the animal). An example for such an evaluation is shown in figure 9.15.

9.2.4 Remediation concept

To eliminate the risk presented during the risk assessment of the contaminated site, measures of remediation and usage restrictions are proposed within the scope of a remediation concept. Generally speaking, the main goal is to find a remediation solution, which eliminates as many contaminants from the site as possible whilst causing no further harm to the environment at the lowest possible cost. The result must be optimized for the future usage of the site.

The remediation concept includes a variation study of different remediation projects. Based on the technical investigation and the risk assessment, a measurable accountable remediation target of the polluted area is defined with respect to the protection of the environment. For example, in consideration of highly polluted soils, a remediation target of the contaminated site is set at 2000 ppm Pb. This would require all soil with a content of > 2000 ppm to be excavated. These values generally refer to values defined in environmental laws (e.g. soil protection, water protection and contaminated sites). It is possible to suggest different target values for different areas and depths depending on the exposed environmental compartment.

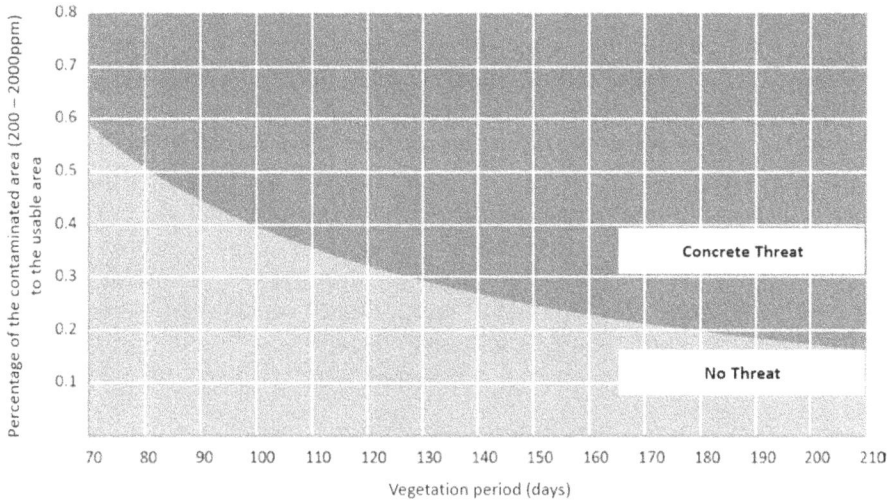

Figure 9.15. Risk assessment of shooting ranges used as pastureland applicable for all livestock. Adapted from [1]. Copyright, Federal Authorities of the Swiss Confederation.

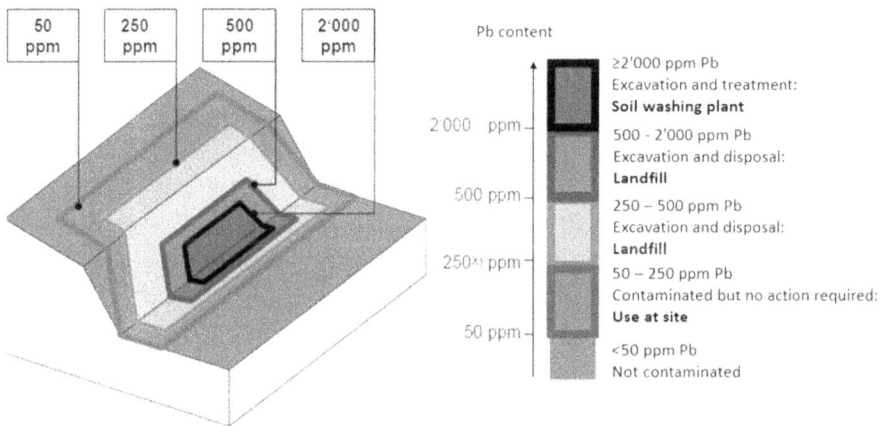

Figure 9.16. Classification of material groups and their disposal measures according to Swiss guidelines. Adapted from [1]. Copyright, Federal Authorities of the Swiss Confederation.

To describe the different measures required, material classes of the contaminated site are classified according to their pollutant content. An example for this according to Swiss guidelines is shown in figure 9.16. Soils < 50 ppm Pb have soil Pb levels comparable to normal background concentrations and are thus not considered influenced by the shooting activity. Material between 50–250 pp Pb are considered contaminated but do not pose an environmental risk, so no action with this material class is required. Between 250–2000 ppm Pb, depending on the usage of the site, it may be necessary to clean it up and dispose of the soil in a landfill. Landfills usually also have their own guidelines as to the pollutant content of the soil to be stored, so the material classes should be divided accordingly to landfill standards in the country. In Switzerland, soils $\geqslant$ 2000 ppm Pb cannot be deposited in a landfill and must be treated especially in a soil washing plant, where the soil is separated mainly by grain sizes and density and the metals recovered. Gravel and sand can be used as secondary building materials while the fine material, which contains the mobilized heavy metals, is disposed of separately. The classification into material classes allows a framework of the different disposal measure required. In a next step, the cubage of the material classes are estimated. Figure 9.17 shows excavation works in the scope of remediation of a contaminated shooting range. It demonstrates that excavation works can be a great intervention into the environment. In addition, excavation works usually include a great amount of removed soil, which is thereafter waste and can no longer be used. Thus, alternative remediation procedures other than excavation works should always be considered, such as prohibition of use, or a permanent change of use.

With remediation measures, all identified risks should be eliminated. However, there are special cases where higher interests are involved and so certain risks accepted. This is the case if the contaminated target areas are within the perimeters of protected natural reserves (e.g. dry grasslands, moorland). For this, a case by case approach is needed to balance the interest of the different protected environmental compartments and the preservation of the protected habitats. As a framework for this discussion, maps on valuable species and habitats are generated. In addition, restoration and compensatory measures for the protected habitats are to be prepared and verified for by experts. An example for such an area in Switzerland is shown in figure 9.18, where the contaminated target area is within the perimeters of a protected landscape, especially known for its high diversity of orchids. Within the scope of the remediation concept, also the interest of the protection of the area within the protected landscape has to be included in the discussion. For the

Figure 9.17. Excavation works within the scope of the remediation of a contaminated shooting range.

Figure 9.18. An orchid field (protected habitat) within the perimeter of the contaminated target location (left picture shows the remains of a fixed target).

evaluation of the different conflict of interest it is important to work according to principles and that guidelines are provided.

In addition to different variants of remediation, also a list of factors to support the selection is prepared, such as environmental protection measures, the amount of Pb excavated per area, costs, etc. Thus, the report concludes with a suggestion of the best variant for remediation of the site with regard to the environment. The suggestion is complete with a project organization, disposal concept, triage, restoration verification concept, restoration concept and all other relevant information concerning the execution works. It is very important that the works are carried out correctly as not to cause any more harm to any environment during remediation works, such as soil compaction or an increased discharge of the contaminated soils into the groundwater. Also, an initial estimation of time planning and costs are calculated with regard to preparing the final plans (remediation project), project and construction management, construction work, transport, disposal/treatment of excavation materials, restoration of the landscape and possible recurring compensations for restrictions of agricultural use of the land. The Swiss Federal Department of Defence, Civil Protection and Sport has released an example for a table of contents of a report for the remediation concept for all military shooting ranges [2].

9.3 Pollutant management techniques

The aim of current military training practices regarding its pollutant management is to release the minimum amount of contaminants into areas with the lowest environmental impact. Thus, any technique should consider at least one of the following principles:

- Reduction of pollution load;
- Limit exposure of the environment;
- Prevent the spread of contaminants to soil or water.

Other important aspects include the effectiveness of pollutant management, the safety of the construction as well as the availability of the technique. As each range is unique, site-specific conditions are the basis to evaluate the selection of appropriate pollutant management techniques. The following section gives a short introduction into available techniques. Discussions in this section does not include 'green ammunition' or the reduction of military training.

9.3.1 Improve construction properties of artificial backstops

The construction of backstops should always fulfill maximal safety requirements because possible safety shortcomings could include more ricochets, which could lead to a higher environmental influence of the shooting activity. In order to minimize the environmental impact of target areas, artificial backstops have become a standard over recent years. There are two construction principles:

- Backstops are isolated from the natural environment (e.g. boxes, compare figure 9.19);
- The whole target area (figure 9.20) is sealed off from the environment with a water management system for the infiltrating water (cf. subsequent section).

Figure 9.19. 300 m shooting range in Switzerland. Artificial backstop from the front (left) and the back (right). Adapted from [1]. Copyright, Federal Authorities of the Swiss Confederation.

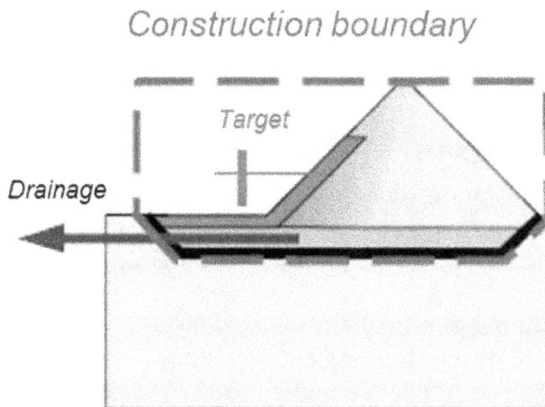

Figure 9.20. Construction boundary of a sealed-off target area. Adapted from [1]. Copyright, Federal Authorities of the Swiss Confederation.

Backstop boxes are cheap in construction and maintenance and the fired bullets easily collected. The system is completely closed off from the environment. However, this type of construction cannot be used for all sorts of military field training, as is the case for sealed-off target areas. However, the construction of a proper water management is of a more complex and costly manner. It is very important that it is performed and maintained correctly, as not to let contaminated leakage into the environment.

9.3.2 Drainage systems

When the target area is sealed off from the environment (figure 9.21), the infiltrating water has to be collected. One option is to collect the water in an underground water storage (i.e. underground gravel basin), with a shaft for sampling. When the basin is full and water samples show no signs of contamination, the water is released to the surrounding environment (i.e. to flowing waters or to a canalization). Another option is to release the drainage of the target area via a cleaning and infiltration system to the underground as shown in figure 9.22. The cleaning can be done by amendments or filtration through a soil or sand column. In addition, an inspection shaft should be in place to control the contamination status of the soil water before it is released into the environment.

Under certain environmental conditions, it can also make sense to implement water management for large target areas that are not sealed off. For example, if the underground of the target area permits good conditions for the groundwater to be collected (e.g. crystalline underground) and in a further step treated so that the water complies with legal guidelines, this can be a possibility.

Figure 9.21. Installation of a waterproofing foil for a sealed-off target area. Adapted from [4]. Copyright, Federal Authorities of the Swiss Confederation.

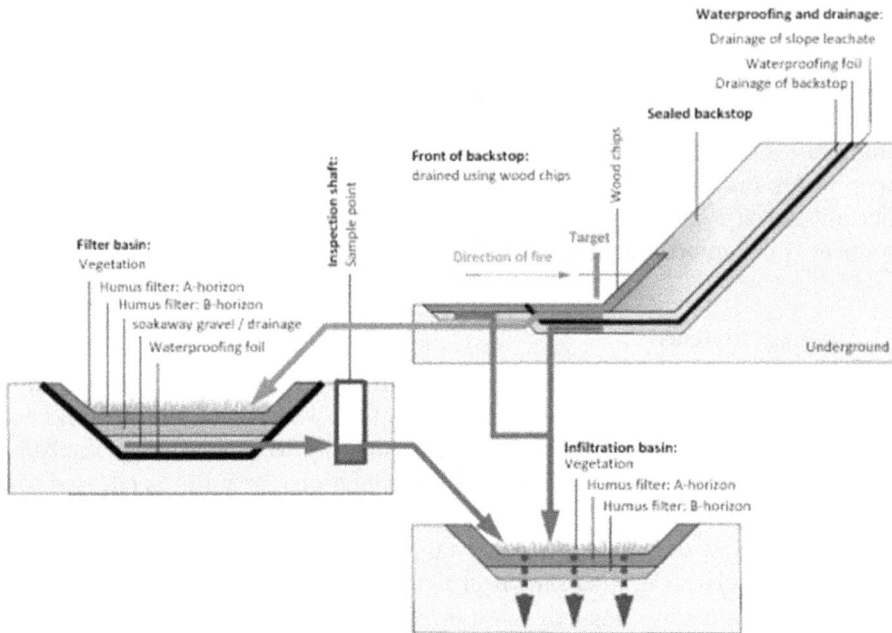

Figure 9.22. Schematically sealed target area with controlled infiltration using a filter- and infiltration basin. In red, potentially polluted water from training exercises. Adapted from [4]. Copyright, Federal Authorities of the Swiss Confederation.

Figure 9.23. Protected areas (e.g. moorlands) in the perimeter of the Swiss shooting range 'Glaubenberg'.

9.3.3 Avoid protected areas

On every shooting range, areas under environmental protection should be identified, documented and can so be included in the discussion of further usage of a site (figure 9.23). No matter if it is groundwater protection or endangered birds brooding in the heath, if these areas are avoided during military training exercises, then the harmful influence of the shooting practice is minimized from the beginning. To minimize the exposure of water bodies, target areas should be far from surface waters or exceed a certain depth to the natural groundwater level.

There are four important steps for the assessment of the situation. Firstly, all existing data on natural and landscape assets are summarized. For this, different environmental inventories are used and a map of all protected areas and valuable habitats in the perimeter of the range is generated. If not much information on environmental values are available for a certain site, then also specialized environmental offices can be commissioned to acquire further information in the field. Secondly, all areas used for military activities are assessed, described and mapped out. Current practices and future practices are described in a legend and are verified with someone knowledgeable on the activities within the perimeter of the shooting range. Thirdly, if there is third party usage of the area, the activities are to be documented and the used area mapped out. For example, apart from leasing the land to local farmers, areas in Swiss shooting ranges are also used for tourism activities (e.g. walking trails and for cross-country ski trails). The next step is to overlay these three maps and spot areas with potential conflict of interest. These areas are then listed, a weighting of interest carried out and a strategy for a solution drawn up including an action plan. Generally, as environmental values have received more and more attention over recent years, most conflict of interests' concern current environmental protection versus military needs, or environmental protection versus third party use. Based on this evaluation, the future use of the area is defined.

In theory, the lowest environmental impact through shooting exercises can be deduced from such evaluations and already a small shift of military practices in areas can achieve a significant change in favor for the environment. When successfully implemented, the communication of concepts of protected areas on military ground help with the acceptance of the public towards military training. In addition, it can also be used to sensitize military personnel for the protection of the environment.

9.3.4 Adapt military training exercises to the terrain

Already, small adaptations to training exercises allow a reduction of the pollution potential of a specific environment. For example, tracer rounds should not be used under dry conditions, as they could ignite a fire. Another example are soluble contaminants, such as Sb and Cu, which should not be used in areas with a high precipitation rate to prevent washout.

Seasonal restrictions can also counter environmental problems resulting from shooting practices. For example, snow influences the activity of explosive shells. Thus with more snow, the dud rate rises. For small arms shooting ranges, frozen ground or ice can cause more ricochets. This is not only a safety problem; it can also spread contaminants over a wide surface area.

Practice ammunition contains fewer contaminants. Therefore, by using practice munition or simulators for shooting practices the overall amount of pollutants released to the environment is reduced. Within reason, these adaptations should be favorably implemented. Even though acquisition costs for these adaptations may be expensive, it most likely will result in lower costs for the management of the shooting range over its entire lifetime.

9.3.5 Correct maintenance and appropriate use of shooting ranges

Technical specifications for the correct maintenance and for the appropriate use of shooting ranges must be developed and its implementation ensured. Military training should only be carried out in designated areas. Correct maintenance is the key for the security of a shooting range as well as controlled environmental pollution. For example, an area designated for training with small arms should not be used for heavier arms, as it could possibly destroy technical installations for the reduction of the environmental impact such as drainage systems.

9.4 Case study shooting range, Flumserberg, Switzerland

The shooting range Flumserberg (figure 9.24) had been used intensively for military training since 1920 before it was given up in 2002. The shooting range is situated between 1400–1500 m above sea level and covers a surface area of 13 ha. Within the target areas are no major constructions, but there is a ski region in the nearby surroundings. The military leased the area from the civil community of Flums–Grosberg and was to be returned in a safe to use, remediated state.

In 2005, within the scope of the historical investigation, it could be determined that the shooting range consisted of three spatially separated areas. For these areas all military activities, the location of their target areas, and the geological and hydrogeological conditions were documented. Within all shooting areas, there were

Figure 9.24. Shooting range Flumserberg (CH). View from the target area before remediation works began.

spring catchments that were being used for either drinking water or watering troughs for cattle. Surface streams traverse the target areas and the property was being used as grasslands for grazing animals during early summertime and autumn. In addition, one target area was within a moorland landscape of national importance with considerable diversity of species.

With the pollutant content of the used ammunition and an approximation of the shooting activity, it was concluded that four target areas have the potential of pollution. But only for the target area for small arms Madils, which is also located in the moor landscape, was the potential contamination with heavy metals considered high. For the others the potential of contamination was classified as low due to the low environmental impact of the described activities. Thus, the target site Madlis needed a further investigation (i.e. technical investigation). The results from the historical investigation are demonstrated in a suspicion matrix in table 9.4. The estimation of the shots fired is 30–40 000 per year.

In 2012, the effective heavy metal pollution of the shooting range was assessed with the completion of a technical investigation of the site Madils. Water samples from the two spring catchments and from surface water within the perimeter of the target area were analyzed and results demonstrated no sign of Pb, Cu or Sb in any water body. However, results from soil measurements proved that there was indeed a large-scale pollution of the soil. Lateral Pb distribution demonstrated local hot-spots and when the edge of the measuring grid was adjusted, another target location with high Pb content was discovered. Deduced from the generated pollution map, it could be concluded that giant boulders within the target area were used as targets. The bullets must have shattered whilst hitting the boulders, thus explaining the high pollution content surrounding them. To illustrate this, figure 9.25 shows a boulder used as a target from another Swiss target site. The vertical distribution showed that after a depth of 0.2 m, the Pb levels decrease drastically.

The risk assessment for this site considered the exposure of soil, groundwater and grazing animals. In Switzerland, soils are not allowed to be used for agricultural purposes when soil Pb concentrations exceed 2000 ppm. This is the case for large sections of the target area and for them a remediation target of 2000 ppm was defined. Even though no traces of contamination were found in the groundwater samples, the risk for one spring water catchment, located downstream from the contaminated site (Pb > 2000 ppm) was considered. The underground of the area consists of rubble and coarse grains, which have insufficient retention property

Table 9.4. Suspicion matrix of the pollution potential of the shooting target site Madils.

	Time period	Used ammunition	Pollutants	Location of pollutants	Potential of pollution	Distribution pathways
Shooting range Madils	1920–2002	GP 11, GP 90	Pb, Sb (Cu, Ni)	Soil, underground (groundwater)	Locally high	Moor water, stream water, leachate, grazing animals

Figure 9.25. Boulder with shot holes that was used as a target for military shooting exercises.

materials. Thus, the risk of contaminants being mobilized from the contaminated site is increased and the exposure of groundwater justified. To eliminate the potential exposure of groundwater a remediation target of 1000 ppm Pb was considered. For these areas a detailed investigation with a narrower sampling grid was suggested as a further measure to better evaluate the risk. The whole area of the shooting range is used as grassland for dairy cows during early summer and autumn. The milk is caseated on the same alp. The risk assessment for the exposure of the grazing cows was relativized by the criteria of duration of exposure (i.e. seasonal, thus short) and percentage of the contaminated area to the whole area (i.e. <10%). It was concluded that remediation values of exposed soils (remediation target 2000 ppm Pb) would be sufficient to eliminate the risk to the dairy cows.

Figure 9.26 shows the maps generated as a result of the technical investigation and the risk assessment. Figure 9.26(top) demonstrates the map of measurement, which shows the corrected XRF results of the samples. Figure 9.26(middle) shows the pollution map, where the area of the same pollution categories (here according to Swiss guidelines) are demonstrated. The plan for action (figure 9.26(bottom)) shows the areas that require the same measures. There are certain areas that need to be decontaminated, but also areas where a detailed investigation is required to assess the risk of exposed groundwater.

The aim of the remediation was to restore previous usage of the land before the military used the area for shooting training, which is pasturing of cattle in early summer and autumn. The remediation target for the contaminated site was set for 2000 ppm Pb and within the perimeter, where the groundwater is potentially exposed, a remediation target for 1000 ppm Pb was set. Thus, the risk for exposure of soil, groundwater and grazing animals was to be eliminated.

The project organization, the triage, how the excavated material is to be transported and all environmental protection measures were defined and handed in to the responsible authorities for approval. The excavation and triage works were carried out during dry conditions in order to protect the soil from further damage.

Figure 9.26. Illustrations of the measurement results of Madils in three maps: (top) plan of measurements, (middle) pollution plan, and (bottom) action plan.

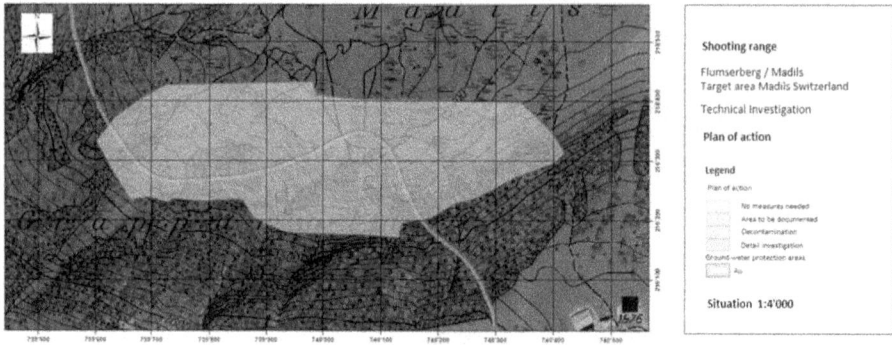

Figure 9.26. (Continued.)

Table 9.5. Material categories and estimates of cubage for excavation works.

Category	Pb content [ppm]	Pb content (average) [ppm]	Quantity excavated [t]	Pb excavated [kg]
Excavation and treatment	>2000	2400	629.04	1510
Excavation and disposal	1000–2000	1400	553.62	775
Total	**>1000**		**1182.66**	**2285**

Furthermore, the water protection was ensured by collecting all water leachate and controlling the pollutant content before it was disposed of. Remediation verification was done on-site with the m-XRF, where continuous measuring and analysis were carried out. Only when all Pb contents were under the defined remediation target for the site, were the excavation works completed.

The following material categories and cubages that were carried out are shown in table 9.5. In total, 2.3 tons of Pb could be excavated.

As the site is located in a mountainous region within a protected moor landscape, there are certain challenges with the performance of the remediation work that increased the resulting remediation costs. Table 9.6 shows a listing of the estimated costs drawn up in the scope of the remediation concept. The total costs are still ongoing as the restoration work takes up to four years for such a site.

The exposed surface was restored with a humus ground cover, which was grassed with site-specific vegetation. The areas are fenced off for grazing animals during the 3–4 years after restoration in order to aid recovery. Figure 9.27 shows pictures from the remediation works.

The Swiss Department of Defence released a short film in 2017 on YouTube describing the management of its contaminated sites. For further information, visit the following link: https://youtu.be/8PtYfpeLfJI.

Table 9.6. Remediation costs of the target site Madils.

Performance	Costs [Swiss Francs: CHF ≈ $]
Marking of the area	2052.00
Remediation works and landfill	504 013.25
Ground cover	9000.00
Technical planning of remediation	12 936.15
Realization remediation measures	60 425.45
Eco-monitoring of construction activities	27 255.00
Pedological site support	1124.30
Compensation and maintenance for the owner	22 000.00

Figure 9.27. Remediation works for the shooting range Flumserberg CH: (a) target area Madils; (b) shows physical soil protection measures (i.e. no direct driving on moor) in preparation of the construction work and (c)–(d) show the excavation works in the target area.

9.5 Discussion and conclusion

As the different management steps of contaminated sites build upon each other, it is very important to work transparently, uniformly and professionally. To ensure this, responsible authorities should provide technical guidelines from the investigation phase to the remediation phase and qualified personnel should collect the required data (preparation, fieldwork and data processing) and write the reports. The remediation of a contaminated site should be conducted in a targeted and problem-solving fashion, considering environmental benefits, technical advances and financial viability.

To ensure a healthy environment on shooting ranges, not only are legally binding guidelines to be implemented, but also communication and environmental education should be encouraged for the persons in charge. Protecting the environment can be considered an issue of national importance and/or security and all persons involved in the management of military shooting ranges should understand this message. The military use and the environmental protection of a site do not exclude each other, but both have very legitimate demands on an area that have to be met. This can most effectively be achieved by promoting an understanding of why lead and other substances are an environmental and public concern, by pointing out which laws and regulations apply and what the benefits of applying good management practices are. An understanding of the shooting range issue at hand is the base of greater cooperation and coordination of all involved persons. A platform to involve everyone and allow the exchange of information should be integrated in pollutant management plans. In addition, successfully implemented measures to minimize the environmental risk from military shooting exercises can be promoted and publicized to improve public relations.

Further research is needed with regard to other contaminants than lead. The major focus of shooting range contamination on lead is justified because of the amount discharged to the environment and because research (most notable, development of guideline values) and measuring techniques are readily available. However, the distribution of lead is only an approximation of the distribution of antimony, which due to its higher mobility in soils must be distributed differently upon weathering of spent bullets. Legal guidelines for antimony in all environmental compartments are still scarce. Information on the environmental behavior of all pollutants discharged into the environment (e.g. also for any new material) is important. Thus, with further research the knowledge gaps on the distribution and environmental behavior of the pollutants released into the environment during military training exercises are becoming smaller.

Even if range activities currently do not cause adverse public health and environmental impacts, they can profit from the proposed environmental management techniques by reducing the potential of lead exposure and contamination to humans, animals and the environment. Precaution instead of aftercare is very important and is best achieved in the planning and maintenance of the range design. The fewer contaminants that are discharged into the environment and the more that is done to prevent the off-site migration of lead and other contaminants to the subsurface and surrounding water bodies will result in costs being avoided in the future for lengthy remediation activities; this saved money can be applied for other military activities.

In conclusion, it is important to know the environmental risk potential of all small arms shooting ranges, as we do not want harmful levels of Pb, Sb or Cu in the environment. Through military shooting practices, their introduction to the environment cannot be completely avoided. However, efforts to minimize pollution should be implemented and appropriate remediation efforts conducted.

References

[1] Philipp R, Keiser R and Walcher C 2017 Untersuchung der Belastungen auf Schiessplätzen und Schiessanlagen des VBS (Eidgenössisches Departement für Verteidigung Bevölkerungsschutz und Sport VBS)

[2] Philipp R, Keiser R and Walcher C 2017 Erarbeitung eines Sanierungsprojekts (Eidgenössisches Departement für Verteidigung Bevölkerungsschutz und Sport VBS)

[3] Rupflin C and Krebs R 2016 Gefährdungsabschätzung auf militärischen Schiessplätzen mit Graslandnutzung (Eidgenössisches Departement für Verteidigung Bevölkerungsschutz und Sport VBS)

[4] Philipp R, Keiser R and Walcher C 2018 Emissionsfreie Kugelfänge auf Schiessplätzen des VBS (Eidgenössisches Departement für Verteidigung Bevölkerungsschutz und Sport VBS)

Chapter 10

Metal and energetics survey of the Borris shooting range, Denmark

Philip de Lasson

In 2012 it was decided that a survey of the Danish shooting ranges was needed. The Borris range (figure 10.1) was picked as the first range to be examined because of a number of reasons. It is one of the most heavily used ranges being utilised by the army for training with weapon systems from small arms to artillery. Moreover it is situated on a sandy plain with little natural protection of the aquifer and it contains a stream well known for its fishing potential.

During 2013–2014 a survey was done using MULTI INCREMENT[1] sampling. The survey showed that metals are present in slightly heightened concentrations in waterways and lake sediment in the range. Energetics have been found in the topsoil in varying concentrations at different sites on the range, but were not detected in the aquifer below the range. Only at the anti-tank range was any significant leaching to groundwater detected.

During 2017 a follow-on survey was done on the stream running through the range to ascertain if harmful amounts of metals were leaching into it. It was concluded that this was not the case even though heightened levels of some metals were detected in the stream and sediment.

The survey shows that the range, as must be expected, is being affected by its use. However, the contaminants are confined to the range and therefore they do not pose a risk towards fishing or extraction of drinking water. There have been no adverse effects on plant or wildlife reported and the range is one of the best known sites of significant natural value in Denmark, housing many rare plant and wildlife species.

---

[1] MULTI INCREMENT® is a registered trademark of EnviroStat, Inc.

Figure 10.1. Borris shooting range—image courtesy Danish Defence.

10.1 Introduction

The Borris shooting range is situated in the south-western part of Jutland in a relatively sparsely populated part of the country (figure 10.2).

The topography of the area is flat (12–20 m above mean sea level) with only small sandy mounds up to around 5 m above the terrain surface (figure 10.3). The surface of the range consists of a glacial meltwater deposit comprised mainly of sand and gravel. These sediments were washed out of a glacier front during the latest ice age approx: 20–100 thousand years ago. The aquifer under the range has a groundwater table around 0 to 15 m below the surface and is at some sites in direct contact with surface water like trenches and lakes.

Through the area a stream (Omme Å) has eroded a valley around 3 m below the terrain (figure 10.4).

The climate is temperate with mean temperatures around 0 °C in January and 15 °C in August. Precipitation is around 800 mm per year with the highest amounts during the months September–November.

The range consists of 4743 ha and is by far the largest in Denmark. It has a modern camp that is able to house approx. 1 battalion of troops at a time. The range is being used extensively for training of all kinds of weapon systems ranging from small arms to the largest artillery types. Shooting activities are common for more than 200 days every year and are conducted mainly by Army and National Guard units. Because of the long history of usage there is quite a lot of unexploded ordnance buried in the soil at the range. The entire range is encircled by a 30 km long ring road that enables easy access to firing points along the entire circumference of the range.

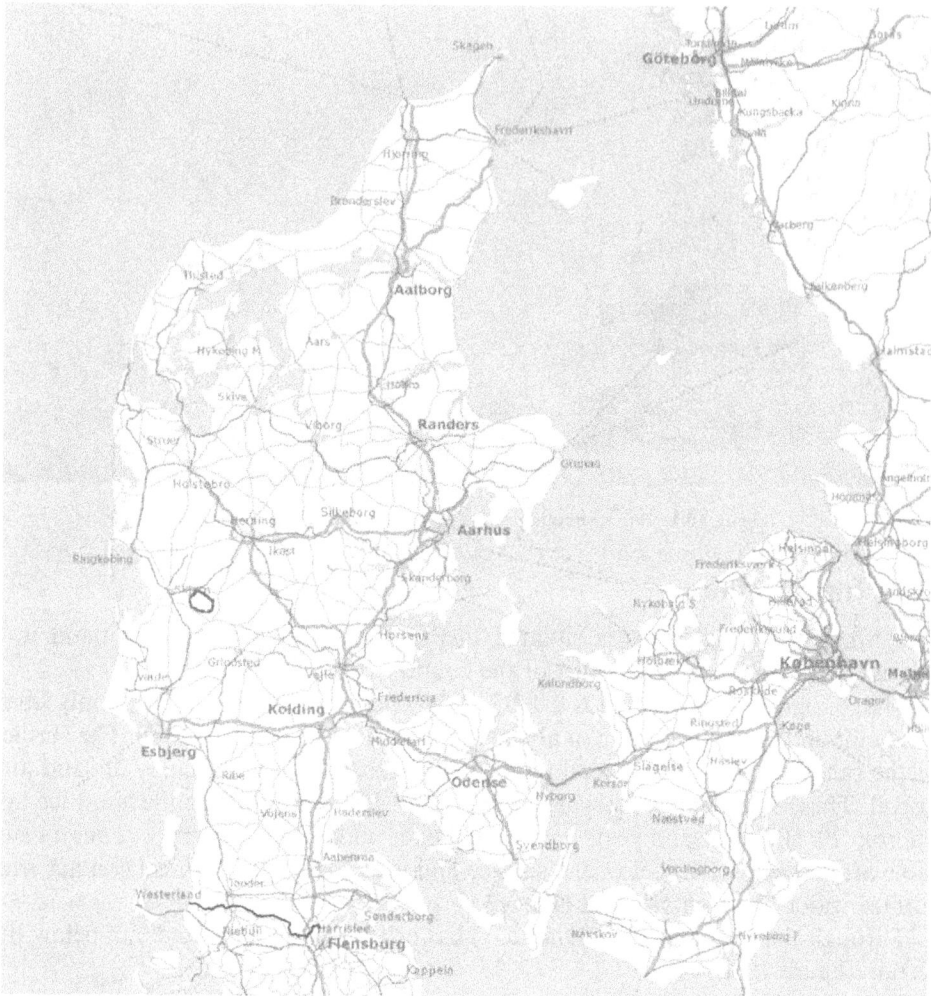

Figure 10.2. Borris shooting range (marked with red)—image courtesy Danish Defence.

10.2 History of Borris shooting range

On the first maps from 1803 the area is marked mainly as heath and moor. Every few years a major blaze would sweep across the heath during the dry summer months thereby preventing the area from becoming forested.

The area has not been used intensively for farming because of its sandy soil. An exception is the area along the stream Omme Å where farming was conducted along the banks of the stream. Not much had changed by the time parts of the area were converted into a military shooting range in 1902. In the beginning the area was mainly used during summer time (July–September) and the soldiers were billeted in tents. During the 1930s permanent structures where erected and the area was used by the German occupation forces during the Second World War.

Figure 10.3. Typical scenery at Borris range—Image courtesy by Niras consultants.

Figure 10.4. Omme stream meandering through the range—image courtesy by Danish Defence.

During the next expansion of the range in 1953 things were much different. At that time only 550 of the 3000 ha expansion was in the form of heath. The rest had been converted into farmland during the late 1800s and early 1900s. The expansion was necessary because of the dramatic development of weapon systems during the

Second World War and the period following the war. The new weapon systems had a much longer range and destructive potential, which required larger areas for training.

10.3 Conducting the survey

10.3.1 Background

After having participated in a NATO working group regarding munitions-related contamination it was decided that a risk assessment of the Danish shooting ranges was needed. The Borris range was picked as the first range to be examined for a number of reasons. It is one of the most heavily used ranges, being used by the army for training with weapon systems from small arms to artillery. Moreover, it is situated on a sandy plain with little natural protection of the water aquifers and a river well known for its fishing potential intercepts the range.

10.3.2 Purpose

The purpose of the survey was defined as an investigation into the occurrence of explosives and metals in soil, groundwater and open water recipients (streams, lakes, etc) at the range. The assumption being that explosives and metals were deposited directly into the soil as a result of training activities and that potential leaching would transport these contaminants into the groundwater and into open water recipients.

10.3.3 Scope

There are many fixed ranges for small arms at Borris, but since these normally do not contain explosives and because several have already been surveyed for metals the focus of the survey was on the effect from heavy weapons. The following sub-sites were chosen for sampling:
- Artillery firing position (propellants);
- Impact area artillery (explosives and metals);
- Anti-tank weapons range (propellants/explosives/metals);
- Burn site surplus artillery propellant (propellants);
- Hand grenade range (explosives/metals);
- Omme Å stream (explosives and metals).

Artillery firing position
The artillery firing position (figure 10.5) is located outside the main range at Borris because of the very long trajectory of modern artillery. The site consists of an open heath dominated area approx. 8.5 km from the impact area on the range. It has not been possible to determine exactly how long this area has been used as a firing position, but it is at least for the last 20 years and probably much longer.
    The weapon systems that have been fired from the site are mainly:
- 105 mm artillery;
- 155 mm artillery;
- 203 mm artillery.

Figure 10.5. Artillery firing position—image courtesy Morten Barker.

It is not known how many rounds have been fired from the site, but during the period 2007–2014 approx. 2970 grenades were fired. Mainly from the M109 Self Propelled Artillery Vehicle fielding a 155 mm gun.

The site is normally used for single battery shooting, which means that six guns are set up and fired simultaneously. The most commonly used propellant is of the variant 'M1 propellant', which consists of 84.2% nitrocellulose, 9.9% dinitrotoluen, 4.9% dibutylphthalat and 1.0% diphenylamin.

Impact area artillery
The impact area for artillery, mortars and anti-tank missiles at Borris is located approx. in the middle of the range and consists of a sandy heath area (figure 10.6) with several small lakes.

The weapon systems that have been used to fire into the site are mainly:
- 105 mm artillery;
- 155 mm artillery;
- 203 mm artillery;
- 60 mm mortar;
- 81 mm mortar;
- 120 mm mortar;
- Anti-tank missile (figure 10.7).

But because of the long history of usage a lot of other kinds of weapon systems have been used to shoot into the area.

Figure 10.6. Impact crater in the sandy soil in the process of being revegetated—image courtesy of Niras consultants.

Figure 10.7. TOW missile remnant—image courtesy by Niras consultants.

Figure 10.8. Unexploded artillery shell in impact area—image courtesy Niras consultants.

The typical type of grenades used are:
- High explosive (HE);
- High explosive anti-tank (HEAT);
- Smoke;
- Illumination;
- White/yellow/red phosphorous.

It is not known how many rounds have been fired into the site, but during the period 2007–2014 approx. 23 000 artillery and 53 000 mortar rounds were fired into the impact area. The dud rate of these rounds is not known, but a considerable number of rounds will be present in the area as either unexploded shells (figure 10.8) or as low-order detonations.

Anti-tank range
The range is mainly used for infantry-issued anti-tank weapon systems. The area is dominated by heath and has been used for an unknown number of years (but at least 25 years). This site is somewhat unusual compared to the other sites investigated, because it potentially contains propellants both in the back-blast area and in front of the firing line from rocket engine ignition. At the target zone it is expected that explosives and metals will be present.

The weapon systems that have been used to fire into the site are mainly:
- 84 mm recoilless rifle;
- 106 mm recoilless rifle;
- 66 mm anti-tank rocket;
- 84 mm anti-tank rocket.

The typical type of grenades used are:
- High explosive (HE);
- High explosive anti-tank (HEAT).

The most commonly used propellant during the last 20 years is of the variant 'AKB propellant', which consists of 61% nitrocellulose, 37.5% nitroglycerine and 1.5% ethyl centralite.

It is not known how many rounds have been fired at the site, but during the period 2007–2014 approx. 11 500 rounds were fired on the anti-tank range. The dud rate of these rounds is quite high with numbers like 5% not uncommon. When possible they are rendered harmless by blow-in-place detonations.

Burn site surplus artillery propellant
In this area surplus artillery propellant is burnt directly on the ground. The burning is done where vegetation has been cleared to construct firebreaks. The sandy soil makes it easy to ensure that the fire does not spread from the burn site.

It is not known how much propellant is being burnt at the site, but the maximum amount of propellant in a single burn is restricted to 1000 kg.

Hand grenade range
In this area hand grenade training is undertaken. The grenades are thrown directly onto an area with sandy soil free from vegetation. The area has been in use since the 1950s and a significant number (>1 000 000) of hand grenades have been detonated at the site since.

The Danish Army has been using the same type of hand grenade since 1954. This grenade contains 147 g of TNT as the explosive charge inside a steel casing.

Omme Å stream
The stream meanders through the southern part of the range area with a flow direction from southeast towards northwest (figure 10.9). The stream is known because of its excellent fishing and at least one fish farm is located downstream of the range.

Potential surface discharge from the range to the stream was mapped from a boat while travelling from the point where the stream enters the range until the stream exits the range (figure 10.10). Based on the field observation from this expedition 10 sampling points were located along the north side of the stream.

There are no major pathways for surface water exiting the range other than Omme Å. Because no detectable concentrations of explosives have been found in the run-off from the range the survey of Omme Å did not consider this group of compounds, but focused only on metals. Since no elevated levels of metals and explosives were found in the aquifer on the range the influx from groundwater into the stream was not considered in this survey.

Figure 10.9. Typical setting Omme stream—image courtesy Niras consultants.

Figure 10.10. Exit point for Omme stream from Borris range—image courtesy Niras consultants.

Table 10.1. Groups of compounds.

Energetics (explosives)	Energetics (propellants)	Metals
HMX (Octogen)	Ammonium	Aluminium
Nitroglycerine	DEGDN (diethylene glycol dinitrate)	Antimony
PETN	Hydrazine	Bismuth
Picric acid (2,4,6-trinitrophenol)	Nitro-cellulose	Lead
RDX (Hexogen)	Nitroglycerine	Copper
Tetryl	Nitroguanidine	Chrome
TNT (2,4,6-trinitrotoluene)	Perchlorate	Mercury
TNB (trinitrobenzene)	2,4-DNT (2,4-dinitrotoluene)	Nickel
1,3-DNB (1,3-dinitrobenzene)		Zinc
2-AT (2-Amino-4,6-dinitrotoluene)		
4-AT (4-Amino-2,6-dinitrotoluene)		
2,4-DNT (2,4-dinitrotoluene)		
2,6-DNT (2,6-dinitrotoluene)		

10.4 Compounds

The compounds of interest were grouped into three groups so that each area of interest could be analysed for the compounds that are relevant to that particular sub-site (table 10.1). The compounds were chosen based on prior experience with similar surveys and from NATO STO publication 323/(AVT-197)/TP/668 'Munitions-Related Contamination—Source Characterisation, Fate and Transport' and other literature sources.

No official guideline values for acceptable concentrations of energetics exist in Denmark. Guideline values from other NATO countries are typically used when doing a survey involving energetics.

10.5 Sampling

10.5.1 Artillery firing position

Soil

Soil samples were taken during the installation of a well. The well was installed with a regular drill rig using a 6″ auger.

Two MULTI INCREMENT samples were taken from an area of 70 × 100 m covering the uppermost 5 cm of soil. The sampled area is situated on the typical setup site of the artillery firing from the site. Each sample was taken with 50 increments and triplicates for quality assessment.

Water

Two separate water samples were taken from the well approx. one month apart from each other. Groundwater level in the area is approx. 1.5 m below surface level.

10.5.2 Impact area artillery

Soil

Soil samples were taken during the installation of ten wells (B1–B10). Because of the danger of unexploded ordnance all the driving and drilling was done with a remote controlled drill rig with a 6″ auger (figure 10.11).

The exact spots where the wells were situated were screened to a depth of 0.8 m with a metal detector beforehand to ensure that no grenades were present in the topmost layer. It was deemed possible, but unlikely that grenades were present at greater depths.

Two MULTI INCREMENT samples were taken from a decision unit (DU) measuring approx. 50 × 100 m covering parts of impact area (green square figure 10.12). The two samples were taken, respectively, at the depth intervals 0.0–0.1 and 0.3–0.4 m below surface. Each sample was taken with 100 increments and triplicates for quality assessment (figure 10.13).

Six sediment samples were collected from various streams and lakes in or near the impact area. These are taken from a depth of 0.1–0.2 m below the water/bottom interface.

Water

A water sample was taken from each of the ten wells. Groundwater level in the area is between 0.5–3.0 m below surface level depending on local topography. Water samples were taken in various streams/lakes at the same six positions where sediment samples were collected (figure 10.14).

Figure 10.11. Remote control with screen for operating drilling rig—image courtesy Niras consultants.

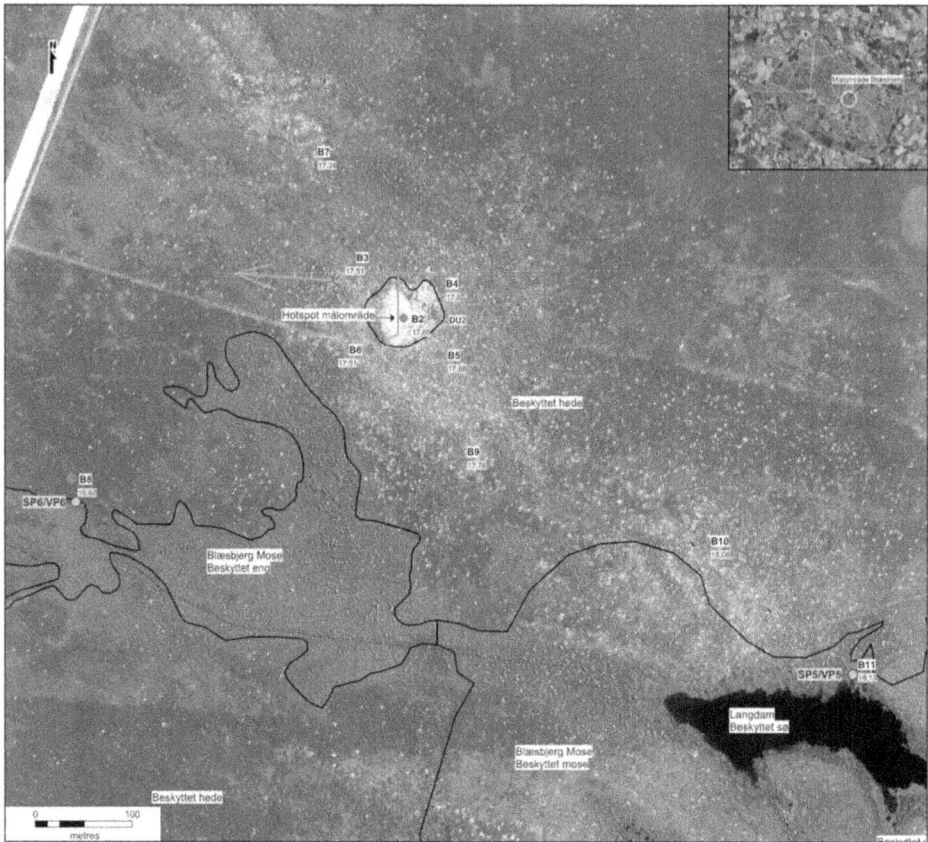

Figure 10.12. Sampling at the impact area—image courtesy by Danish Defence.

10.5.3 Anti-tank range

Soil

Four wells were installed at the site. Because of the danger of unexploded ordnance all the driving and drilling was done with a remote controlled drill rig with a 6″ auger.

Five MULTI INCREMENT samples were taken from the back-blast area behind the firing line with four samples covering the depth interval 0.0–0.05 m below surface and one sample covering the depth interval 0.05–0.15 m below surface (figure 10.15). Each sample was taken with 50 or 100 increments. One of the samples was taken in triplicate for quality assessment.

Two MULTI INCREMENT samples were taken from the rocket ignition area in front of the firing line covering the depth interval 0.0–0.05 m below surface (figure 10.16). Each sample was taken with 50 or 100 increments.

Three MULTI INCREMENT samples were taken from the target zone with two samples covering the depth interval 0.0–0.1 m below surface and one sample covering the depth interval 0.1–0.2 m below surface (figure 10.17). Each sample was taken with 50 or 100 increments.

Figure 10.13. MULTI INCREMENT sampling of the topsoil—image courtesy by Niras consultants.

Water

A water sample was taken from each of the four wells. Groundwater level in the area is between 4 and 5 m below surface level.

10.5.4 Burn site surplus artillery propellant

Soil

Soil samples were taken during the installation of a well. The well was installed with a regular drill rig using a 6″auger.

A MULTI INCREMENT sample was taken from an area of 65×80 m covering the uppermost 5 cm of soil. The sample was taken with 48 increments.

Water

A water sample was taken from the well. Groundwater level in the area is approx. 2 m below surface level.

Figure 10.14. Typical lake in the impact area—image courtesy Niras consultants.

Figure 10.15. Back-blast area—image courtesy by Niras consultants.

10.5.5 Hand grenade range

Soil

Soil samples were taken during the installation of two wells. The wells were installed with a regular drill rig using a 6″ auger.

Two MULTI INCREMENT samples were taken from the target area. One centrally with a diameter of 10 m and the other as a donut-shaped area around the

Figure 10.16. Rocket ignition area—image courtesy by Niras consultants.

Figure 10.17. Target zone—image courtesy by Niras consultants.

central zone with a width of 10 m. The centrally placed MULTI INCREMENT sample covered the uppermost 5 cm of soil and was taken with 100 increments. The donut-shaped MULTI INCREMENT sample covered the uppermost 10 cm of soil and was taken with 49 increments.

Water

Water samples were taken from the wells. Groundwater level in the area is approx. 0.5 m below surface level.

Figure 10.18. Water sampling at the entry point to the range—image courtesy Niras consultants.

10.5.6 Omme Å stream

Water and soil

A sampling transect was established across the stream at the point of entry (figure 10.18) and at the point of exit of the range (figure 10.10). At these transects five rounds of measurements were conducted during the summer and fall of 2017. During each round of measurement water and sediment samples were taken. The water samples were collected centrally in the stream using an electric pump and five sediment samples were taken with a piston sampler across the stream bed. The piston sampler collected the topmost 15 cm of the sediment. The velocity of the stream was measured at each sampling and the water table was logged continually during the entire period with an automated logger. The following parameters of the water samples were measured in the field: temperature, pH, oxygen, conductivity and redox potential.

At each of the 10 previously defined tributaries to the stream a water sample was collected centrally in the stream using an electric pump and a sediment sample was taken with a piston sampler (figure 10.19). The piston sampler collected the topmost 15 cm of the sediment.

10.6 Results

10.6.1 Artillery firing position

Soil

Metal content in the soil samples did not deviate from background values.

Figure 10.19. Typical tributary entering Omme stream—image courtesy Niras consultants.

The soil samples contained heightened average concentrations of ammonium 1140 mg kg^{-1} and nitrocellulose 703 mg kg^{-1} (table 10.2).

Water
The groundwater samples from the well did not contain any elevated concentrations of metal or propellant compounds.

10.6.2 Impact area artillery

Soil
Metal content in the soil samples from the wells did not deviate from background values.

In the two MULTI INCREMENT samples low concentrations of TNT, HMX and RDX were measured.

Table 10.2. Propellants in soil at artillery firing position.

| | | Laboratory analysis, soil | | | | | | | |
| | | Propellants (mg kg^{-1}) | | | | | | | |
ID[1]	Sample depth m b.s.	Nitro-glycerin	2,4-DNT	Ammonium	Perchlorate	Hydrazin	Diethylen-glykol-dinitrat (DEGN)	Nitro-guanidin	Nitro-cellulose
DU1b.1	0.05	—	—	920	—	0.15	—	—	480
DU1b.2	0.05	—	—	1340	—	0.15	—	—	870
DU1b.3	0.05	—	—	1160	—	0.20	—	—	760
Average value DU1b.(1-3)		—	—	**1140**	—	0.17	—	—	**703**
Relative standard deviation (RSD)				15%		14%			23%
Detection limit		0.10	0.050	10	0.10	0.05	0.5	0.5	10

Note:
Green marking: RSD-values <35%, considered an indicator that samples are replicable.
-: Below detection limits.

Table 10.3. Metal content in sediment samples.

ID	Sample depth m b. sea bed	Laboratory analysis, sediment sample Metals (mg kg^{-1})									
		Aluminium	Antimony	Lead	Barium	Chrome	Copper	Mercury	Nickel	Zinc	Bismuth
SP1	0.1–0.2	863	—	3	14	1.3	1.1	—	2.6	15	—
SP2	0.1–0.2	802	—	2	30	1.3	0.7	—	4.5	24	—
SP3	0.1–0.2	594	—	2	4.8	1.3	1.7	0.01	1.1	6.3	—
SP4	0.1–0.2	658	—	36	6.3	0.9	6.3	0.02	1.2	13	0.0 209
SP5	0.1–0.2	2.810	1	**42**	76	8.6	46	0.09	7.4	73	0.0 366
SP6	0.1–0.2	133	—	1	2.1	0.8	2.0	—	1.0	4.7	—
Threshold limit		n. l.	80	40	100	500	500	1	30	500	n. l.
Detection limit		10	1	1	0.5	0.2	0.01	0.1	0.1	0.4	0.02

Note:
–: Below detection limits.
n. l.: no legal threshold limit.

In two of the six sediment samples that were collected from various streams and lakes in or near the impact area heightened concentrations of antimony, lead, barium, chrome, copper and mercury were detected. Only one single sample verges on the legal threshold limits (lead 40 mg kg^{-1}), with a content of 42 mg kg^{-1} lead (table 10.3).

Water
The groundwater samples from the wells did not contain any elevated concentrations of metals (apart from zinc) or explosives compounds.

Surface water samples from various streams/lakes contained heightened concentrations of antimony, lead, cadmium, chrome and copper. Of these, lead, cadmium and copper exceeded legal threshold limits (table 10.4).

The same water samples did not contain any traces of explosives.

10.6.3 Anti-tank range

Soil
In the back-blast area very high concentrations (max. 26 000 mg kg^{-1}) of propellants in the form of nitrocellulose were measured. Lower levels of ammonium were also measured in the soil samples (table 10.5).

Metal content in the soil samples did not deviate from background values except cadmium, which was measured in concentrations above expected levels (maximum concentration measured 0.75 mg kg^{-1}).

Table 10.4. Metal content in surface water samples.

	Laboratory analysis, surface water								
	Metals (μg l^{-1})								
ID	Antimony	Lead	Cadmium	Chrome	Copper	Mercury	Nickel	Zinc	Bismuth
VP1	0.0 746	0.92	0.072	0.77	0.76	—	**4.1**	32	—
VP2	0.0 647	0.20	0.076	0.85	0.63	—	3.9	**6.4**	—
VP3	—	**1.2**	0.047	0.90	**2.2**	—	0.54	17	—
VP4	—	**10**	0.072	0.69	**11**	—	0.71	35	—
VP5	0.25	**3.6**	0.046	1.1	**6.2**	—	0.55	22	—
VP6	—	**3.1**	**0.11**	0.90	33	—	0.87	22	—
Threshold limit	113	1.2	0.08–0.25	4.9	1	0.07	4	3.1	n. l.
Background level	n. v.	n. v.	n. v.	n. v.	0.66	n. v.	0.82	1.5	n. v.
Detection limit	0.2	0.025	0.003	0.2	0.03	0.2	0.03	0.3	0.05

Note:
–: Below detection limits.
n. l.: no legal threshold limit.
n. v.: no official value.

Table 10.5. Propellants in soil back-blast area.

| | | Laboratory analysis, soil | | | | | | | |
| | | Propellants (mg kg^{-1}) | | | | | | | |
ID	Sample depth m b. s.	Diethylen-glykoldinitrat (DEGN)	Nitro-guanidin	Nitro-cellulose	Nitro-glycerin	2,4-DNT	Ammonium	Perchlorat	Hydrazin
DU4	0–0.05	—	—	**26 000**	**	—	**171**	—	—
DU6	0–0.05	—	—	**8800**	**	4.3	**167**	—	—
DU12	0–0.05	—	—	280	4.3	—	n. a.	n. a.	n. a.
DU13	0.05–0.15	—	—	8400	210	—	n. a.	n. a.	n. a.
DU5.1	0–0.05	—	—	**22 000**	**	—	**167**	—	—
DU5.2	0–0.05	—	—	**22 000**	**	—	**377**	—	—
DU5.3	0–0.05	—	—	**13 000**	**	—	**155**	—	—
Average value DU5 (1–3)				**19 000**			**233**		
Relative standard deviation (RSD)		0%	0%	22%	0%	0%		0%	0%
Detection limit		0.5	0.5	10	—	0.5	0.050	0.1	0.10

Note:

n. a.: Not analysed.

Green marking: RSD-values <35%, considered an indicator that samples are replicable.

: RSD-values >35%, considered an indicator that samples are of doubtful quality.

** Not possible to analyse because of interference from another compound.

In the rocket ignition area high concentrations (max. 4600 mg kg^{-1}) of propellants in the form of nitrocellulose were measured. Lower levels of ammonium were also measured in the soil samples (table 10.6).

Metal content in the soil samples did not deviate from background values.

At the target zone low concentrations of TNT, HMX, RDX, Nitro-glycerin, 2-AT and 4-AT were measured. Elevated concentrations of aluminium, bismuth, lead, cadmium, chrome and copper were found (table 10.7).

Water
In the back-blast area elevated concentrations of cadmium and aluminium were measured in the groundwater samples from the two wells (table 10.8).

In the target zone HMX, RDX and 4-AT was measured in the groundwater sample from the single well (table 10.9).

10.6.4 Burn site surplus artillery propellant

Soil
In the burn area heightened concentrations of lead, 2,4-DNT, ammonium and hydrazine were measured (tables 10.10 and 10.11).

Water
The water sample from the well did not reveal any compounds with concentrations above normal background levels.

10.6.5 Hand grenade range

Soil
The soil samples from the range did not reveal any compounds with concentrations above normal background levels.

Table 10.6. Propellants in rocket ignition area.

			Laboratory analysis, soil						
			Propellants (mg kg^{-1})						
ID	Sample depth m b. s.	Diethylen-glykoldinitrat (DEGN)	Nitro-guanidin	Nitro-cellulose	Nitro-glycerin	2,4-DNT	Ammonium	Perchlorat	Hydrazin
DU3 0–0.05		—	—	**4600**	**	—	**359**	—	—
DU6 0–0.05		—	—	**1100**	40	—	n. a.	n. a.	n. a.
Detection limit		0.5	0.5	10	—	0.5	0.050	0.1	0.10

Note:
n. a.: Not analysed.
** Not possible to analyse because of interference from another compound.

Table 10.7. Metal content in the target zone.

	Sample depth m b. s.		Laboratory analysis, soil Metals (mg kg^{-1})								
ID		Aluminium	Antimony	Bismuth	Lead	Cadmium	Chrome	Copper	Mercury	Nickel	Zinc
DU8	0–0.1	6.340	2	1.79	**80**	n. a.	18	**929**	0.01	9.1	166
DU9	0.1–0.2	4.270	—	0.855	33	**0.61**	88	441	0.01	5.9	86
DU10	0–0.1	2.430	—	0.740	22	0.44	8.5	190	0.01	3.6	44
Threshold limit		n. l.	80	n. l.	40	0.5	500	500	1	30	500
Detection limit		10	1	0.02	1	?	0.2	0.01	0.1	0.1	0.4

Note:
–: Below detection limits.
n. a.: Not analysed.
n. l.: no legal threshold limit.

Table 10.8. Metals in groundwater in back-blast area.

| | Laboratory analysis, groundwater | | | | | | | | | |
| | Metals (μg l^{-1}) | | | | | | | | | |
ID	Antimony	Lead	Cadmium	Chrome	Copper	Mercury	Nickel	Zinc	Bismuth	Aluminium
B14	—	0.049	0.32	0.28	0.62	—	3.0	77	—	1.200
B15	—	0.041	**0.61**	0.22	0.97	—	1.8	76	—	1.200
Threshold limit	n. l.	1	0.5	25	100	0.1	10	100	n. l.	n. l.
Background level	n. v.	0.1->1	0.005->0.5	0.04–10	0.1->50	n. b.	0.1>10	0.5->100	n. v.	n. v.
Detection limit	0.2	0.025	0.003	0.2	0.03	0.2	0.03	0.3	0.05	3.0

Notes:

n. l.: no legal threshold limit.

n. v.: no official value.

Table 10.9. Explosives in groundwater in the target zone.

| ID[1] | Laboratory analysis, groundwater | |
| | Explosives (µg l^{-1}) | |
	Target zone B13	Detection limit
2,4,6-Trinitrotoluen (TNT)	—	0.20
Octogen (HMX)	180	0.20
Hexogen (RDX)	13	0.20
Nitro-glycerin	—	0.40
1,3-Dinitrobenzen	—	0.20
1,3,5-Trinitrobenzen	—	0.20
2,4-Dinitrotoluen (2,4-DNT)	—	0.20
2,6-Dinitrotoluen (2,6-DNT)	—	0.20
2-Amino-4,6-Dinitrotoluen (2-AT)	—	0.20
4-Amino-2,6-Dinitrotoluen (4-AT)	0.43	0.20
Tetryl	—	0.20
PETN	—	0.40
Picric acid	—	0.050

Note:
–: Below detection limits.

Water
Apart from slightly elevated concentrations of zinc the water samples from the wells did not reveal any compounds with concentrations above normal background levels.

10.6.6 Omme Å stream

Soil
No elevated concentrations of metals were measured in the sediment samples at the inflow and outflow transects or in the 10 sampling points along the stream course.

Water
The pH-value in the stream fluctuates between 7.5 and 8.0, which excludes metal mobilisation from low pH. Elevated concentrations of copper (max. value 3.4 µg l^{-1}), mercury (max. value 0.7 µg l^{-1}) and zinc (max. value 32 µg l^{-1}) were measured in water samples from both the inflow and outflow transect and in all 10 of the sampling points along the stream course.

The measured concentrations do not indicate a rise in average content of these metals in the stream from the point of entry of the range compared to the point of exit. It seems that the regional background levels for these metals are slightly elevated compared to 'normal' levels. This could potentially be a result of the very sandy low pH geology in the region being conducive to a heightened mobility of certain metals.

Table 10.10. Metal content in soil at burn site.

		Laboratory analysis, soil Metals (mg kg^{-1})										
ID	Sample depth m b. s.	Aluminium	Antimony	Barium	Lead	Bismuth	Chrome	Copper	Mercury	Nickel	Zinc
DU7	0–0.05	685	2	15	**229**	0.0584	4.8	6.1	0.03	1.8	10
Threshold limit		n. l.	80	100	40	n. l.	500	500	1	30	500
Detection limit		10	1	100	1	0.02	0.2	0.2	0.01	0.1	0.4

Note:
n. l.: no legal threshold limit.

Table 10.11. Propellant content in soil at burn site.

		Laboratory analysis, soil							
		Propellants (mg kg^{-1})							
ID	Sample depth m b. s.	Diethylen-glykoldinitrat (DEGN)	Nitro-guanidin	Nitro-cellulose	Nitro-glycerin	2,4-DNT	Ammonium	Perchlorat	Hydrazin
DU7	0–0.05	—	—	—	—	0.45	278	—	0.03
Detection limit		0.5	1.0	10	0.10	0.05	0.050	0.1	0.02

Note:
–: Below detection limits.

10.7 Assessment

Risk assessment with regards to polluted soil and water in Denmark is typically done for the following receptors:
- Humans (ingestion/inhalation/skin contact);
- Groundwater;
- Surface water;
- Fauna/flora.

10.7.1 Artillery firing position

The topmost soil at the firing site is slightly polluted with lead and nitrocellulose. These compounds are not found in the groundwater sample.

As long as the area continues to be used as a firing position covered by vegetation for the entire year there is *little or no risk towards any receptor.*

10.7.2 Impact area artillery

Even though there are many shrapnel pieces and other remains from grenades in the target area the metal content in the surrounding soil is not elevated above background values (figure 10.20).

The soil is polluted with explosives TNT, HMX and RDX as a result of low-order detonations and duds. The explosives are found in the topmost 0.5 m, which seems reasonable considering the source. The groundwater below the impact area does not seem to be affected by shooting activities. The surface water bodies in the impact area seem to be lightly affected with metals, but not explosives.

There could be a *potential risk towards a person when digging in the exposed soil or drinking from the lakes in the impact area.* This is deemed very theoretical since humans should not be digging or drinking water in the area. It is uncertain if the fauna or flora in the area are affected in any way, but no ill effects have been observed.

Figure 10.20. Metal fragments in the sandy soil—image courtesy by Niras consultants.

10.7.3 Anti-tank range

The back-blast area is polluted with propellants (nitrocellulose, nitro-glycerin and ammonium). There could be a *potential risk towards a person digging in the exposed soil in the back-blast area*. This is deemed very theoretical since humans should not be digging in the area. Another potential concern is *the risk towards inhalation of polluted dust when shooting at the range*. This is deemed negligible since the shooters and instructors only spend a few hours each year at the site. It is uncertain if the fauna or flora in the area are affected in any way, but no ill effects have been observed. A bare patch of earth is deemed to be a result of the shock wave from the back-blast.

The rocket ignition area is polluted with propellants (nitrocellulose). There could be a *potential risk towards a person digging in the exposed soil in the rocket ignition area*. This is deemed very theoretical since humans should not be digging in the area. Another potential concern is *the risk towards inhalation of polluted dust when shooting at the range*. This is deemed negligible since the shooters and instructors only spend a few hours each year at the site. It is uncertain if the fauna or flora in the area are affected in any way, but no ill effects have been observed. A bare patch of earth is deemed to be a result of heat from the ignition of the rocket engines.

The target zone is polluted with metals (lead and copper) and explosives (TNT, HMX, RDX and nitro-glycerin). There could be a *potential risk towards a person when digging in the exposed soil in the target zone*. This is deemed very theoretical since humans should not be digging in the area. It is uncertain if the fauna or flora in the area are affected in any way, but no ill effects have been observed.

10.7.4 Burn site surplus artillery propellant

The burn site is polluted with metals (lead). Since the area where the propellants are being burnt is very barren (firebreak) there could be a *potential risk towards a person digging or inhaling dust at the site.* This is deemed negligible since the shooters and instructors only spend a few hours each year at the site. It is uncertain if the fauna or flora in the area are affected in any way, but no ill effects have been observed.

Of greater concern are the toxic fumes being produced during the burning. These contain harmful substances and an alternative way to get rid of excess propellant that does not involve burning would be preferable.

10.7.5 Hand grenade range

The groundwater below the hand grenade range is slightly polluted with zinc. This is deemed as a result of unusually high background values in the area and does not stem from the grenade range. The conclusion is therefore that there is *no risk towards any receptor at the grenade range.*

10.7.6 Omme Å stream

Eight of the ten places where surface water enters Omme Å are slightly polluted with metals (lead, mercury or zinc) and there could be a *potential risk towards aquatic life, and ultimately towards ingestion of fish caught in the stream.*

Because the pollution is only just above the guidance values for open water bodies and because of the almost instantaneous dilution (the influx is very small compared to the stream) in the stream the risk is deemed negligible. It is uncertain if the fauna or flora in the area are affected in any way, but no ill effects have been observed.

10.8 Conclusion

It seems that even though there has been intensive military usage of the Borris shooting range for more than half a century, the impact caused by leaching explosives and metals is very limited. Heightened concentrations were found in the topmost soil layers and in surface water on the range. But in such low concentrations no significant danger towards any receptor can be defined. Should the area ever be converted to some other usage than a military range the site should be investigated in greater detail. For now and into the foreseeable future the range will continue to serve the armed forces of Denmark as one of its major training and shooting areas.

IOP Publishing

Global Approaches to Environmental Management on Military Training Ranges

Tracey J Temple and Melissa K Ladyman

Chapter 11

Mitigation of the environmental footprint of a munition

S Thiboutot, R Martel, S Brochu and Michael R Walsh

In all nations, the military forces operate and train over vast areas of lands, at home and abroad. In Canada alone, the Department of National Defence manages approximately 2.2 million hectares of land, out of which a large portion is dedicated to military live-fire training ranges. Therefore, the Armed Forces have a responsibility to show leadership in environmental sustainability and to manage these assets efficiently on behalf of the population. Military readiness is ensured by live-fire training with a wide variety of munitions that, if not managed properly, can have considerable impacts on human health and the environment, which can jeopardize the existence of these training ranges. Range contamination has also led to groundwater contamination, resulting in some cases in the loss of ranges, large monetary penalties, and significant clean-up costs. The site characterization methodology described in chapter 4 of this book allows a better understanding of soil and water contamination resulting from live-fire training. To enhance range sustainability, the next logical step is the implementation of mitigation methods or permanent solutions to reduce and prevent environmental contamination. By investing in the understanding of munitions' footprint, tailored solutions can be applied to minimize or eliminate their adverse impacts. This chapter will present a few solutions that were undertaken to mitigate contamination that was identified through a thorough characterization of various training ranges across North America. Surface soils contamination with high concentration of munitions' residues were identified at artillery firing positions, in small arms range stop butts and in grenade and demolition ranges. To mitigate or remediate the identified problems, various tools were developed such as demilitarization and containment tools, use of reactive membranes and the application of remediation measures.

11.1 Introduction

The intimate knowledge gained by characterization of military live-firing ranges allows the measurement of their environmental health. This has been done extensively in North America in major live-fire training ranges, which has led to a good knowledge of areas of concern, where contamination poses a threat to range sustainability [1–7]. The next logical step is to transform the knowledge gained in munitions constituents (MC) dispersion into design and management practices that will mitigate and attenuate range contamination from live-fire training. Many types of mitigation measures can be implemented in range and training areas (RTAs) to either remediate or mitigate the problems identified. Physical measures such as contaminant containment are suitable whenever the combustion/detonation processes are not too intense. Chemical or biological treatment of contaminated soils is another option, when levels of contaminants are above accepted guidelines or thresholds, or when the contamination lies over a sensitive geological/hydrogeological area. Ranges can therefore be designed in a sustainable manner, which minimizes the environmental footprint of live-fire training. The implementation of mitigation methods or permanent solutions to reduce and prevent environmental contamination of ranges is now being actively studied in many North Atlantic Treaty Organization (NATO) nations [8–11]. A NATO task group started in 2017, AVT-291, focuses on innovative technology and design practices that will ensure that the accumulation of munitions constituents does not occur in concentrations that will result in adverse effects on the environment. The measures and concepts undertaken are anticipated to result in a set of technical recommendations that will enable the minimization of live-fire training environmental impacts on operational RTAs. New solutions are currently being evaluated through research and development to mitigate negative impacts on training ranges on which energetic compounds or heavy metals are used as well as prevention and mitigation of contaminant migration of ranges. This chapter will cover a few efforts that were made to mitigate the adverse environmental impacts of live-fire training with munitions in North America.

11.2 Development of field demilitarization methods in Canada and the USA for the destruction of the excess artillery gun propellant

11.2.1 Introduction

Canada and the United States follow a training doctrine known as 'train as you fight.' For artillery units, part of this training is field burning of excess propellant charges during and following field training exercises. In the past, this consisted of burning the charges on the ground in a prescribed manner. The field expedient open burning (OB) of propellant was a regular activity associated with the firing of artillery guns such as the 105 and the 155 mm howitzers. Both guns use a propelling charge system composed of multiple charge bags to fire a projectile to the required distance. Firing is typically conducted using the fewest charge bags to minimize the wear and tear on the gun barrel, as well as to avoid ricochet of the projectile. Following gun firing operations, unused propelling charge increments were burned near the gun positions as part of the unit's training. This procedure was done by piling unused bags on the surface soil or on the snow/ice surface during winter and by igniting them (figure 11.1). Solid combustion residues presented a very high risk to the environment and human health, as it leads to the build-up of toxic and persistent contaminants in the surface soils, such as 2,4-dinitrotoluene (2,4-DNT) and lead (figure 11.1) [12].

Amongst the environmental issues identified in North American RTAs, propellant burning has been shown to be a major source of soil and groundwater contamination [7]. As an example, the US Army Cold Regions Research and Engineering Laboratory (CRREL) characterized soils surrounding a fixed propellant burn pan at site Observation Point 7 on the Donnelly Training Range, Ft. Wainwright, Alaska [13]. Soil samples were taken around the burn pan using multi-increment sampling after the removal of unburned propellant grains. The samples were analyzed for energetic compounds, primarily nitroglycerine (NG) and 2,4-dinitrotoluene (2,4-DNT). High concentrations of these compounds were recovered in the samples, ranging from 3 to 35 mg kg^{-1}. In addition, the soils were analyzed for lead, and concentrations varied between 360 and 5100 mg kg^{-1}.

Small-scale tests (<1 kg) for residues from burning of excess propellants on soil were also conducted. Single-base artillery propellant containing 2,4-DNT was burned on strips of clean sand. Charges were lit and burned linearly along the sand (figure 11.2). Following cooldown, all the sand containing residues was collected for analysis and the experiment was repeated using wetted sand. Several tests were also conducted on clean snow surfaces to determine the effect of snow on burning efficiency [14].

11.2.2 Field demilitarization method for excess gun propellants/Canadian fixed tool

A team led by DRDC and involving various munitions' stakeholders as well as CRREL scientists designed, developed and deployed a fixed field demilitarization method for Canada for the burning of excess artillery propellants [16]. As the burning of propellants generated very high temperatures and violent reactions, many

Figure 11.1. Excess propellant burning over snow surface (left) and post-combustion residues (right).

Figure 11.2. Charges were lit and burned linearly along the sand.

Figure 11.3. The final design of the demilitarization unit.

iterations of various designs were needed to achieve a safe and efficient process. Therefore, a tailored solution was developed with the invention and development of a field demilitarization method for surplus powders that is now durable, efficient and safe. It consists of a burn table, totally adapted to the specific scenario of field demilitarization of propellants, involving high-burning rate and temperatures. The design of the demilitarization method was complex and required the consideration of many factors, such as material resistance to extreme flame temperature, violence and corrosiveness, resistance to a wide variety of weather conditions, and several aspects related to safety and ergonomics. Consultations were held with users as well as with explosives and ammunition neutralization experts to integrate their sound advice. In 2010, several burning tests were carried out in the field using various prototypes and iterations of the burning table. Following each of these trials, the team made significant and innovative improvements to better meet customer needs. The final design of the demilitarization unit is illustrated in figure 11.3. A great effort was put into the design of the burning shields to optimize the burning while minimizing the projection of propellant grains in the burning process.

11.3 Introduction

In Canada, the characterization of small arms sand butts showed that significant amounts of Cu, Zn, Sb and Pb accumulated in the soil of the small arms backstop berms, often above the Canadian Council of Ministers of the Environment (CCME) industrial criteria [5]. The transport of these heavy metals can occur either through surface water run-off, migration towards the aquifer and groundwater, or wind erosion. Therefore, a method to contain the contaminated soils was developed to avoid further dispersion of toxic heavy metals. At that time, a thorough review of the commercially available containment systems revealed that none fulfilled the Canadian Armed Forces (CAF) needs. Defence R&D Canada (DRDC) collaborated with several partners to develop environmentally-friendly bullet catchers that greatly reduce the risk of soil contamination at small arms firing ranges, and that considerably reduce maintenance cost [8].

11.3.1 Canadian bullet catcher

After numerous trials and iterations, the final design comprehends a self-healing membrane over a catcher filled with sand equipped with a water exit pipe and a treatment system to remove any heavy metal that could leach out of the catchers. After four years of intensive use, the catchers have proven to be highly efficient and cost-effective. A study by an independent firm showed that the bullet catchers are 54% less expensive than conventional sand butts, can be adapted to a variety of weather conditions, and can be installed by engineering and construction teams on military bases using plans and drawings. The bullet catchers are a 'green' solution for managing Canadian Armed Forces' small arms ranges because their structure prevents seepage from contaminating the groundwater. This in turn helps avoid considerable maintenance costs related to soil decontamination. The catcher is illustrated in figure 11.4, and drawings are available to any nation interested in building them.

The small arms green range has been in service since November 2012. So far, approximately 320 000 projectiles have been fired and results suggest that this

Figure 11.4. Self-healing bullet catcher from a small arms range's target impact area.

technology shows great promise as a method of mitigating environmental contamination due to small arms training in Canada. The use of electronic targeting enhances the longevity of the membranes by lowering the frequency of tumbling rounds.

11.3.2 Summary

It is strongly believed that these bullet catchers are part of the solution to mitigate and solve the lead and antimony issues encountered in Canadian small arms ranges. It is thought that this approach will provide more flexibility and potentially lower costs that will facilitate the implementation of this technology throughout the CAF. The proposed green small arms range is economically viable and will decrease the maintenance cost and frequency when compared to conventional sand butts. It will greatly decrease the environmental footprint of small arms training and the self-healing membrane also reduces significantly the noise of live-fire training.

11.4 The development of reactive membranes for adsorption of heavy metals and energetic materials

11.4.1 Introduction

For many years, INRS conducted field and laboratory experiments to understand the fate and transport of MC [20–24]. In 2015, INRS was tasked by DRDC Valcartier to study reactive barriers that would prevent MC from reaching sensitive receptors. The objective was to develop a reactive barrier that would adsorb contaminants moving with soil water under grenade, anti-tank or demolition ranges.

11.4.2 Methodology

Reactive materials were selected using many columns studies and an outdoor pilot test began in 2015 with the most promising materials. The objective was to determine the efficiency of reactive barriers under real field conditions (freezing-thawing, rain and snowmelt infiltrations). The tests were made into 2.6 m^3 water-tight containers buried into the soil (figure 11.5). Four containers were installed in 2015: three containing different reactive barriers bone char (BC) over activated carbon (AC), AC over BC and AC over brown peat moss (BPM) mixed with coconut coir (CC) and the fourth one had no reactive material to act as a control. The 1182 ± 3 kg of contaminated soil put in each container was made by mixing a demolition range surface soil with 160 kg of contaminated soil from a small arms range (to add metals contamination), 16 g of ammonium perchlorate and 80 g of crushed composition B. Two other containers were installed in 2016 to replicate the highest metal concentrations measured in the soil of Canadian grenade ranges. A synthetic contaminated soil was produced by mixing clean natural soil with metallic powders and 20 g of crushed composition B. In November 2016, a solution of 1 l of dissolved ammonium perchlorate (AP) (16 g l^{-1}) was spread over the surface of all containers except BC treatment. The water quality at the outflows of the containers was evaluated 23 times in the first four containers and 14 times in the two others.

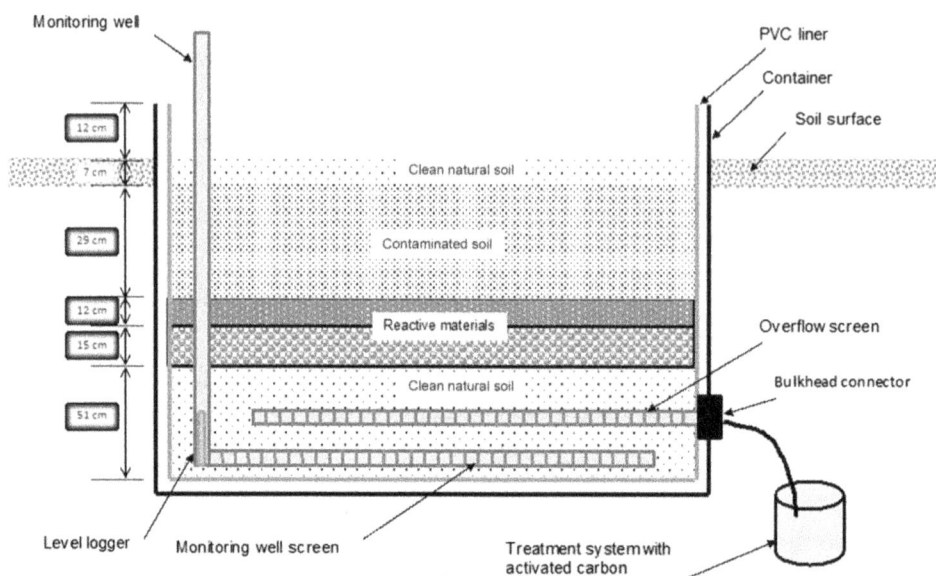

Figure 11.5. Testing set-up for reactive barriers.

Energetic materials, metals, perchlorate and physicochemical parameters (pH, Eh, temperature and electrical conductivity) were followed. The analytical results from water samples were compared to the drinking water guidelines from Health Canada and to the aquatic life criteria. Two flowmeters were installed in 2016 to measure the water outflow to establish a mass balance of the contaminants.

11.4.3 Results

Results obtained for the first 28 months of the experiment indicated that no problem was encountered by scaling-up reactive materials from the column experiments to the field conditions. The three reactive barriers of the containers installed in 2015 performed well to decrease perchlorate and energetic materials concentrations below the most severe of the two water guidelines. Despite the high concentrations measured in the control for perchlorate (up to 43 000 µg l^{-1}), RDX (640 to 40 340 µg l^{-1}), HMX (267 to 2860 µg l^{-1}) and TNT (15 to 5300 µg l^{-1}), all treatments (AC/BC, BC/AC and AC/BPM+CC) succeeded in removing more than 99.9% of the contamination and to lower the concentrations below the guidelines (see figure 11.6 for perchlorate results). For metals, the low concentrations measured in the control are caused by the presence of clean sand between the contaminated layer and the sampling point, which causes a delay on the arrival of metals at the sampling points. Antimony took 20 months to migrate to the bottom of the control. A mass balance calculated from measured concentrations in water and measured or simulated water flow rates show that containers with treatments released 99.9% less perchlorate and energetic materials than the control. For metals, the mass balance calculation indicated that more than 97% of antimony was removed. The treatment AC/BPM +CC leached some zinc. The low concentrations of some metals (close to the limit of

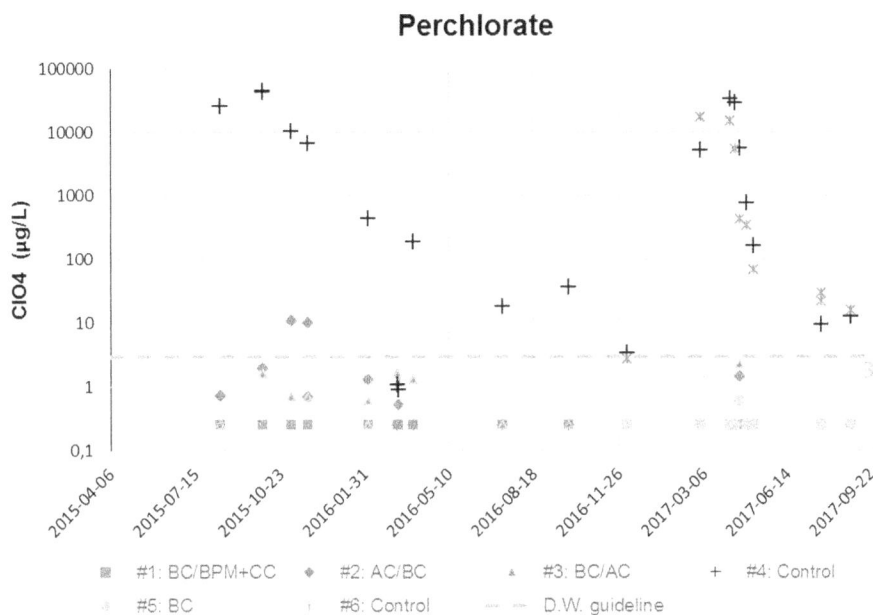

Figure 11.6. Evolution of perchlorate concentrations over time in six containers.

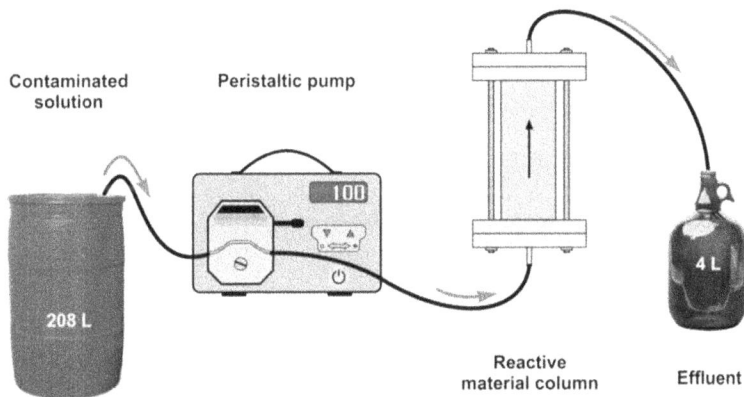

Figure 11.7. Schematic representation of one test column.

detection) in the control and in the treatments resulted in a not reliable mass balance estimate for cadmium, copper and lead.

In parallel to the field work, eight saturated column tests (figure 11.7) were conducted to evaluate the efficiency and the lifespan of reactive materials. The experiments were realized with the same three reactive materials (in duplicate) as in the field tests, mainly activated carbon and bone char combinations and an empty column for control. A contaminated water, produced by the leaching of the same stock of contaminated soil used in the containers test, was injected into the columns at an average 20 mL h^{-1}, representing the equivalent of 330 years of infiltration for the 479 days of the experiment. Water samples were collected twice a week at the outflow of each column and a total of 1064 samples have been collected. Samples were analyzed for metals, perchlorate and energetic materials to determine when guidelines were exceeded. Not all samples were analyzed, only a selected number at specific dates to follow outflow concentrations of contaminants. Results were compared to an empty column used as a control.

Column tests with 15 cm thick reactive materials work extremely well to adsorb dissolved energetic materials (EM), perchlorate and metals from the contaminated water injected. In fact, most of the contaminants tested were measured below the selected water quality guidelines at the outlet of the treatment column during a simulated period of infiltration of over 300 years on average. Antimony (Sb) and perchlorate (ClO_4) had a breakthrough (observed concentrations above guidelines) at the outflow of the treatment columns after a reasonable period of, respectively, 50 years (55 y for AC/BC or BC/AC and 43 y for AC/BC mixture but 0 y for AC/BPM +CC) and 175 years (150 y for AC/BC, 175 y for BC/AC and AC/BPM+CC, and 215 y for AC/BC mixture). The only treatment that does not work is AC/BPM+CC treatment for antimony with breakthrough right at the beginning of the contaminated water injection. Using the above breakthrough for antimony and perchlorate and assuming that the breakthrough was reached at the last available chemical analysis below the guidelines for the other parameters tested (EM at 314–343 y; cadmium, copper, lead and zinc at 308–335 y) it is possible to evaluate the lifespan of the reactive materials and their adsorption capacity. The above number of years was considered more than enough for the design of a reactive barrier and the extra years normally needed to reach the breakthrough account for a security factor. For all treatment tested, the lifespan of EM (HMX, RDX, TNT) is on average 652 years, it is between 314 and 436 years for perchlorate, 637 years for Cu, Zn, Cd and Pb and on average 100 years for Sb. For all treatment, the adsorption capacity of the reactive material calculated by the outflow concentrations method is on average 1820 mg RDX kg^{-1}, 480 mg HMX kg^{-1}, 215 mg TNT kg^{-1}, 300–400 mg, ClO_4/kg, 37 mg Cu kg^{-1}, 160 mg Zn/kg, 4 mg Cd kg^{-1}, 15 mg Pb/kg and 25 mg Sb kg^{-1}. A mass balance was calculated on solid reactive material and resulted in values in the same order of magnitude as the outflow concentrations method, excepted for Pb and Sb in the AC/BPM+CC treatment. Mass balance calculation in column tests indicated that the precision was better with the outflow concentrations method than the solid reactive material method. The outflow concentration method does not necessitate additional extraction and the mass balance for energetic materials and perchlorate are possible. However, the mass balance realized on reactive material allowed one to determine where the adsorption occurred in the columns.

11.4.4 Summary

The implementation of a reactive barrier is site specific. The thickness of the barrier should be based on infiltration rates evaluated with meteorological data, water quality data from pan lysimeters installed in the range over at least a 6 month period and measured adsorption capacity from column tests. This study showed the high potential of reactive material technology for treating various types of munition-related contaminants in ranges and training areas. A few examples of application would be:

1) A vertical permeable reactive barrier buried into the saturated zone to prevent dissolved contaminants to reaching a surface water body;

2) An horizontal permeable reactive barrier buried in the unsaturated zone below a limited extent training area such as grenade, demolition, or small arm ranges to avoid dissolved contaminants reaching an aquifer;

3) A permeable reactive zone built within a dam in a river to treat water as it flows through it;

4) As a filter placed at the outflow of a contaminated water collecting system in grenade, demolition or small arm ranges, or used in a pump and treat system.

This project allowed the development of a protocol for the scale-up from the laboratory to field applications. Many efficient reactive materials were identified for munition-related contaminants made of organic and inorganic chemicals. Their life spans were established for eastern Canadian conditions while they could be evaluated for any other climatic condition based on the amount of infiltration water and the concentration of contaminants of concern in the infiltrating water. The strength of our approach is that it could be tailored for any type of mixed contamination either in an *in situ* horizontal or vertical configuration.

11.5 Investigations on the efficiency of remedial methods for energetic materials: dithionite and lime

11.5.1 Introduction

Several soil remediation processes for energetic materials contaminated soils have been investigated by the international community. They can be classified into four main categories:

1. Bioremediation (biodegradation, phytoremediation, constructed wetlands, composting, etc);

2. Physical methods (adsorption, permeable reactive barrier, etc);

3. Chemical methods (oxidation, reduction, basic hydrolysis, etc);

4. Thermal methods (incineration, onsite thermolysis, etc).

Some of these processes are commercially available, while others are still in their infancy. An extensive review of all these technologies is beyond the scope of the current work but additional information can be found in [25, 26]. Soil remediation of active RTAs is not frequently undertaken unless the source term must be removed, or controlled. Typically, this point is reached in cases where energetic material guidelines in soils or water are exceeded, or when energetic materials have migrated to nearby groundwater or surface water, or when adverse impacts on sensitive fauna, flora or human receptors are observed. Remediation strategies for RTAs must be considered on a site-by-site basis. The optimal remediation strategy for active RTAs depends on numerous factors such as the type and granulometry of soil, the type and depth of contaminants, as well as the desired end result. In addition, remediation techniques must be inexpensive (because of the large size of RTAs), easily applied to remote locations, non-intrusive, because of the presence of unexploded ordnance (UXO), and of short duration, to avoid disturbing military training as much as possible, and must prevent further environmental issues. Consequently, *in situ* remediation methods are usually preferred.

In addition, the selection of an appropriate remediation method for a specific RTA must include mid-scale testing, because laboratory experiments are not representative enough of the actual field environment. The success of the remediation is closely related to the representativeness of a mid-scale test site, which must replicate as closely as possible the contaminated site to be remediated. A thorough knowledge of the contaminants' environmental fate, concentration and particle size as well as of their dispersion on the surface and sub-surface of the soil is therefore the key for a successful remediation.

Energetic materials are known to be heterogeneously dispersed as particles smaller than 1 mm, generally on the top 2.5 cm of the soil [5]. Moreover, these particles are commonly made of a mixture of energetic materials, of composition similar to Composition B (RDX:TNT 60:40), Octol (HMX:TNT 70:30) or propellants (various mixes of NG, 2,4-DNT and NG). According to Taylor *et al* [27], the presence of RDX and HMX in Comp B and Octol, respectively, slows the dissolution rate of TNT due to their lower water solubility. Moreover, according to the same authors, the dissolution rate of large energetic material particles produced by a detonation is slower than that of small particles due to the surface area effect. Because the solubility and dissolution rate of a particular energetic material probably affect its reaction with a remediation reagent (the reaction does not occur in the solid state), the energetic materials used for mid-scale testing should be ground to particles smaller than 1 mm. For best results, soil spiking should be performed by dispersing ground energetic materials' particles on the surface of the soil, rather than spiking with energetic materials already dissolved in an organic solvent.

11.5.2 Testing of remediation technologies

Representative mid-scale testing can be conducted using containers exposed to weathering. As an example, a 7 m^3 container needs to be filled with clean soil (approximately 7000 kg) from the RTA to be remediated, compacted and covered by a thin layer (approximately 2.5 cm) of contaminated soil of precisely known concentration. The contaminated soil must be made by spiking the same clean soil as the one used to fill the container with ground energetic formulations particles (<1 mm) to a concentration representative of that of the RTA to be remediated. The contaminated soil has to be mechanically mixed before its spreading on the clean soil, using a barrel mixer, for example. The remediation reagents must be added to the container using the same process as one would in the field (raking, tilling and watering may be used if necessary). Composite sampling must be used to collect soil samples. The reaction can be followed by taking samples at regular time intervals in the course of several weeks. If desired, lysimeters can be added to collect 'aquifer' water for analysis. A control container (with no remediation reagents) is of absolute necessity to assess the natural attenuation and downward migration of the contaminants due to weathering. Two remediation strategies, which rely on the basic hydrolysis of energetic materials with hydrated lime [28–34] or on their chemical reduction with dithionite [35, 36] have been assessed using laboratory experiments and container studies (figure 11.8). These methods were selected because of their usage increase in recent years, due to their relatively low cost, their short duration, their ease of use and their efficiency for the degradation of energetic materials. Dithionite is a strong reductant that also reacts with organic constituents and living microorganisms in the vicinity of the remediation site, leaving behind an easily eroded soil with zero organic content. Metallic compounds are claimed to react with dithionite to form insoluble sulfur salts that are not bioavailable. Lime works in a slightly different manner, by elevating the pH of the soil up to 10. Consequently, every organic compound in lime's path undergoes basic hydrolysis. Most metallic compounds are claimed to become insoluble in such a high pH environment, but this is not the case of amphoteric elements, such as arsenic and antimony. Both remediation processes require a thorough mixing of the soil and remediation reagents using, for example, a tiller, as well as the addition of water to start the degradation reaction. Tilling soils in impact areas is a risky process, because of the presence of UXOs. A surface clearance must be performed before the beginning of operations, which increases the remediation cost.

Both remediation processes are initiated by the addition of water to the mix of soil and remediation reagent, which is a logistic burden, especially for dithionite for which warm water is needed. In addition, the large quantities of water necessary for the reaction may trigger the migration of metallic and energetic residues deeper in the aquifer, which could lead to the false belief of degradation on the surface of the soil and to a higher environmental risk.

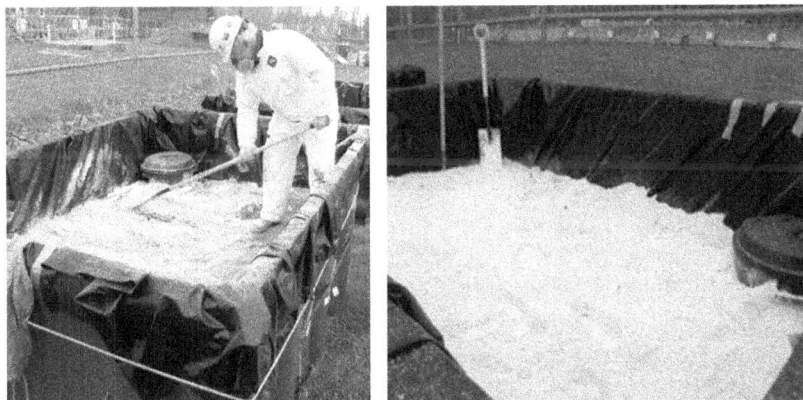

Figure 11.8. Dithionite (left) and lime (right) container after addition to their respective container.

The monitoring of the reaction to determine the remediation efficiency represents an analytical challenge. The first report on this issue came from Larson *et al* [31], who indicated that the extraction and analysis of soils samples containing energetic materials and hydrated lime led to the 'report of false degradation as legitimate results', and consequently to overestimate efficiency. Without proper lime quenching, the degradation reaction goes on in the extraction solvent, leading thus to underestimated concentrations that are not representative of what remains in the soil, and consequently to overestimate efficiency.

The same observations were reported by Turcotte-Savard and Brochu [37], who evaluated the effects of sample preparation and analysis on the overall degradation of energetic materials exposed to sodium dithionite or hydrated lime in soil. Samples of known TNT, RDX and HMX concentrations were prepared by adding Comp B to general purpose sand, which was then extracted and analyzed as would be an energetic material contaminated soil sample using the SW-846 EPA method 8330b [38]. The reaction occurred in a vial and the contact time between soil, energetic materials and remediation reagent was less than 1 min The soil mixture was then extracted 18 h with acetonitrile, as prescribed by EPA method 8330b. Given the short-contact time between the remediation reagents and the energetic materials, no reaction should have occurred. Instead, variable degradation levels were observed. TNT exposed to hydrated lime of buffered dithionite was completely degraded during the extraction process. RDX was slightly more resistant, with losses of 90% and 30% to 35% when exposed to lime and buffered dithionite, respectively. These findings indicate that TNT and RDX were degraded during sample extraction and analysis, due to the presence of residual remediation reagents in the soil samples, which may react once the extraction solvent is added to the sample. This leads to a systematic underestimation of the actual residual energetic material concentrations in soils and to an overestimation of remediation efficiency, leading to the false belief of energetic material degradation during a chemical remediation treatment. This issue could be resolved by the use of a proper quenching agent that would neutralize the unreacted dithionite and lime before the extraction. Research still needs to be done in that area.

The degradation reactions with dithionite and lime were also followed in outdoor containers using the method described above. Lime made most of the TNT (90%) disappear in less than a day, while RDX and HMX never got below 40% of their initial concentrations even after three weeks. With dithionite, TNT completely disappeared in less than a day, while RDX and HMX never got below 60% and 70% of their initial concentration, even after three weeks.

Another important element to consider in a remediation process is the degradation products. Ideally, the remediation processes should lead to complete mineralization of all contaminants to innocuous constituents. However, the production of an intermediary, potentially toxic compound is seldom checked, because the efficiency of the reaction is generally followed by monitoring the disappearance of energetic materials of interest by SW-846 EPA method 8330b.

11.5.3 Summary

Remediation of sites contaminated with energetic materials constitutes a complex operation, because of the presence of UXOs, of the remoteness of RTAs and of the large size of the areas. The need to avoid disturbance of military training as much as possible also represents a challenge. In addition, the follow-up of the remediation process using EPA method 8330b leads to overestimated efficiency, because a significant part of the degradation reaction occurs during the extraction process. Research is underway to find a solution to this important issue.

11.6 Conclusions

The objective of this chapter was to stress the importance of developing solutions to mitigate munitions-related contaminants dispersion, fate and impacts in military live-firing ranges. If their concentration reaches levels of concern in RTA environmental samples, it represents a high risk in the heart of our priceless training assets, and ultimately it might trigger their closure. Losing these assets would be a catastrophe for most nations, as opening new training ranges would be almost impossible when considering the encroachment with civilian developments.

Twenty years ago, there was no awareness of the risk associated with the accumulation of munitions constituents in ranges and the methods and tools for representative assessment had to be developed. The journey to range sustainability was not straightforward and a lot of resistance had to be overcome to convince the military users of the benefits of land assessment to promote risk management. Sampling for MC from multiple sources of munitions in a variety of range types represented a great challenge. Their small and large scale fate and transport were studied, and in Canada, the numerous benefits of hydrogeological characterization of RTA were clearly demonstrated. The priceless information obtained from all these efforts allows the design of an action plan to mitigate or eliminate the identified problems. We presented here a few of the efforts dedicated to minimizing or eliminating the risk of their dispersion and we are very proud of the numerous achievements towards military range sustainability due to hard work, strong leadership and to years of successful collaboration with allied countries. Many countries already begun to implement solutions in RTA to achieve sustainable training. The process applied to MC of our conventional stockpile must be applied to new emerging constituents and formulations to avoid repeating mistakes from the past and ensuring the viability of our ranges. Leadership and vision towards sustainable training ranges have risen in importance in most NATO nations and shall allow successful continuous training of our allied forces.

References

[1] Jenkins T F 2008 *Characterization and Fate of Gun and Rocket Propellant Residues on Testing and Training Ranges* TR-08-1 Final Report ERDC Technical Report

[2] Pennington J C 2006 *Distribution and Fate of Energetics on DoD Test and Training Ranges* TR-06-13 Final Report ERDC

[3] Sunahara G, Lotufo G, Kuperman R, Hawari J, Thiboutot S and Ampleman G 2009 *Ecotoxicology of Explosives* (CRC Press, Taylor and Francis Group) pp 1–325

[4] Thiboutot S, Ampleman G and Alan H 2002 *Guide for Characterization of Sites Contaminated with Energetic Materials* TR-02-1 Report CRREL

[5] Thiboutot S *et al* 2012 Environmental characterization of military training ranges for munitions-related contaminants *Int. J. Energetic Mat. Chem. Prop.* **11** 15–57

[6] Thiboutot S, Ampleman G, Brochu S, Diaz E, Walsh M R and Walsh M E 2013 Canadian Programme on the Environmental Impacts of Munitions – Deposition Rate of Explosives, Propellants and IM Explosives *Proc. of the European Conf. of Defence and the Environment* (Helsinki May 2013) https://www.defmin.fi/files/2608/Conference_proceedings_web_2013.pdf

[7] Walsh M R 2010 *Characterization and Fate of Gun and Rocket Propellant Residues on Testing and Training Ranges* TR 10-13 Interim report 2 (U) SERDP Project ER-1481ERDC

[8] Ampleman G, Thiboutot S, Diaz E, Brochu S, Martel R and Walsh M R 2013 New range design and mitigation methods for sustainable training *Proc. of the European Conf. of Defence and the Environment* (Helsinki May 2013) https://www.defmin.fi/files/2608/Conference_proceedings_web_2013.pdf

[9] Humer T 2015 Precautionary soil and water protection by the German Ministry of Defence—Vulnerability assessments on military training areas *Proc. of the European Conf. of Defence and the Environment* (ECDE) (Helsinki, June 9-10, 2015)

[10] Svanström T and Warsta M 2015 Technical and practical solutions for environmental protection of heavy weapons ranges *Proc. of the European Conf. of Defence and the Environment* (ECDE) (Helsinki, June 9-10, 2015)

[11] The Interstate Technology & Regulatory Council 2005 Environmental Management at Operating Outdoor Small Arms Firing Ranges ITRC guidance document https://itrcweb.org/GuidanceDocuments/SMART-2.pdf

[12] Diaz E, Brochu S, Poulin I, Faucher D, Marois A and Gagnon A 2011 Residual dinitrotoluenes from open burning of gun propellant *Environmental Chemistry of Explosives and Propellant Compounds in Soils and Marine Systems: Distributed Source Characterization and Remedial Technology American Chemical Society Symposium Series* vol 1069 p 401

[13] Walsh M R, Walsh M E and Hewitt A D 2012 Energetic residues from field disposal of gun propellants *J. Hazard. Mater.* **173** 115–22

[14] Walsh M R, Walsh M E, Ramsey C A, Rachow R, Zufelt J, Collins C, Gelvin A, Perron N and Saari S 2006 *Energetic Residues Deposition from 60- and 81-mm Mortars on Snow* ETR-06-10 Technical Report ERDC/CRREL U.S. Army Corps of Engineers Cold Regions Research and Engineering Laboratory (Hanover, NH, USA)

[15] Ampleman G, Thiboutot S, Marois A, Gagnon A, Walsh M R, Walsh M E, Ramsey C and Archambeault P 2011 Propellant Residues Emitted by Triple Base Ammunition Live Firing using British 155-mm Howitzer Gun at CFB Suffield, Canada. ch 3 in Technical Report ERDC/CRREL TR-11-13, U.S. Army Corps of Engineers Cold Regions Research and Engineering Laboratory

[16] Thiboutot S *et al* 2011 Development of a table for the Safe Burning of Excess Artillery Propellant Charge Bags DRDC Valcartier TR 2010-254

[17] Thiboutot S, Ampleman G, Pantea D, Whitwell S and Sparks T 2012 Lead emissions from open burning of artillery propellants *Air Pollution XX, WIT Trans. Ecol. Environ.* **157** 273–84

[18] Walsh M R 2016 *A Portable Burn Pan for the Disposal of Excess Propellants* MP-16-2 Technical Report ERDC/CRREL (U.S. Army Corps of Engineers Cold Regions Research and Engineering Laboratory, Hanover, NH, USA)

[19] US Army National Guard 2014 *Best Management Practices for Army National Guard Operational Ranges—Fact Sheet: Burn Pans* (Washington, DC: US Army National Guard Bureau)

[20] Bordeleau G, Martel R, Ampleman G, Thiboutot S and Poulin I 2012 The fate and transport of nitroglycerin in soils and groundwater at active and legacy anti-tank firing positions *J. Contam. Hydrol.* **142–143** 11–21

[21] Lewis J, Martel R, Thiboutot S and Ampleman G *et al* 2009 Trépanier. Quantifying the transport of energetic materials in unsaturated sediments from cracked unexploded ordnance, Québec, Canada *J. Environ. Qual.* **38** 1–8

[22] Lewis J *et al* 2013 The effect of subsurface military detonations on vadose zone hydraulic conductivity, contaminant transport and aquifer recharge *J. Contam. Hydrol.* **146** 8–15

[23] Martel R, Mailloux M, Gabriel U, Lefebvre R, Ampleman G and Thiboutot S 2009 Behavior of energetic materials in groundwater at an anti-tank range *J. Environ. Qual.* **38** 75–92

[24] Mailloux M, Martel R, Gabriel U, Lefebvre R, Ampleman G and Thiboutot S 2008 Hydrogeological study of an anti-tank range *J. Environ. Qual.* **37** 1468–76

[25] Kalderis D, Juhasz A L, Boopathy R and Comfort S 2011 Soils contaminated with explosives: Environmental fate and evaluation of state-of-the-art remediation processes (IUPAC Technical Report) *Pure Appl. Chem.* **83** 1407–84

[26] EPA Contaminated Sites Clean-Up Information, available at https://clu-in.org/remediation/ (accessed October 11, 2018)

[27] Taylor S, Bigl S, Vuyovich C, Roningen J, Wagner A, Perron N, Daly S, Walsh M and Hug J 2011 *Explosives Dissolved from Unexploded Ordnance.* Hanover, VA, Cold Regions Research and Engineering Laboratory, US Army Engineer Research and Development Center (Report DTIC- ADA565446)

[28] Davis J L, Brooks M C, Larson S L, Nestler C C and Felt D R 2006 Lime treatment of explosives-contaminated soil from munitions plants and firing ranges *Soil Sediment. Contam.* **15** 2006

[29] Davis J L, Brooks M C, Larson S L, Nestler C C and Deborah R F 2006 Lime treatment of explosives-contaminated soil from munitions plants and firing ranges *Soil Sed. Contamin.: Inter. J.* **15** 565–80

[30] Hansen L S, Larson S L, Davis J L, Cullinane J M and Nestler C C 2003 Lime Treatment of 2,4,6-Trinitrotoluene Contaminated Soils: Proof of Concept Study. ERDC/EL-TR-03-15. Vicksburg, MS: U.S. Army Corps of Engineers, Engineer Research and Development Center, Environmental Laboratory, 2003. Available at: http://wes.army.mil/el/elpubs/pdf/trel03-15.pdf

[31] Larson S L, Davis J L, Martin W A, Felt D R, Nestler C C, Brandon D L, Fabian G and O'Connor G 2007 Grenade Range Management Using Lime for Metals Immobilization and Explosives Transformation Treatability Study. ERDC/EL-TR-07-5. Vicksburg, MS: U.S. Army Corps of Engineers, Engineer Research and Development Center, Environmental Laboratory. Available at: https://el.erdc.dren.mil/elpubs/pdf/trel07-5.pdf

[32] Larson S L, Davis J L, Martin W A, Felt D R, Nestler C C, Fabian G, O'Connor G, Zynda G and Johnson B A 2008a Grenade Range Management using Lime for Dual Role of Metals

Immobilization and Explosives Transformation. Field Demonstration at Fort Jackson, SC. ERDC/EL-TR-08-24 *Vicksburg, MS: U.S. Army Corps of Engineers, Engineer Research and Development Center, Environmental Laboratory* Available at: http://acwc.sdp.sirsi.net/client/search/asset/1003187

[33] Larson S L, Davis J L, Martin W A, Felt D R and Nestler C C 2008 Grenade Range Management Using Lime for Metals Immobilization and Explosives Transformation *ESTCP Cost and Performance Report ER-0216* (Arlington, VA: Environmental Security Technology Certification Program) Available at: https://frtr.gov/costperformance/pdf/Grenade-Range-Managemnt-ETSCP-2008.pdf

[34] Martin W A, Larson S, Nestler C, Fabian G, O'Connor G and Felt D 2012 Hydrated lime for metals immobilization and explosives transformation: treatability study *J. Hazard. Mater.* **215–16** 280–6

[35] Boparai H, Comfort S, Shea P and Szecsody J 2008 Remediating explosive-contaminated groundwater by *in situ* redox manipulation (ISRM) of aquifer sediments *Chemosphere* **71** 933–41

[36] Nzengung V A 2014 Sulfur-based bulk reductants and methods of using same *U.S. Patent* 8,722,957 B2

[37] Turcotte-Savard M O and Brochu S 2019 Degradation of RDX, TNT, and HMX during EPA 8330B sample processing and analysis of soils under hydrated lime or dithionite-based chemical remediation *Soil Sed. Contamin.: Inter. J.* **28** 245–57

[38] USEPA 2006 *Method 8330B: Nitroaromatics, nitramines, nitrate esters by high performance liquid chromatography (HPLC), Revision 2* (Washington, DC: Environmental Protection Agency) Available at: https://epa.gov/sites/production/files/2015-07/documents/epa-8330b.pdf

IOP Publishing

Global Approaches to Environmental Management on Military Training Ranges

Tracey J Temple and Melissa K Ladyman

Chapter 12

Environmental assessment at a Brazilian Army site

M E S Marques, E B F Galante, M M Reis and M C Barbosa

Military activities require ammunitions, which are manufactured, stored, used and eventually disposed of. In the Brazilian Army storing and disposing of ammunitions is a mission for a few depots placed in the country. Considering that continuous disposal of explosives can impact the environment, this work involved carrying out an environmental assessment of the overall process at one of the Army depots located in the southwestern part of the country. The assessment begins with a hierarchical analysis method to determine what parameters are relevant, followed by a survey of what contaminants are to be expected, which informed soil and groundwater analysis. Then the fate and transport of heavy metal within the site are assessed using vegetation as a possible indicator of environmental pollution. For that analysis, a grid of sampling points was created based on cardinal directions, using linear transects and circular sampling units, following the effect of shock waves caused by the detonations. The whole array of results showed that the site might be contaminated by cadmium, copper and lead in soil as the vegetation showed toxicity by the elements manganese, chromium, zinc, cadmium, copper and lead. The vegetation proved to be a good indicator of environmental pollution. Correlation coefficients between soil and vegetation revealed a direct relationship between nickel, chromium, zinc, copper and lead site contamination and the local source of pollution. Moreover, the results were plotted showing that the contamination is well established at the military site, therefore very unlikely to affect the surrounding areas.

12.1 Introduction

The disposal of ammunition and explosives with expired shelf life is an assignment of the Brazilian Army, according to Presidential Decree No. 3.665 (November 2000), which approves 105 Brazilian Army regulation (R-105) [1]. This decree establishes that products controlled by the Army and also those that are seized by the competent authorities, that is, by the police authorities, must be sent to the warehouses of the Army Units for final disposal, one of them being destruction. The disposal of these materials is carried out in the Army's own lands, in the units designated for this purpose, arranged in distinct areas in the seven Military Regions of the national territory. Article 224 of this decree specifically describes the minimum safety conditions for carrying out open detonation.

One of the activities that the Army carries out and which causes great environmental impact is the destruction of ammunition and unserviceable explosives. This procedure is regulated by the Army's Technical Manual T9 [2] and Explosives Field Manual [3], but due to the great impact that this activity causes to the environment, it is necessary to adopt techniques to monitor the contaminated areas and plan how to recover them.

In Brazil, the Army undertakes the open burning of explosives and ammunition technique on its own land, complying with safety standards. However, from an environmental point of view, there is still little discussion about which region would be most suitable for the destruction of the material within the areas that the Army itself already has for this purpose and what impacts these activities will have.

The open burning technique consists of the combustion of any material without control of the combustion gases, without containment of the combustion reaction in a closed device, nor the certainty that the combustion will be complete. Open pit detonation is a chemical process by which an explosive charge ignites the remaining munitions to be detonated (figure 12.1).

According to the Federal Remediation Technologies Roundtable (FRTR) [5], the factors that limit the applicability and effectiveness of the detonation/burning process are: minimum distance requirements for safety purposes requiring large areas; the emissions of gases generated are difficult to capture in the treatment; in

Figure 12.1. Preparation for open detonation. Reproduced from [4].

flaring or blasting operations the winds can carry sparks, flame and toxic gases to neighboring facilities; risk of premature detonation when the process is being prepared and the difficulty in treating the waste and contamination because of the cost and risk associated with this task.

According to the United States Environmental Protection Agency (USEPA) [6] data, there are around 1 000 000 potentially contaminated areas in the United States, of which 69% are from heavy metals from different sources of contamination. In Europe, according to the Sustainable Management for Trace Elements Contaminated Soils (SUMATECS) report [7], potentially contaminated sites also reach this value, but with a 31% lower percentage of soil contaminated by metals. In Brazil, unlike the USA and Europe, the amount of destroyed ammunition and explosives is relatively small, which brings some questions about the feasibility of demilitarizing (recycling) [8, 9] these materials.

Studies on techniques for the remediation of explosive contaminated soils have been described by USEPA [6, 10] as early as 1993, however, in Brazil it is necessary to develop a methodology for the diagnosis of contamination of destruction areas, considering the characteristics of the climate and the formation of tropical soils, and the variability of the geotechnical and morphological characteristics of the areas used for this purpose in Brazilian territory.

The difficulties of geotechnical investigation at this stage are mainly associated with safety issues, since field investigations are carried out in the destruction area itself, with drilling, sampling of soil, surface water and installation of wells, etc, which are areas where there may be remnants of ammunition (figure 12.2). The ammunition disposal process used does not totally destroy the ammunition, which could lead to contamination by organic products and the metal residues should be removed after destruction. In order for this process to be carried out, it must be ensured that all explosives and ammunition have been completely detonated. Another difficulty for the research is that the areas of destruction are of restricted military security.

The Brazilian Army, within the scope of its Environmental Management Systems (EMS), has been concerned with preserving and recovering possible contaminated

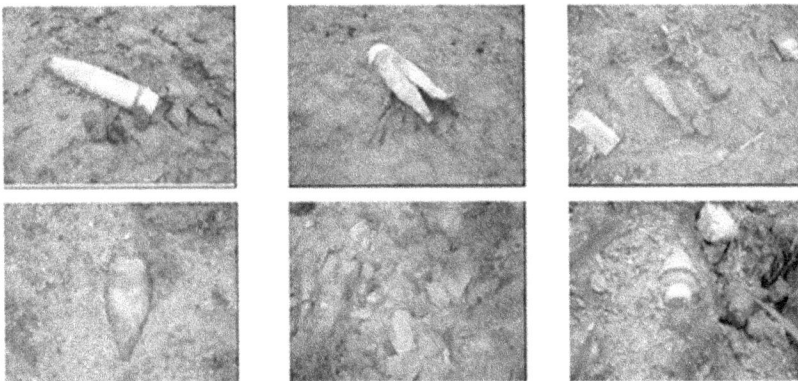

Figure 12.2. Remnants of ammunition from open destruction. Reproduced from [4].

areas. It is necessary to know the life cycle of the munitions, to carry out diagnosis, monitor and catalog the contaminated areas and then develop environmental remediation techniques for each area [4]. Therefore, a team from the Military Engineering Institute (IME) [11], in conjunction with the Federal University of Rio de Janeiro (COPPE/UFRJ), implemented a series of environmental studies on soil contaminants [4, 12], vegetation [13] and groundwater [14], and the results of which, in summary, are presented in this chapter.

12.2 Case study

The Brazilian Army disposes of ordnances through open burning and open detonation in eleven sites throughout Brazil, which cover the Army's needs in every region.

This particular survey was carried out in the southeast region of Brazil, where the Brazilian Army maintains a large ammunition depot, which is located within a 22.5 km^2 estate. Due to Brazilian military protocols, the depot is responsible for storing and distributing ordnances, as well as destruction and disposal; the latter began in the 1960s. Moreover, the depot area case study (figure 12.3) is located in the state of Rio de Janeiro, Brazil, and is imbedded in a river basin already affected and impacted by domestic sewage and other industrial pollutants. This same hydrographic area is of great importance for the entire area as a water supply, as well as enabling the operation of hydroelectric and thermoelectric plants and the production process of hundreds of industries [15].

The topology of the area (figure 12.4) can be represented by level curves 20 m apart, which show the inclination of the site. For the research, a steeper relief may represent a positive point when present in the vicinity of the area, as it may serve as a physical barrier to retain the aerial plume generated at the time of the explosion. Flatter areas may be closer to surface water bodies or groundwater, accelerating possible transport and contamination.

The regional climate is essentially rainy and hot tropical ('type A' according to Koeppen classification), ranging from 'almost hot' to 'hot' and from 'humid' to 'superhumid' with a dry season taking place during winter. Most of the yearly rainfall (85%) occurs between October and March. The proximity of the region to the coastal strip makes it subject to periodic and irregular maritime influence [12]. The average rainfall is 1225 mm, the driest months being in between May and September, which increases awareness and safety precautions in avoiding and controlling fire outbreaks. The entire area is part of the 17% remaining Atlantic Forest biome, which includes several associated ecosystems, such as dense ombrophiles forest, semideciduous forest, mangroves, altitude fields, swamps and plains [16].

The overall vegetation distribution in the area is shown in figures 12.5 and 12.6 (top left), which can be divided into dense or sparse. The dense area is determined by the presence of several large trees, whilst the sparse area is mainly undergrowth, predominantly grasses. Dense vegetation can retain both soil contaminants and prevent aerial plume dissipation, whilst ground vegetation may allow the boom to move away from the ammunition destruction area but may retain leachate.

Figure 12.3. Case study boundaries (Source: IBGE orthophoto).

Figure 12.4. Topographic assessment of the disposal area (scale 1:25 000) (image courtesy of 2M Topografia, 2013).

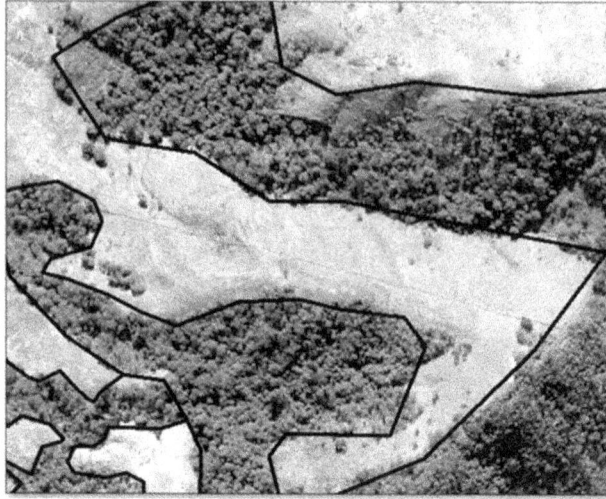

Figure 12.5. Vegetation classes: dense and sparse.

The Brazilian Forest Code determines how and where the Brazilian land can be explored and how the protected areas of native vegetation should be preserved. This legislation in Brazil complements guidelines for range management providing a list of mandatory safety distances (table 12.1). The array of safety distances helps secure the land, covered (or not) by native vegetation, with the environmental function of preserving water resources, landscape, geological stability and biodiversity, facilitating the genetic flow of fauna and flora, protecting the soil and ensuring the well-being of human populations.

Moreover, the Brazilian legislation also encourages the creation of environmental preservation units in some areas (known as 'Environmental Protected Area'), where the basic objective is 'to reconcile nature conservation with the sustainable use of a portion of its natural resources'. In particular for the case study, we can identify the Guandu Environmental Protected Area [15] (figure 12.6, top right) that was created to guarantee the quality and quantity of water from the Guandu River Basin, which supplies the whole metropolitan region of Rio de Janeiro. The top-right of figure 12.6 shows a topographic map that represents all land surface and what is on it through level curves every 20 m. This map can help to determine whether or not vegetation can be considered as a physical barrier in the dispersion of pollutants.

The slope for the case study has been plotted and is shown in figure 12.6 (bottom left). The greater the slope and the height of the relief, the more its importance around the area of ammunition destruction. In the case study we observed mainly two types of soil, which are classified by the Brazilian Soil Classification System [17, 18] as fluvial neosols for alluvial soils and red-yellow argosols for red-yellow podzolic (soils are shown in figure 12.6, bottom right).

In general, the fluvial soils occur in the floodplain, fluvial plains and alluvial terraces, along drains of the main hydrographic basins. These are non-hydro-morphic mineral soils of recent Quaternary sediments. In general, they present

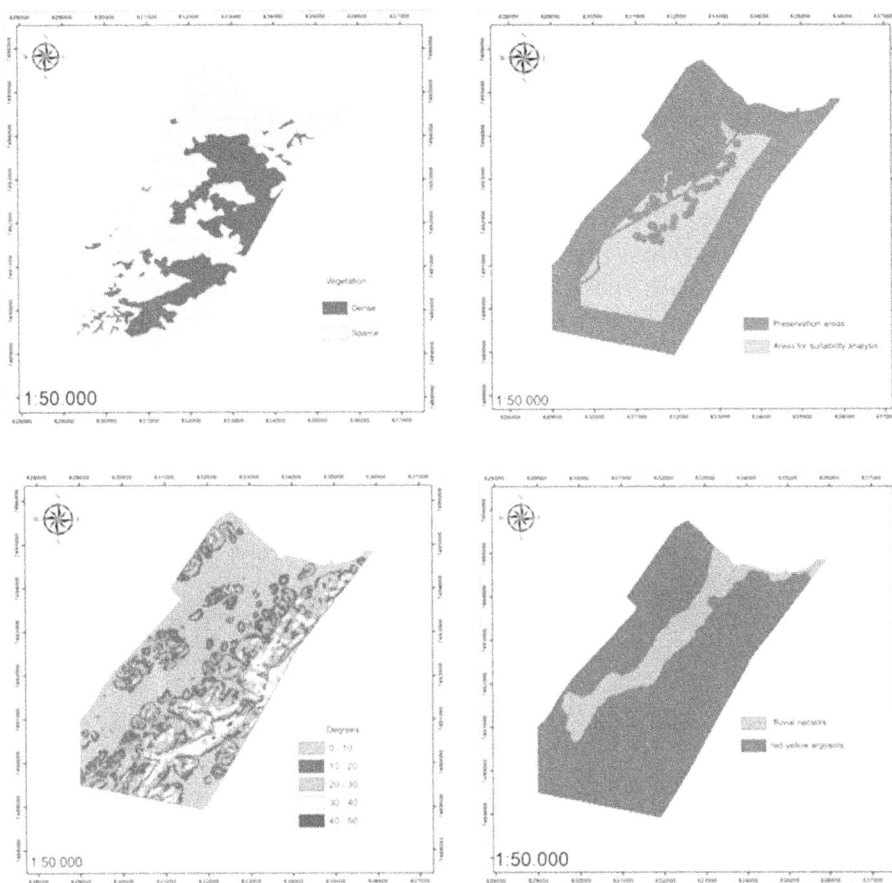

Figure 12.6. Case study sub-area maps: vegetation distribution map (top left), Guandu Environmental Protected Area (top right), declivity map (bottom left) and soil map (bottom right).

Table 12.1. Safety limits regarding environmental preservation.

Definition	Limits	Legislation
Water course less than 10 m wide	30 m	Brazilian Law n° 12.651, May 2012
Water course less than 20 m wide	100 m	Brazilian Law n° 12.651, May 2012
Ordnance warehouses	100 m	R-105 [1]
Houses, railways, roads and warehouses	700 m	R-105 [1]
Estate borders	700 m	R-105 [1]
Hills and mountains peaks	2/3 of the maximum height	Res INEA N° 93 from 24th October of 2014

diversified thickness and granulometry, along the soil profile, due to the diversity and the forms of deposition of the originating material. The differentiation between the layers is generally quite clear [17, 18]

The soil of the current ammunition destruction area [13] is found in red-yellow argosols. It has a general average of about 60% sand, 19% silt and 21% clay at 0–20 cm depth. At a depth of 20–40 cm the results were approximately 58% sand, 19% silt and 23% clay. In general, the soil of the area presents an average of 59% of sand and 22% of clay. Soils with this high percentage of the sand fraction usually present high permeability and leaching potential of pollutants; low water holding capacity; good aeration; low organic matter content; rapid decomposition of organic matter; low capacity to store nutrients; and low resistance to pH change.

Hydrogeological maps were elaborated in two ways: as a product of a specific mapping, in which the data are generated from that study, or as a thematic synthesis of pre-existing data such as geological and hydrological maps and reports, among others [19].

The essential data that the map should present for the study are depth, velocity, flow and direction of flow, as well as installation of wells to monitor the flow underground [14]. This information is considered very important and it is necessary to include it in other Army sites, since the available maps in the databases are not adequate and present poor image quality, due to the small-scale study.

In summary, the total estate area is of 22.5 km^2, in which 15.8 km^2 are considered preservation area and security land, and thus are not available for military activity. Therefore, only the 6.7 km^2 remaining area is suitable for disposing of ordnance material.

12.2.1 Methodology

The environmental assessment was carried out in steps: Firstly a hierarchical analysis method was applied to determine what variables should be considered, secondly we conducted a survey of possible contaminations, thirdly we assessed the fate and transport of contaminants in soil, groundwater and vegetation, and finally we ran calculations to predict how soil is affected by shock waves from the constant detonations.

The hierarchical analysis method was proposed by Thomas Saaty [20]. It assists the decision-making process and consists of facilitating human reasoning by comparing the elements of a dataset, i.e. comparing each element with the superior element using a scale with values from 1 to 9, standardizing and facilitating an understanding of all comparisons. Together with the analytic hierarchy process [20] and a variation of the hazard preliminary matrix analysis method [21–23], some spatial analysis procedures were adopted to generate results with the geographic component. The collected data were analyzed using ArcGIS 10.2.2 software [24].

This initial assessment was followed by an initial survey of possible contaminations [4] that informed soil and groundwater evaluations. The assessment of the water course contamination of the water courses was carried out through the installation of three main groundwater wells in the study area (figure 12.7). From each well a set of groundwater samples was gathered following a specific Brazilian environmental guideline [25]. The samples were analyzed for Al, Sb, Ba, Cd, Cu, Cr, Pb, Mn, Ni, Fe through mass spectrometry and high-performance liquid chromatography (HPLC). The samples were monitored from March 2014 to March 2016.

The flux and transportation of contaminants within the area were assessed indirectly, sampling and analyzing vegetation. For that purpose, a grid of sampling points was created based on cardinal directions, using linear transects and circular sampling units, following the effect of shock waves caused by the detonations. The leaves collected were extracted by the use of a Soxhlet solvent extractor [26–29], followed by chemical analysis in mass spectrometry, HPLC and gas chromatography mass spectrometry (GCMS).

The effect of shock waves from repeated detonation though the soil was assessed by a series of numerical calculations [30] carried out using VISED discrete element method software developed at IME [11].

12.2.2 Results and reports

From the initial hierarchy assessment and hazard preliminary analysis, it has been determined that variables such as soil (35.9%) and hydrogeology (29.6%) were the most important for the selection of sites to destroy unmanageable ammunition, followed by vegetation cover (16.5%), declivity (8, 4%) and topography (5.3%). For the case study it was verified that the area that is currently used to destroy unsuitable ammunition is located in an appropriate place and there is no requirement for change, but continuous monitoring is needed.

Figure 12.7. Groundwater wells' location within the area of study.

12.2.3 Polluting materials from ammunition detonation

An initial assessment of the disposal of ordnances by open burning and open detonation within the area of study [4] indicated a probable contamination by metals other than organic components, which are shown in table 12.2. The toxicity of these explosives can cause a range of symptoms from mild headaches or dermatitis, or even cause serious damage to internal organs.

This survey [4] reported the presence of heavy metals in the soil, the origin of which was traced back to ammunition components, such as ammunition kits, and in the initiators, which contain lead and probably mercury in the oldest ammunition. The mobility of metals from soil detonations varies with several factors, such as, for example, the Zn^{+2}, Cu^{+2}, Pb^{+2} and Cd^{+2} cations are highly hydratable. In acid and well drained soils, the relative mobility of these elements may vary. In general, cadmium, zinc, manganese and nickel present greater mobility than lead, copper and chromium. To further assess mobility of contaminants, a survey protocol of groundwater, soil and vegetation was implemented.

12.2.4 Soil survey

The soil assessment was initiated by a geotechnical, geophysical and chemical evaluation to assess the expected retention of heavy metals in the soil, by determining values for distribution coefficient (Kd) and other soil attributes such as pH. The results of the geophysical tests were efficient in the detection of

Table 12.2. Consolidated list of contaminants generated from open burning.

Explosives	Munitions
	Nickel alloys
Lead styminate ($C_6H_3O_9N_3Pb$)	Copper
Lead azide ($Pb(N_3)_2$)	Nickel and zinc
Mercury azide ($Hg(N_3)_2$)	Copper and zinc
mercury fulminate ($Hg(ONC)_2$)	Zinc
Black powder (KNO_3, C, S)	Tin
Calcium resinate	Lead
Barium peroxide	Antimony
Strontium nitrate	Hardened steel with chromium alloy and tungsten
Carbon tetrachloride	(Cr–W)
Polyvinyl chloride	Magnesium and molybdenum alloy
Toluidine red	Tombac (Cu and Zn)
Zinc magnesium stearate powder	Iron oxide
Torpex	Aluminum
TNT and aluminum	Brass 70/30
Magnesium and potassium perchlorate	Copper and zinc alloy
Barium nitrate	As well as organic polymers (plastics)
	Cast iron-tungsten

conductive anomalies in the most contaminated areas. The behavior and mobility of Cu and Pb metals were prioritized (Brazilian limits shown in table 12.3).

Moreover, the survey [12] confirmed the presence of copper with high transport risk while lead was trapped in the soil. We also observed that geophysics can assist in the delimitation of the impacted area; however, the electrical geophysical soil characteristic survey did not correlate with contamination levels, as shown in figure 12.8.

12.2.5 Vegetation survey

The main objective here was to evaluate the impact of detonations in terms of the transport of heavy metals through the air, by mapping this impact through the collection and analysis of vegetation at different distances from the burning and detonation area. Based on the analysis of different parts of the plant species and also the soil at the site the translocation of metals in plants was evaluated. The tests on plants were carried out at the Brazilian Agricultural Research Corporation (EMBRAPA) and the results were compared against the legal thresholds for contaminants in vegetation (table 12.4).

Table 12.3. Guideline values for soil contamination (mg kg^{-1}) [31, 32].

Metal	Reference values	Prevention limits	Safety threshold values		
			Agriculture	Residence	Industry
Lead	17	72	180	300	900
Copper	35	60	200	400	600
Cadmium	0.5	1.3	3.0	8.0	20.0
Chromium	40	75	150	300	400
Iron	—	—	—	—	—
Manganese	—	—	—	—	—
Nickel	13	30	70	100	130
Zinc	60	300	450	1000	2000

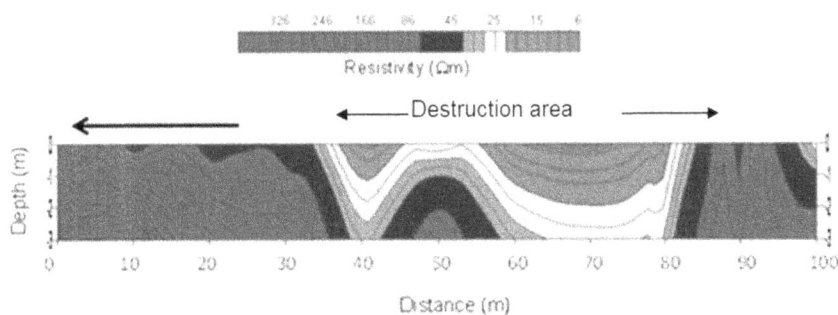

Figure 12.8. Electrical geophysical method section [12] COPPE/UFRU 2012.

The results produced a map (figure 12.9) of the spatial variability of lead, copper, cadmium, zinc, chromium and manganese. The Army site is distant to urban areas, without industries nearby, which could influence/mask the results of the tests, hence the results are only related to soil susceptibility, erosion processes, and atmospheric scattering of contaminants from the burning and detonation area.

The study suggests that contamination is restricted to a 'small' area of 500 × 500 m within the Army site. An important partial conclusion of this work is that the spatial location of the contaminants may vary from site to site because the environmental variations and conditions vary greatly across the different geographic conditions in Brazilian regions.

12.2.6 Propagation of shock waves

Another aspect that can impact the environment in Army units is the effect of shock waves from repeated detonation through the soil. Therefore, a series of numerical calculations [30] were carried out in an attempt to model the destruction of

Table 12.4. Heavy metals toxicity in vegetation [33].

Chemical element	Normal levels (ppm)	Toxic levels (ppm)
Lead	10	30
Copper	20	20
Cadmium	0.2	5
Zinc	100	100
Chromium	0.5	5
Nickel	5	10
Manganese	300	400
Iron	xxx	xxx

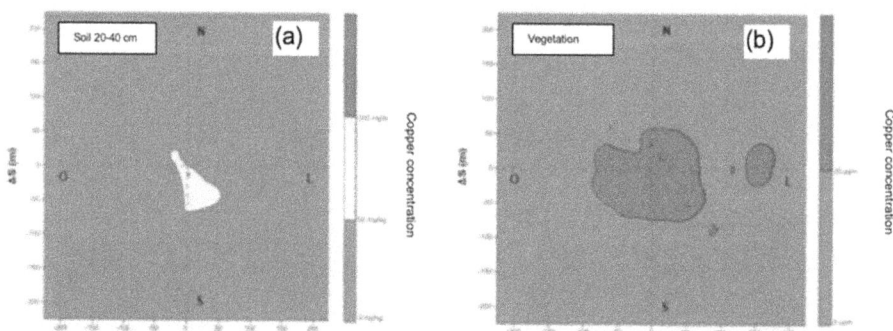

Figure 12.9. Spatial variability of copper metal: (a) in the soil − 20 to 40 cm depth and (b) in the vegetation. Reproduced from [13] COPPE/UFRU 2012.

ammunition in the soil. VISED software provides an estimation of the size of the crater created by the explosion, as well as stresses and deformations of the soil mass around the crater. The code used enables one to determine how the soil is physically affected in each detonation.

VISED software was used for the modeling of detonation waves [12], since for the explosion characteristics we were required to use the discrete elements method (DEM) instead of the finite element method (FEM). It was possible to qualitatively simulate the depth of the explosion (figure 12.10), allowing visualization of the whole process.

A partial conclusion of this numerical analysis is that it still needs to be further developed coupling DEM and FEM (to the deeper layers of soil and further from the detonation zone). Through this geophysical study, the need to understand the propagation of shock waves in the soil and its comprehensiveness is reinforced.

12.3 Conclusions

This study aimed to evaluate the level of heavy metal pollution in soil and surrounding vegetation at a particular military site in Brazil that is used for both storage and disposal of ammunition. It was implemented through a series of surveys, which included groundwater and vegetation analysis as a possible indicator of environmental pollution, followed by detonation shock wave effect calculations. Results showed that the site is contaminated by heavy metals cadmium, copper and lead in soil and that the vegetation showed toxicity by the elements manganese,

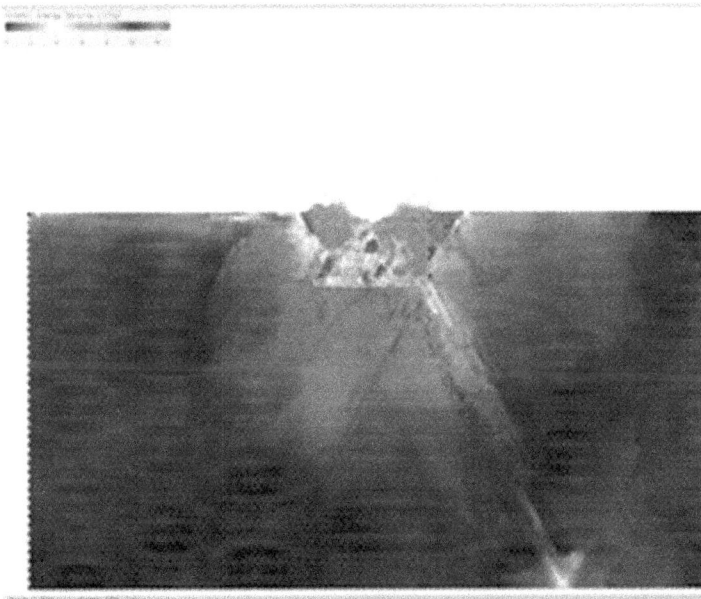

Figure 12.10. Crates formed by 1 g of TNT −1.5 m. Reproduced from [30] COPPE/UFRU 2013.

chromium, zinc, cadmium, copper and lead. The vegetation proved to be a good indicator of environmental pollution.

Maps of spatial variability of metal contamination in the study area, related to the total amount of trace elements found in soil and vegetation per linear transect, indicate a strong relationship between atmospheric dispersion of the pollutants and local topography, as well as with the orientation of winds at the site. Previous studies showed similar results regarding soil contamination, but the vegetation site was investigated for the first time.

The whole set of data was plotted (figure 12.11), as well as the current location of the ammunition destruction location. We can observe that the contaminants are well established at the military site. We were also able to stabilize contamination levels (indicated by the colors red, yellow and green in figure 12.11), which show that the contamination is contained within the Army site and it is very unlikely to affect the surrounding areas.

By analyzing the figure we can determine whether or not the disposal area must be transferred, as well as determine where to install monitoring points for soil and groundwater. Moreover, the survey studies [4, 11, 12, 14, 34] showed high

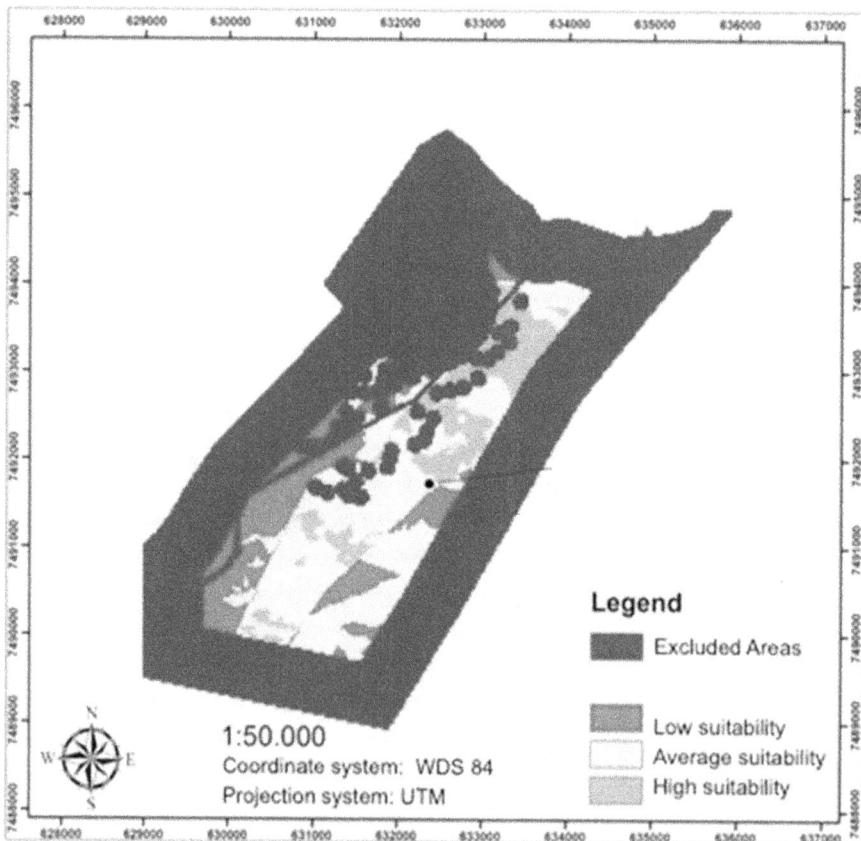

Figure 12.11. Suitable areas for disposal in the case study site.

contamination levels of lead and average levels of copper and cadmium inside the destruction area, hence it is not recommended to transfer the burning area to avoid spreading contamination; however, the site should monitor its groundwater and record meteorological conditions using stations. The body of evidence reported here also showed that the mass flux and transportation of metals in the soil and in the air can be measured through vegetation.

Finally, the data reported here was gathered between 2010 and 2018 through a series of Brazilian Army projects assessing this same site [4, 11–14, 34]. Further work could be implemented in other Army sites that conduct similar missions.

References

[1] Brazil, Regulamento para a Fiscalização de Produtos Controlados (R-105) Brasilia, Brazil: Brazil 2000

[2] USDoD (US Department of Defence) 1990 *TM 9-1300-214 - Technical Manual* (Washington, D.C: USA Department of the Army)

[3] US Army US Army Field Manual 5–250 https://www.bits.de/NRANEU/others/amd-us-archive/fm5-250%2892%29.pdf

[4] Brum T 2010 Environmental remediation of areas contaminated by explosives (in portuguese) MSc dissertation in Defense Engineering, Military Institute of Engineering, Rio de Janeiro, Brazil

[5] FRTR Federal Remediation Technologies Roundtable 2008 [Online] Available: http://frtr.gov/matrix2/section4/4-24.html [Accessed: 23-Mar-2019]

[6] USEPA Environmental Protection Agency 2014 [Online] Available: http://epa.gov/ [Accessed: 01-Jan-2014]

[7] SUMATECS 2015 Sustainable management of trace element contaminated soils— Development of a decision tool system and its evaluation for practical application - Project No. SN-01/20. Available at: https://snowmannetwork.com

[8] USA 1982 TM 9-1300-277 - General Instructions For Demilitarization/ Disposal Of Conventional Munitions Washington, USA http://miscpartsmanuals2.tpub.com/TM-9-1300-277/

[9] Shapira N I, Patterson J, Brown J and Noll K 1978 EPA-600/2-78-012 - State of the Art Study: Demilitarization of Conventional Munitions (Cincinnati, Ohio) Available at: https://babel.hathitrust.org/cgi/pt?id=ucl.31210012666564;view=1up;seq=1. [Accessed March 23, 2019]

[10] USEPA 1996 *U.S. Environmental Protection Agency - Soil Screening Guidance: User's Guide— EPA 540/R-96/018* (Washington, DC: Office of Solid Waste and Emergency Response)

[11] IME 2019 IME - Instituto Militar de Engenharia Rio de Janeiro, Brazil, Brazil

[12] da Silva A A D 2010 Study of heavy metals contamination in ammunition destruction area (in portuguese) Dissertation, Program of Civil Engineering, COPPE/UFRJ

[13] Xavier R B L 2012 Impact of the activity of ammunition destruction in surrounding vegetation - case study for heavy metals (in portuguese) Dissertation, Program of Civil Engineering, COPPE/UFRJ

[14] Castilho C de O 2015 Study of contamination by metals in water bodies in a field of destruction of ammunition and explosives (in portuguese) MSc dissertation in Defense Engineering, Military Institute of Engineering, Rio de Janeiro, Brazil

[15] INEA 2012 Bacia Hidrográfica dos Rios Guandu, da Guarda e Guandu-Mirim - Experiências para a gestão dos recursos hídricos [Online] Available: http://comiteguandu. org.br/conteudo/livroguandu2013.pdf [Accessed: 23-Mar-2019]

[16] INEA-CONSELHO-DIRETOR 2014 Resolução INEA no 93 de 24 de outubro de 2014. Estabelece a metodologia a ser utilizada para delimitação de área de preservação permanente de topo de morro no estado do Rio de Janeiro. Available at: http://www.inea.rj.gov.br/cs/ groups/public/documents/document/zwew/mdyy/~edisp/inea0062360.pdf. [Accessed March 23, 2019]

[17] Empresa Brasileira de Pesquisa Agropecuária - EMBRAPA 2015 Levantamento de Reconhecimento de Alta Intensidade dos Solos das Bacias Hidrográficas dos Rios Guapi-Macacu e Caceribu Report. Available at: https://www.embrapa.br/solos/busca-de-publicacoes/-/ publicacao/1039563/levantamento-de-reconhecimento-de-alta-intensidade-dos-solos-das-bacias-hidrograficas-dos-rios-guapi-macacu-e-caceribu [Accessed March 23, 2019]

[18] Empresa Brasileira de Pesquisa Agropecuária - EMBRAPA 2015 Neossolos Flúvicos Report [Online] Available: http://agencia.cnptia.embrapa.br/gestor/territorio_mata_sul_pernambu cana/arvore/CONT000gtq6c11e02wx7ha087apz24wkllak.html. [Accessed: 20-Sep-2001]

[19] CPRM 2015 Mapa Hidrogeológico do Brasil ao Milionésimo [Online] Available: http://cprm. gov.br/publique/cgi/cgilua.exe/sys/start.htm?infoid=756&sid=9 [Accessed: 01-Jan-2015]

[20] SAATY T L 1991 Método de Análise Hierárquica (São Paulo: MC Graw-Hill)

[21] USDoD 2000 (US Department of Defence), MIL STD 882-D - Standard Practice for System Safety, no. January 1993 (Washington, DC: USA)

[22] USDoD 2012 (US Department of Defence), MIL STD 882-E - Standard Practice for System Safety, no. February 2000 (Washington, DC: USA)

[23] USADoD 2017 Mil-Std-882-D - Standard Practice for System Safety no. February 1–17

[24] ESRI 2015 ArcGIS for Desktop Available at: http://www.esri.com/software/arcgis/arcgis-for-desktop. [Accessed March 23, 2019]

[25] CONAMA 1986 CONAMA No. 001 - Dispõe sobre critérios básicos e diretrizes gerais para a avaliação de impacto ambiental [Online] Available: http://www2.mma.gov.br/port/conama/ legiabre.cfm?codlegi=23 [Accessed: 23-Mar-2019]

[26] Luque de Castro M D and Priego-Capote F 2010 Soxhlet extraction: Past and present panacea *J. Chromatogr.* A **1217** 2383–89

[27] Koning S, Janssen H-G and Brinkman U A T 2009 Modern methods of sample preparation for GC analysis *Chromatographia* **69** 33–78

[28] Jensen W B 2007 The origin of the Soxhlet extractor *J. Chem. Educ.* **84** 1913

[29] USEPA 1995 EPA Method 3540C, Soxhlet Extraction USA

[30] Firmo T 2013 An initial study of the modeling of the destruction of unseen ammunition in the soil by the method of the discrete elements (in portuguese) Dissertation, Program of Civil Engineering, COPPE/UFRJ

[31] MMA 2014 MMA - Ministério do Meio Ambiente [Online]. Available: http://mma.gov.br/ conama/ [Accessed: 01-Jan-2014]

[32] CETESB, "CETESB" 2014 [Online]. Available: https://cetesb.sp.gov.br/normas-tecnicas-cetesb/ [Accessed: 23-Mar-2019]

[33] Kabata-Pendias A and Pendias H 1985 *Trace Elements in Soils and Plants* (Boca Raton, FL: CRC Press) 2001

[34] Marangoni C 2015 Methodology for the selection of areas of destruction of ammunition and inservable explosives (in portuguese) MSc dissertation in Defense Engineering, Military Institute of Engineering, Rio de Janeiro, Brazil.

IOP Publishing

Global Approaches to Environmental Management on Military Training Ranges

Tracey J Temple and Melissa K Ladyman

Chapter 13

Bushfire management (Australia)

L Brennan

The Australian Department of Defence manages 150 parcels of land that are prone to bushfires. These areas require careful management to ensure human life and property are protected in an environmentally sustainable way while maintaining defence capability. This chapter outlines the Australian Department of Defence approach to bushfire management, which must ensure safety, comply with governmental and Defence policy and consider public perception. It provides examples of management tools that are utilised by Defence to make informed decisions to protect life and property and when it should undertake training activities. The chapter also provides case studies of wildfire incidents given to demonstrate how improvements in land management can reduce environmental and safety risks. The chapter highlights some keys strategies and decision making methodologies that Defence uses to manage the bushfire threat on its Estate.

13.1 Background

Fire as a land management technique has been used in Australia for centuries by the Aboriginal people and more recently by European settlers. Aboriginal people use cultural burning to enhance the wellbeing of the land and its people. Generally this would mean supporting increased biodiversity, promoting mosaics of vegetation, hazard reduction, and ceremonial purposes and increasing amenity for people. 'Burning Off' may be implicated in the extinction of some fire-sensitive species of plant and animals dependent upon infrequently burnt habitats, and is widely believed to have maintained structurally open vegetation such as grassland and also extended the range of fire-adapted species such as Eucalyptus into environments climatically suitable for rain forests [1].

Australia is characterised by dry vegetation which is easily burnt. The weather in Northern Australia is quite different to that of Southern Australia. In Northern

Australia the climate is characterised by the annual monsoon season, which extends from November/December through to March/April, the wet season. The remainder of the year is characterised by warm dry conditions, the dry season. The fire season in the north is defined by the period during which the savannah, the predominant biome, becomes fully cured and will burn readily and generally lasts from April/May to the commencement of the wet season [2].

In Southern Australia the peak fire season is in the summer months as it experiences cool winters and warm to hot summers, with rainfall predominantly in winter and spring. In some parts of Australia there is pronounced season summer/autumn drought, while other parts of the country such as Tasmania experience rain all year round [2]. Unusual rainy periods or droughts can alter the timing and severity of the fire season [3]. As a result of the climatic conditions of Australia, the timing of the fire seasons varies greatly across the country (figure 13.1).

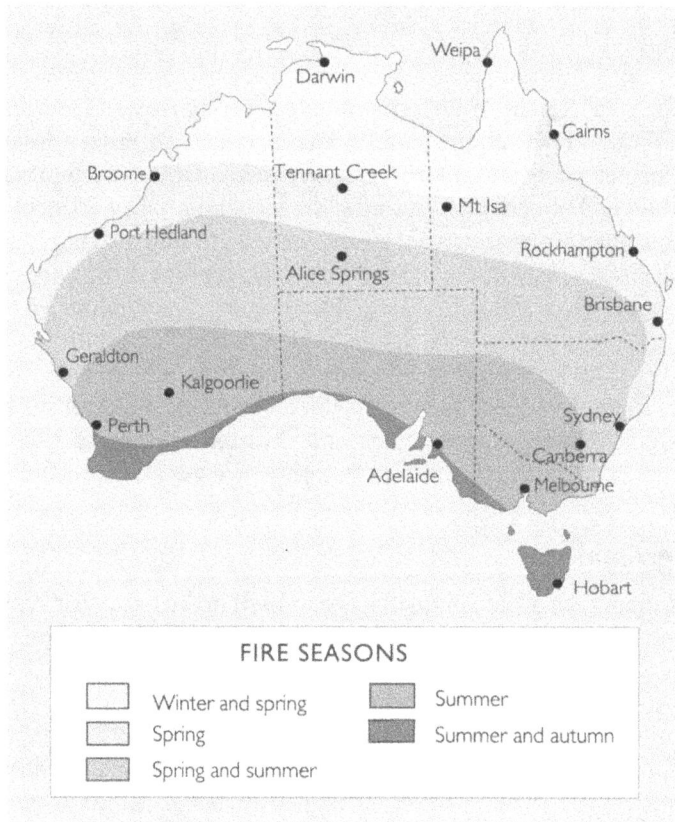

Figure 13.1. Map of fire seasons across Australia. Reproduced by permission of Bureau of Meteorology, © 2018 Commonwealth of Australia.

13.2 Outline of the Defence Estate

In Australia the Department of Defence (Defence) is in a unique situation where it owns and leases large and small parcels of land in isolated areas and abutting urban development. One hundred and fifty of those parcels of land are bushfire-prone properties. As such, Defence has small and large landholder responsibilities, employer responsibilities, and high-risk activity manager and capability management responsibilities.

Defence approach to risk management is mandated by a Joint Directive by the Chief of Defence Force and Secretary of Defence to be integrated into all planning, approval, review and implementation processes, at all levels, to ensure that risk is one of the major considerations in decision making.

Defence's approach to risk management from bushfire hazards has been developed to establish a clear line of accountability, a clear decision making process and an ability to effectively identify and manage corporate and operational level risks.

Environmental risk assessments including any proposed controls or treatments shall be critically reviewed by the Risk Management Governance Board (RMGB) to ensure that risk assessment processes reflect Defence policy, procedures and guidelines. A challenge which Defence faces is urban encroachment that increases the consequences if a bushfire was to leave a Defence base or training area and cause severe damage to property or cause severe injury or death to people. Furthermore, urban encroachment also poses significant risk to Defence whereby its infrastructure and personnel could be put at risk as a result of a bushfire entering a Defence base or training area from neighbouring land.

Given this management paradigm there is a heightened risk that the way Defence undertakes bushfire management will be increasingly criticised by the surrounding community and this criticism may affect Defence's reputation. Therefore, a key consideration for Defence is how will routine bushfire management measures such as burning off be perceived by the local community. Consequently, Defence gives consideration to the proposed bushfire mitigation measures that it intends to implement and determines how Defence's actions will affect surrounding neighbours or be perceived by the public/local community and where possible Defence aims to minimise any impact on the local community.

There have been times where bushfires have left a Defence Base or Training Area and impacted on the local community. There have also been occasions where bushfires have entered a Defence Base or Training Area due to actions undertaken outside of the Defence Estate. An example is where a live-fire exercise at the Marrangaroo Training Area (MTA) resulted in a bushfire on the Defence Estate and was not able to be contained within the MTA and impacted an area of approximately 56 590 ha on and off the MTA in 2013. The fire is known as the State Mine Fire and further information is provided on this fire in a case study later on in this chapter. Subsequently, Defence has been constantly reviewing and improving the way it manages bushfire to optimise the best outcome for the people, assets, Defence capability and the environment both within and adjoining Defence property.

13.3 Defence bushfire management policy

Bushfire-prone areas expose Defence personnel, capability, contractors and local communities to varying degrees of risk. Defence processes and procedures aim ensure that these risks are managed to be as low as is reasonable practicable to minimise or prevent the detrimental impacts of bushfire, while recognising the fundamental role that bushfire plays in the Australian environment. Defence's Bushfire Management approach is aligned with the Australian Government's National Bushfire Policy Statement for Forests and Rangelands (2014) [4]. The vision of the Policy is that:

> *'Fire regimes are effectively managed to maintain and enhance the protection of human life and property, and the health, biodiversity, tourism, recreation and production benefits derived from Australia's forest and rangelands.'*

This policy statement sets out principles, strategic objectives and national goals for Bushfire Management. The principles establish protection of human life as the highest consideration, and emphasise the importance of bushfire risk awareness, preparedness and proactive planning for bushfire survival. The notion of 'shared responsibility' is emphasised, whereby fire and emergency services, bushfire-prone property owners, and individuals living, working or visiting bushfire-prone areas all have important parts to play in managing the bushfire risk. The need to actively and adaptively manage the land with fire is highlighted to maximise environmental benefits to ecosystems' resilience and reduce the adverse impact of severe bushfires.

Defence has also adopted the philosophy of 'Prepare, Act, Survive' bushfire awareness, a warning and safety framework that is used by Australia's land and fire management agencies. A critical element of the 'Prepare, Act, Survive' framework is that communities (Defence and Civilian) understand local bushfire risks, know how to mitigate those risks, prepare for the fire season, and respond to bushfire incidents in accordance with pre-planned bushfire survival plans.

In Australia every employer and employee is required to abide by the Workplace Health and Safety (WHS) Act 2011 (legislation). The WHS Act requires that hazards are identified and all reasonable practical actions are taken to eliminate or minimise risks to ensure a safe work environment. Consequently, Defence has developed a bushfire policy that establishes a framework for assessing and managing risk arising from bushfire hazards. Defence implementation of appropriate mitigation activities is consistent with Australian codes of practice; How to Manage Work Health and Safety Risks [5], Managing the Work Environment and Facilities [6] and Work Health and Safety Consultation, and Cooperation and Coordination [7].

Defence objectives for bushfire management are to:
  (a) Protect human life;
  (b) Protect Defence and civilian property and assets;
  (c) Support Australia Defence Force training;
  (d) Promote proactive, environmentally sustainable management of bushfire.

The Defence bushfire management approach incorporates an assessment of each property owned or leased by Defence to determine if it is bushfire prone.

A Defence property is designated as bushfire prone if:

(a) It includes areas mapped as bushfire prone by a local government or local fire authority;

(b) If the property has not been assessed by local authorities, it includes areas that would meet the definitions of class 1, 2 or 3 bushfire-prone land as set out in the NSW Rural Fire Service Guide for Bush Fire Prone Land Mapping (2015) [8].

All bushfire-prone properties are subject to an overall site bushfire risk (OSBR) assessment to determine the degree of bushfire management planning required for the property. Once assessed, the OSBR is recorded and then only requires review when a substantive change occurs in the parameters contributing to the assessment. The criteria to determine OSBR are:

Low OSBR properties, which include:

(a) Small arms ranges where the range, safety template and immediate surroundings constitute the entire property and are the sole function of the property;

(b) Properties where live-fire activities do not occur, and either; there are less than 5 built assets in fire prone vegetation of any class, or only class 2 vegetation (grassland) occurs, or urban fringe buildings where evacuation or fire response are entirely controlled by civilian authorities.

Medium OSBR properties, which include:

(a) Training areas on which Defence live-fire activities occur, but which only contain basic assets such as camp, range control and firing ranges;

(b) Bases with no attached training area that only include restricted areas of class 1 or 3 bushfire-prone vegetation (less than 10 ha total with no patches greater than 2 ha).

High OSBR properties, which include:

(a) Military area incorporating both training areas and cantonments, or bases with significant areas of bushfire-prone vegetation adjoining assets (>10 ha, or with multiple patches 2–10 ha);

(b) For medium training areas, but with medium or high density residential properties or fire-sensitive high-value economic assets such as forestry plantations adjoining boundary.

Those properties identified as having a medium or high OSBR must be included within a bushfire management plan (BMP). The BMP shall include a bushfire risk management plan prepared by an independent external bushfire expert and is to cover all medium and high-risk properties within a base service area. Bushfire management plans are reviewed by a number of key stakeholders but are accepted by the base support manager responsible for that particular property. A Defence BMP is to:

(a) Document the extent and type of bushfire hazard(s);

(b) Identify and remedy issues arising from implementation of previous BMP;

(c) Document fire history since the last BMP was prepared;

(d) Document site-specific roles and responsibilities;

(e) Document and assess the risk associated with bushfire hazard(s);

(f) Identify and document the risk owner for each risk;

(g) Nominate performance standards for mitigation work required to treat risks;

(h) Identify and document the response owner for each risk treatment;

(i) Document a proposed five-year mitigation works programme to implement risk treatments;

(j) For training areas, prepare a bushfire prevention plan that details specific procedural and other mitigations to minimise the risk of Defence activities igniting a fire and impacting identified assets and values at risk;

(k) Develop prepare act service materials;

(l) Provide as relevant to each property a map depicting required bushfire mitigation work;

(m) Provide as relevant to each property a map depicting infrastructure and hazards, including the unexploded ordnance (UXO) risk, relevant to conducting bushfire operations;

(n) BMP shall be reviewed every five years to identify components requiring update;

(o) Any risks identified as carrying a high or very high inherent risk shall be reviewed by the risk owner prior to the commencement of each fire danger period to assess whether treatments are in place and whether residual risk is acceptable, or as low as reasonably practical, given seasonal and other constraints.

13.4 Case study Marrangaroo/State Mine Fire 2013

An example of a live-fire exercise that had a tremendous impact on a military training area and the local community was the Marrangaroo/State Mine Fire of 2013. On 16 October 2013 an explosive demolition serial was conducted on the internal demo-litions range at Marrangaroo Training Area (MTA), (NSW New South Wales, Australia) to dispose of eight surplus 84mm high explosive anti-tank (84mm HEAT) rounds. The demolition serial consisted of two separate stacks each of four 84mm HEAT rounds, which were to be disposed of by initiating plastic explosive/primer sheet. This disposal activity was conducted just before midday. Shortly after, a staff member who was conducting a safety clearance of the demolition site on the internal range noticed a small fire some 25 m to the east of the point of demolition [9].

The member commenced fighting the fire by stomping on it and called for assistance in suppressing the fire. Eventually, five members and a 'Stryker' unit (a utility vehicle with a tank, pump and hose reel along with a number of shovels, beaters and a knapsack) were in attendance at the internal range. Despite the efforts made, the fire continued to grow in intensity and size in the heavily wooded area surrounding the point of ignition [9].

Efforts to extinguish the fire were terminated when UXO present in the vicinity of the internal range began exploding and deemed too dangerous for personnel to remain in the vicinity of the fire. The fire then spread rapidly within and beyond MTA, causing destruction of bushland and property [9].

The local fire service was alerted to the fire and attended MTA. However, they did not deploy any firefighting assets because of the risk of injury from UXO. The fire impacted 56 590 ha on and off the MTA [9].

The Chief of the Defence Force undertook a Commission of Inquiry into the fire at MTA [9]. Some of the recommendations from the inquiry included:

- Range control officer Marrangaroo training area liaise with the local Rural Fire Service unit to develop a map indicating the areas of the range likely to contain unexploded ordnance.
- The development and implementation of a memorandum of understanding (MOU) between Defence and the Rural Fire Service.
- Fire danger ratings are not a suitable mechanism for determining whether live-firing should or should not proceed on the ranges within Marrangaroo training area and that the forest fire danger index was a more appropriate mechanism. It was not possible for the Commission of Inquiry to determine what the exact threshold should be for such practices, nor to recommend how such information should be gathered.
- MTA standing orders should be reviewed to impose a requirement that the Officer in Charge of any live-firing practice ascertain and consider current weather parameters, temperature, humidity, wind strength and direction. The setting of those parameters and their limits should be decided in consultation with the Bureau of Meteorology and Rural Fire Services and inserted into range standing orders.
- Defence engage with both the Bureau of Meteorology and the Rural Fire Service to determine a more suitable index system.

The Recommendations reported above have been implemented resulting in Defence improving the way it manages the bushfire risk on the Estate.

13.4.1 Implementations of automatic weather stations

One of the recommendations from the Marrangaroo Commission of Inquiry was that better decisions could have been made if the range control officer (RCO) had reliable local weather data instead of relying on the weather data from the local fire district to make decisions at the site level.

To help rectify this situation, Defence formalised a partnership with the Australian Bureau of Meteorology (BOM) to provide 21 automatic weather stations (AWS) across 15 Defence training areas. To inform the location of the AWS, Defence undertook a risk assessment to identify which sites were most at risk from bushfire. The agreement between BOM and Defence outlined that Defence would make available the space required for the AWS on the training areas and purchase the automated weather station equipment with the BOM installing stations, maintaining the instruments, analysing historical weather data trends and presenting the weather data on an internet dashboard page in real time.

The AWS provide real time data that includes:

- Wind direction (degrees) by wind vane at 10 m (and direction at peak gust speeds).
- Wind speed (measured in knots but can be displayed in knots and km/h) by cup anemometer at 10 m (and wind gusts).
- Air temperature (°C) (in solar radiation screen at standard height of 1.1 m).
- Humidity (%RH) (in solar radiation screen at standard height of 1.1 m).
- Rainfall (mm) (continuous by tipping bucket rain gauge, at standard 300 mm above ground).
- Pressure (hPA) (digital barometer).

Each AWS has scheduled instrument checks conducted by BOM staff/contractors to ensure they are operating within approved tolerances.

BOM converts the raw AWS data into a weather dashboard with easily viewable screens to provide the RCO with easily interpretable information. This assists the RCO in making better informed decisions about likely bushfire behaviour and subsequent risks for the training area. This helps inform decisions regarding the conduct of live-fire activities. There are three main pages of the Defence weather dashboard site [10] provided by BOM, they include:

1. Fire danger rating—with forecasted wind speed in either km/h or knots, drought factor, maximum forest fire danger index, curing and maximum forest danger index, figure 13.2 below.
2. Meteogram—provides relative humidity (%), temperature (°C) and wind speed (kph) in graphical form, figure 13.3 below. It is a forecasting tool and past data is not shown.
3. Fire danger rating depicted in map format, figure 13.4.

In figure 13.1, there is a table with four values, the drought factor (DF) value 5.6, is the drought factor number, which is a numerical value that estimates the proportion of fine fuels in a forest (such as twigs and leaves) that will burn under current conditions. The higher the DF, the higher the fire danger (value: 0 moist–10 dry). Drought factor values are used in the calculation for the forest fire danger index (FFDI) [10].

The next number to the right of DF is max FFDI with a value of 2. Max FFDI is the maximum forest fire danger index, which is the maximum value for the day (from midnight to midnight). FFDI is a measure of the degree of danger of fire in Australian forests. A relative number denoting the potential rates of spread, or suppression difficulty for the specific combinations of temperature, relative humidity, drought effects, recent rainfall and wind speed (values between 0 and 100 +) [10].

Figure 13.2. Fire danger rating. Reproduced by permission of Bureau of Meteorology, © 2018 Commonwealth of Australia.

Below and left of max FFDI is curing with a value of 62%. Curing describes the annual or seasonal cycle of grasses dying and drying out (or browning). The higher the proportion of cured material in grasslands, the higher the fire danger risks. Curing assessments are based on a combination of visual estimates (updated by field observers) and satellite imagery. Curing values are used in the calculation of the grassland fire danger index (GFDI). Curing values range from 0% (moist)—100% (dry) [10].

The last cell in the table is max GFDI, which is the maximum value for the GFDI for the day (from midnight to midnight). GFDI is a measure of the degree of danger of fire in Australian grassland. This is a relative number denoting the potential rates of spread, or suppression difficult for specific combinations of temperature, relative humidity, wind speed, curing and fuel load, with a value between 1 and 150+ [10].

There are several benefits as a result of the agreement between Defence and the BOM. There is a good working relation between the BOM and Defence and this ensures that issues and obstacles are dealt with in a timely manner and resources can be shared between each department. The BOM already manages the majority of Australia's weather information and therefore it makes sense that Defence would

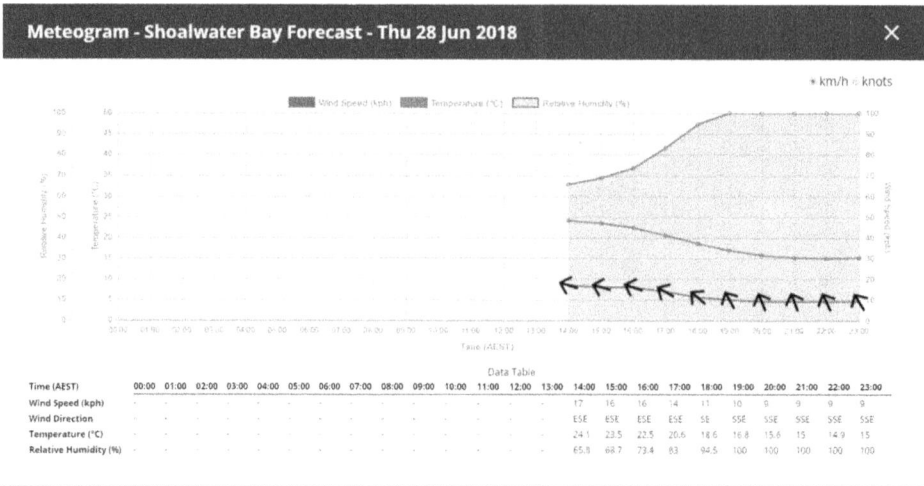

Figure 13.3. Meteogram. Reproduced by permission of Bureau of Meteorology, © 2018 Commonwealth of Australia. (Meteogram is a forecasting tool and past data is not shown.)

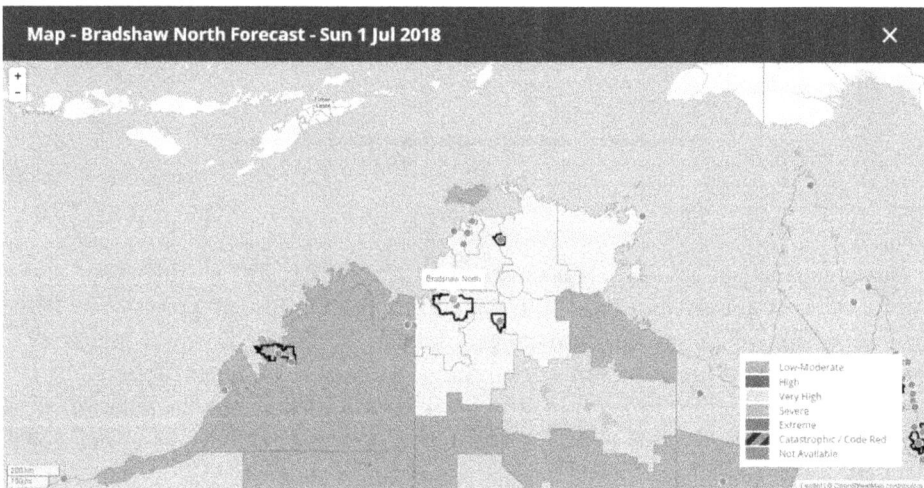

Figure 13.4. Fire danger rating map. Reproduced by permission of Bureau of Meteorology, © 2018 Commonwealth of Australia.

rely on their knowledge and experience. The BOM is able to provide accurate weather data updated every 30 min and this allows Defence to make informed decisions on how military training areas and ranges are managed and what activities can be undertaken at any time throughout the day. The agreement with the BOM also provides the added advantage of having approximately 100 years of weather data that can be interrogated to determine weather trends, aid weather predictions and identify weather that increases the risk of bushfire.

This solid working relationship with BOM enables Defence to access accurate short-term weather forecasts at each site; this enables more efficient management of Defence training areas and assists the range control staff and users to make informed decisions as to when live-firing activities can be conducted safely. The integration of the AWS allows Defence to monitor the local weather conditions whilst an activity is being conducted and to quickly cease the activity if the weather changes and creates an unsafe environment for conduct of that activity. Overall, the AWS combined with the BOM internet dashboard page [10] provide Defence with a crucial tool that supports the safe and efficient operation and management of Defence training areas and ranges.

13.4.2 Memorandum of understanding agreements

In Australia the Fire and Rescue services are operated by the five State and two Territory Governments, who provide these services to their local communities. The Defence Estate (including training areas) is managed by the Commonwealth Government. The Commonwealth Government does not have the necessary Fire and Rescue capability required to manage most of its sites. A majority of the Defence Estate have a basic form of first response firefighting capability provided by a contracted service provider. However, if they can't control the fire they will rely on the State or Territory fire services to assist in controlling and extinguishing a fire, similarly to any other landholder. Therefore, the Commonwealth Government provides funding to the State and Territory Governments for them to provide Fire and Rescue services at Defence Sites. The mechanism under which the Commonwealth Government formalises this arrangement is through a MOU, which is signed by the Commonwealth Government and a representative from each State and Territory responsible for providing the Fire and Rescue Service in their State or Territory.

Defence also uses a Mutual Aid Agreement (MAA) with the local fire authority for particular sites. The MAA formalises a relationship between Defence and the local Fire Authority and aims to document a cooperative framework and agreed commitments between Defence and the Fire and Rescue Service provider in relation to the management of bushfires, bushfire hazard mitigation activities and agreed bushfire response for fires on or threatening a Defence Property.

The key purpose of the MAA is to ensure that each agency understands and accepts local inter-agency arrangements for the operational aspects of bushfire management and the conduct of joint bushfire response exercises at the Defence site. This ensures that both Defence and Fire and Rescue Service provider personnel understand what operational support may or may not be available to manage bushfires and where the conduct joint exercises prior to a bushfire incident bears out this information and understanding.

The MAA with the local Fire and Rescue services are essential for the Department of Defence to maximise the chance of mitigating and a extinguishing a bushfire on the Defence Estate.

13.4.3 Wildfire competency for range control officers

As a result of the Defence Commission of Inquiry into the Marrangaroo Training Area Fire, Defence is committed to providing training to RCO. This will ensure they have sufficient knowledge and skills to manage the day to day operations on a Defence training area in a safe and environmental sustainable way. During their induction training RCO and their assistants are required to gain Competency PUAFIR512 'Develop and Analyse the Behaviour and Suppression Options for a Level 2 Wildfire'. This unit covers the competency required to provide an analysis of the spread and behaviour of an intermediate wildfire and to prepare fire suppression options that are appropriate for the expected fire behaviour. The assessment can be achieved either by the RCO being assessed for the analysis of a hypothetical fire or work undertaken during a real fire event.

The elements of the PAUFIR512, which the RCO is required to perform competently include [11]:

1. Analyse factors impacting on the spread and behaviour of an intermediate wildfire and develop and incident prediction.
   - Includes information on the current and future spread and fire behaviour is collected from a range of sources and recorded.
   - Analysis is conducted using consideration of fuels and fuel assessment, weather analysis, the effects of topography and likely resultant fire behaviour.
   - Fire Prediction tools and references are effectively utilised in the analysis of fire spread and behaviour.
   - Results of the fire behaviour analysis are validated against fire observations as they become available.
2. Develop maps and data, and maintain associated information regarding projected fire spread and behaviour.
   - Necessary map information and data is prepared.
   - Fire spread and behaviour projections are developed in a manner appropriate to the incident.
   - Use of information in planning the control of the incident is facilitated through quality, timeliness and presentation of the information.
   - Fire spread and fire behaviour projections are updated as new weather and fire information becomes available.
3. Analyse and communicate key risks of the projected fire spread and behaviour.
   - Site information is sought from agency databases or experts.
   - Area and timing of potential future impacts of the fire is projected.
   - Key risks of the fire to human, economic and environmental assets are considered.

- Fire and weather are monitored to assess if or when fire danger is likely to suddenly increase.
4. Prepare and analyse a range of fire suppression options consistent with incident objectives.
   - Range of options with an analysis of probable level of success and consequences of failure is prepared for consideration by the incident management team.
   - Time available and the threshold fire behaviour for which each strategy and tactic is likely to be effective are considered.
   - Time available and the threshold fire behaviour conditions (due to fuel, weather, topography, and fire size) are considered.
   - Advice and analysis are provided to the incident management team to assist in development of strategies and fallback strategies.

Participants are required to undertake a formative and summative assessment on their induction course, which is required in order for the RCO to demonstrate competency in the unit. The assessment must confirm the ability of the RCO to provide:

- Accurate analysis and projection of fire spread and fire behaviour, indicating probable and possible scenarios.
- Analysis of a range of appropriate fire suppression options.

As outlined above, the method of assessment of the competency is usually via direct observation in a training environment. Defence provides this training environment through the annual Directorate of Operations and Training Area Management (DOTAM) Range and Training Area Operations Course (RTAOC).

Assessment is completed using appropriately qualified assessors who select the most appropriate method of assessment.

Assessment may occur in an operational environment or in an agency-approved simulated work environment. Forms of assessment that are typically used include:

- Direct observation;
- Interviewing the candidate;
- Journals and workplace documentation;
- Third party report from supervisors;
- Written or oral questions.

The competency defines fire behaviour, sources, fuels and fuel assessment, weather analysis, effect of topography on fire behaviour, fire predictions tools and resources, map information and data, risks, and human, economic and environmental assets in table 13.1 below.

Defence is committed to providing education to its training area and range staff so they are equipped with the tools and knowledge to manage and mitigate the bushfire risk at the training areas and range for which they are responsible.

Table 13.1. Definitions of fire terms for competency (Commonwealth of Australia 2013).

Category	Examples/items
Fire behaviour must include:	• fire perimeter • fire size/growth/shape • fire whirls • flame characteristics (height and depth) • heat output and intensity • junction zones • rate of spread • smoke • spotting
Sources may include:	• air or ground observations • automated weather stations • Bureau of Meteorology websites and/or fire weather experts • fire history maps • fuel type maps • fire ground information, operational situation reports and infra-red scans • geographic information systems (GIS) and agency site-related databases • land managers • persons with local knowledge
Consideration of fuels and fuel assessment may include:	• bark fine fuels • canopy fine fuels • coarse fuels • coarse standing fuels • coarse surface fuels • dead course fuel moisture • dead fine fuel moisture • elevated fine fuels • fine fuels • fuel and fire behaviour • live fuel moisture • moisture content assessment • near surface fine fuels • surface fine fuels • total fuel load
Weather analysis may include:	• atmospheric stability • Bureau of Meteorology products and tools • calculation of fire danger ratings • cold fronts • diurnal cycles • droughts • Foehn winds • Katabatic and Anabatic winds

	• long-term weather cycles
	• relative humidity and dew point temperature
	• sea breezes and land breezes
	• seasonal cycles
	• short-term and local weather effects
	• temperature
	• temperature inversions
	• wind gustiness and directional variation
	• wind speed and direction
Effects of topography on fire behaviour must include:	• acceleration effects
	• dry upper winds—mixing/range effect
	• drought index and drought factor
	• fuel distribution
	• elevation
	• rockiness/continuity
	• land form (channelling)
	• slope and aspect
Fire prediction tools and references may include:	• CSIRO (McArthur) forest fire danger meter
	• CSIRO (McArthur) grassland fire danger meter
	• CSIRO (McArthur) grassland fire spread meter
	• Vesta fire model
	• WA forest fire behaviour tables
	• overall fuel hazard guide (DSE, 1999)
	• other fuel specific fire behaviour prediction systems (such as buttongrass in Tasmania, mallee-heath model, spinifex model)
Map information and data may include:	• maps of fire spread, estimated at time intervals as required by the incident management team, with separate mapping for probable and possible scenarios
	• narrative regarding limitations, assumptions, prediction uncertainties and other comment to assist in the interpretation of the data
	• Victoria fire behaviour estimates
Risks may include:	• operational risk
	• public safety risk
	• risks to public and private assets
	• economic risk
	• environmental risk
	• legal risk
	• technical risk
	• political risk

(Continued)

Table 13.1. (*Continued*)

Category	Examples/items
Human, economic and environmental assets may include:	• areas of environmental or conservation value • areas of tourism value • crops and farm assets • historic sites • indigenous cultural sites • key infrastructure such as a major bridge or power transmission lines • plantations • private or public buildings • towns or settlements • water catchments

13.5 Service delivery model for bushfire management

The Department of Defence in Australia operates in a diverse and complex environment. Consequently Defence governs its environmental footprint by outsourcing that management to skilled contractors. This model allows Defence to engage skilled contractors to focus on the management of bushfire risk on the Estate, including its training areas, while Defence focuses on meeting the operational requirements set by the Government of Australia. The Australian environment and the associated weather which Australia experiences is diverse and can change dramatically from year to year and season to season. These conditions create narrow windows where bushfire management tasks can be completed. Consequently, a key requirement for Defence and its Bushfire Management contractors is the flexibility to adapt and react to the current site conditions rather then follow a schedule of works which does not consider the local environment nor weather conditions. This flexibility enables Defence to make the most of this small time window and allows bushfire management measures to be undertaken to protect the Defence Estate and the local community at the optimum time.

Bushfire Management of the Defence Estate is focused on the following mitigations measures:

(a) Identification of very high and high bushfire risks;
(b) Identification of mitigations works to reduce the bushfire risk;
(c) Timely delivery mitigation works in preparedness for the fire season.

13.6 Bushfire Management Yampi Sound Training Area

In 2016, Defence engaged the Australian Wildlife Conservancy (AWC) to deliver science-based land management at Yampi Sound Training Area. Located in the Kimberley region of Western Australia, 200km north-east of Broome, Yampi Sound Training Area covers an area of 568 000 ha including over 700 km of coastline and is a hotspot for endangered and endemic wildlife. Yampi incorporates three different bioregions: towering escarpment stand guard over long, hidden valleys decorated by pockets of rainforest, tropical streams cascade over waterfalls and through chains of rock pools before meandering through savannah woodland until they reach the coast and rugged sandstone and basalt ranges extend into the ocean, creating an intricate pattern of bays and inlets flanked by dense mangrove forests. Yampi Sound Training Area protects a wide range of endemic, threatened and iconic fauna, such as Northern Guoll, Golden-backed Tree-rat, Monjon (smallest of the rock-wallabies), Western Partridge Pigeon, Black Grasswren, Gouldian Finch and Orange Leaf-nosed Bat [12].

The fire history of Yampi has been diverse, for millennia aboriginal people in northern Australia implemented a regime of frequent, mostly small fires, lit from early in the dry season. This regime resulted in a fine-scale mosaic of vegetation of different ages [13].

In recent decades following the cessation of traditional burning, northern Australia has become dominated by a wildfire (an intensive fire late in the dry season) regime, which is characterised by relatively extensive hot fires late in the dry season. Consequently, late dry season fires impact wildlife by removing sources of food and flora cover, which increase predation rates from feral cats. Wildfire can have a detrimental impact on fire-sensitive vegetation such as rainforest. Under the management of AWC, the Yampi Sound Training Area will change its fire regime to one carried out early in the dry season with an aim to reduce the severity of wildfire [13].

During the first year of AWC managing fire at Yampi they doubled the amount of early burning that was conducted, hence the extent of late dry season wildfire was reduced to approximately half compared to previous years [13].

The minimum burning methodology at Yampi Sound Training Area during 2018 [13] will include:
- 4000 km of aerial incendiary[1] operations;
- Delivery of at least 12 000 incendiaries;
- 10 person days of ground burning operations;
- More incendiaries and or more days on the ground may occur depending on the amount of dry vegetation.

13.7 Conclusion

Bushfire is a common part of the Australian environment and will always be something Defence is required to manage to protect life and property. The examples provided in chapter are just some of the issues/improvements Defence has made to improve how it manages its bushfire risk. Given the nature and consequences of bushfires in Australia, Defence will constantly review how a bushfire occurred and determine if management strategies can be put in place to reduce the severity of the bushfire risk. Defence is committed to determining and applying best practice bushfire management techniques to manage the bushfire risk on its Estate.

---

[1] Incendiaries are devices dropped from an aircraft (usually a helicopter) to initiate controlled burning of grass and undergrowth. The operation involves the injection of a capsule or ball containing potassium permanganate with a quantity of ethylene glycol immediately prior to ejection from the aircraft. Spontaneous combustion occurs within 30–40 s from injection allowing the incendiary time to reach the ground before ignition [14].

References

[1] Bowman D 1998 Tansley Review No. 101 The impact of Aboriginal landscape burning on the Australia biota *New Phytol. J.* **140** 385–410

[2] Sullivan A L, McCaw W L, Cruz M G, Matthews S and Ellis P F 2012 Fuel, fire weather and fire behaviour in Australian ecosystems ed R A Bradstock, M Gill and R J Williams *Flammable Australia: Fire Regimes, Biodiversity and Ecosystems in a Changing World* (Collingwood, Victoria: CSIRO Publishing), pp 51–77

[3] Lucas C, Hennessey K, Mills G and Bathols J 2007 Bushfire weather in southeast Australia: Recent trends and projected climate change impacts. Consulting Report prepared for the Climate Institute of Australia. Bushfire CRC and CSIRO

[4] Forest Fire Management Group 2014 National Bushfire Policy Statement for Forests and Rangelands Prepared by the Forest Fire Management Group for the Council of Australian Governments. Fore Fire Management Group Canberra

[5] Safe Work Australia 2011a *How to Manage Work Health and Safety Risks Code of Practice* (Canberra: Safe Work Australia)

[6] Safe Work Australia 2011b *Managing the Work Environment and Facilities Code of Practice* (Canberra: Safe Work Australia)

[7] Safe Work Australia 2011c *Work Health and Safety Consultation, Co-Operation and Co-Ordination Code of Practice* (Canberra: Safe Work Australia)

[8] NSW Rural Fire Service 2015 *Guide for Bush Fire Prone Land Mapping Version 5b* (Granville: NSW Rural Fire Service)

[9] Department of Defence 2014 *Chief of the Defence Force Commission of Inquiry Report into the Fire at Marrangaroo Training Area 16 October 2013* (Canberra: Department of Defence)

[10] Bureau of Meteorology 2018 *Defence Weather Services (website)* (Canberra: Bureau of Meteorology)

[11] Commonwealth of Australia 2013 *PUAFIR512 Develop and Analyse the Behaviour and Suppression Options for a Level 2 Wildfire Release 2* (Canberra: Commonwealth of Australia)

[12] Australian Wildlife Conservancy 2018a Yampi Sound Training Area (Website) (http://australianwildlife.org/sanctuaries/yampi.aspx). Australian Wildlife Conservancy, Perth

[13] Australian Wildlife Conservancy 2018b *Yampi Sound Training Area YSTA Burn Plan 2018* (Perth: Australian Wildlife Conservancy)

[14] McGuffog T 2001 *The 'how to' of Firebreaks and Aerial Burns Practical Advice from the Bushfires Council NT* (Darwin: Tropical Savannas CRC)

IOP Publishing

Global Approaches to Environmental Management on Military Training Ranges

Tracey J Temple and Melissa K Ladyman

Chapter 14

Greener or insensitive munitions: selecting the best option

Sylvie Brochu

Munitions performance is critical to the conduct of effective Armed Forces operations. While munitions safety, including insensitivity, is also extremely important to prevent catastrophic human, material and financial losses and avoid delays/failure in operations, greener munitions are essential to prevent environmental damage of ranges and training areas and sustain military training. Unfortunately, the concept of insensitive munitions appears to conflict with environmental considerations. Highly insensitive munitions are indeed conceived to not detonate when exposed to external stimuli, and consequently produce a large quantity of residues when demilitarized or blown-in place with the current operational procedures. Until better demilitarization procedures are developed, a trade-off appears necessary to preserve the safety and training capability of the Armed Forces. The successful development of munitions thus relies on an integrated approach considering their performance, their safety and their environmental and health impacts throughout their life cycle, from cradle to grave. With so many factors to consider, how does one make an informed decision when the time comes for developing new munitions or selecting one for a specific need? This chapter will describe a way to select the option that best suits the current requirements while sustaining military training.

14.1 Introduction

Green munitions have been described as *a material designed and manufactured in accordance with the principles of green chemistry, with the minimum requirement to preserve the performance level, and safety of handling of the energetic materials it is intended to replace* [1]. Green chemistry, according to the US Environmental Protection Agency (USEPA) [2] is defined as *the design of chemical products and*

doi:10.1088/978-0-7503-1605-7ch14

processes that reduce or eliminate the generation of hazardous substances. Consequently, green munitions aim to decrease the environmental impacts and potential health hazards resulting from their production and use, while at the same time preserving the performance and safety of the munitions.

Environmental impacts and potential health hazards can occur during every step of the munitions life cycle, from its inception by synthesis and production, during storage and transportation, and finally in operation, training or demilitarization. The worst cases of environmental impacts and occupational health hazards have been observed at production and manufacture facilities. Work is ongoing to develop greener processes. Some have succeeded: the Radford Army Ammunition Plant (RAAP; VA, USA), for example, has conceived and implemented a greener production process for 2,4,6-trinitrotoluene (TNT) based on ortho-nitrotoluene, much less hazardous than the toluene that was used as feedstock in the original process [3]. This new process resulted in the production of an extremely pure TNT, without the production of the red water that was extremely toxic and costly to manage, and in which the toxic air emissions (carbon monoxide, nitrogen oxides, volatile organic compounds and trinitromethane) are captured using a new fume abatement system. As a bonus, the new process generates isotrioil, a mixture of 70 to 90% TNT and 10% to 30% 2,4-dinitrotoluene (2,4-DNT) that can be used in the mining industry. Thus, decreasing the environmental and health hazards in energetic materials (EM) production facilities is a realistic avenue.

The dispersion of EM residues also occurs during their use or their demilitarization, and even in storage, albeit to a lesser extent. EM residues, some of which are constituents of concern because of their potential risk to sensitive receptors, have the potential to adversely impact ranges and training areas (RTAs), and by extension, compromise military readiness [4–6]. For example, the Military Massachusetts Reservation (MMR) and Camp Edwards were shut down by the USEPA in 1997 due to the presence of perchlorate, lead and hexahydro-1,3,5-trinitro-1,3,5-triazine (RDX) in the groundwater, which was the main source of drinking water for the Cape Cod area (MA, USA) [7]. The use of greener munitions, that would control the contamination directly at its source before its dispersion, is seen as a promising avenue to avoid extensive spreading of energetic residues in RTAs. This was successfully demonstrated by the US Armament Research, Development and Engineering Center (ARDEC) Research and Technology Branch and the NSWC Crane, who developed several perchlorate-free pyrotechnic devices that have been transitioned in service [8].

However, it is clear that greener munitions can't be developed at the expense of performance and safety, which are critical factors for military users. Munitions performance is a broad field of expertise. The explosives expert will describe it in terms of blast, precision, lethality and range, which translate into detonation pressure, detonation velocity and fragmentation pattern, while the propellant specialist will speak more of propellant quickness, force and burning rate, muzzle velocity and interior ballistic properties. For obvious reasons, improving munitions performance has been the guiding criterion of munitions development for ages.

Insensitive munition (IM) are munitions which reliably fulfill (specified) performance, readiness and operational requirements on demand, but which minimize the probability of inadvertent initiation and severity of subsequent collateral damage to the weapon platform (including personnel) when subjected to unplanned stimuli [9, 10]. Safety also includes the concepts of ingredients compatibility and stability, ageing characteristics of the formulations, system suitability, and safety and suitability for service, to name a few. Unfortunately, the concept of insensitive munitions appears to conflict with environmental considerations. Highly insensitive munitions are indeed conceived to not detonate when exposed to external stimuli, and consequently produce a large quantity of residues when demilitarized or blown-in-place [11].

In addition, the unit cost has always been, and continues to be, a strong incentive in the choice of munitions. However, it is believed that the whole life cycle cost should be considered instead. Indeed, it has been demonstrated that range remediation accounted for 54%, 70%, 77% and 89% of the total cost for the 81, 120, 105 and 155 mm calibers, respectively [12]. There is a need to shift the current buyers' mind set from unit cost to life cycle cost, which would be more economical and highly beneficial in the long term.

Consequently, a holistic approach is required to develop greener IM munitions that will also be as performant as in-service munitions. However, the use of an integrated approach for munitions development complicates the selection of the best formulation/munitions for a given use. How does one define the best formulation? Defence R&D Canada tried to address this issue within the context of the technology demonstration program (TDP) revolutionary insensitive, green and healthier training technology with reduced adverse contamination (RIGHTTRAC). The aim of this TDP was to prove that green and insensitive munitions had better properties than current munitions, and that it was feasible to implement safer weapon solutions that would ease the environmental pressure on RTAs, and decrease the health hazards for the users [13]. The TDP worked on the three main components (fuse, explosive and propellant) of a 105 mm caliber to reach a near-zero dud rate, eliminate the potential for RDX contamination and the use of toxic and carcinogenic compounds. The vehicle used for this demonstration was a 105 mm army artillery munition (HE M1), currently filled with Composition B (Comp B) and using a single base gun propellant (M1 formulation).

A decision matrix method, also called the Pugh method [14], was used to select the best propellant and explosive candidates to replace the in-service formulations. This team-based selection method is a tool used to compare several concepts, evaluate their pros and cons, and select the best possible option. The process consists in selecting, by consensus of the team, the appropriate criteria, weighting them in order of importance and then scoring the potential options against each criterion. A final result is calculated by adding the scores of each criterion multiplied by their respective weight. Consequently, the Pugh matrix transforms subjective opinions into objective statements, which is a substantial benefit. For the RIGHTTRAC TDP, the criteria were selected by a multi-disciplinary team of experts from Defence R&D Canada (DRDC) and General Dynamics—Ordnance and Tactical Systems in the early steps of the project. Whenever possible, the scoring was made using literature data from official sources (e.g. USEPA [15], Munitions Safety Information Analysis Center (MSIAC) [16], European Chemicals Agency (ECA) [17], Canadian List of Challenge substances [18], Agency for Toxic Substances and Disease Registry (ATSDR) [19], etc). In the absence of reliable data, R&D work was planned to get the information necessary to appropriately score each criterion. The decision matrix served as the basis for the detailed project plan. This chapter provides a brief review of the selection process that has been used to select the explosive candidate.

14.2 Matrix selection criteria

Table 14.1 shows the five main criteria that were selected for the Pugh method: environmental and health impacts, IM, technical feasibility, cost and performance. Each of these criteria were then subdivided in various sub-criteria, each pondered differently according to the scope of the project. The sub-criteria for the explosive and the propellant formulations slightly differed to take into account the specificity of each type of formulation. The score allocated for the performance was rather low because it was a prerequisite that any new green explosive and propellant be at least as good as in-service munition. The suitability score was the following: A: High (or 100%); B: Medium (or 66%); C: Low (or 33%); X: Not suitable (or 0%).

The following sub-sections report a brief description of each of the sub-criterion, as well as the results, only for the explosive formulation. More details are available in the listed references. The scores allocated to each formulation and the rationale behind the process is also provided.

Two explosive candidate formulations were pre-selected for the TDP. The prerequisites were that the main charge explosive candidates were RDX-free and met the current performance criteria of Comp B, which was the current explosive formulation in 105 mm munitions. Moreover, it was decided to select a melt-cast and a cast-cured formulation for comparison purposes. Melt-cast formulations are more common in North America than cast-cured formulations. Consequently, one of the candidates was CX-85, a cast-cured polymer-bonded explosive (PBX) composed of a mix of hydroxyl-terminated polybutadiene (HTPB), 1,3,5,7-tetranitro-1,3,5,7-tetrazocine (HMX) and di(2-ethyl hexyl) adipate (DEHA), a plasticizer. The other candidate developed and patented by DRDC was a melt-cast formulation made of TNT, HMX and an energetic thermoplastic elastomer (ETPE), called green insensitive munition (GIM) [20–22]. TNT was chosen despite its recognized potential human carcinogenicity, because of its limited access to sensitive receptors outside military boundaries due to a high degradability and bonding ability to soil component. HMX was preferred to RDX because of its much lower water solubility and toxicity.

14.3 Insensitive munitions

IM tests, usually performed on filled rounds, are expensive and require a significant quantity of energetic formulation. Consequently, the complete IM testing was

Table 14.1. Explosive formulation selection criteria.

Criteria	Weight	Sub-criteria	Sub-sub-criteria	Sub-weight
IM	25	Bullet impact		20
		Fragment impact		20
		Slow cook-off		15
		Sympathetic detonation		20
		Shape charge jet		15
		Variable confinement cook-off test		5
		Large scale gap testing		5
		Total		100
Environmental impact and health hazards	30	Human toxicity		10
		Environmental toxicity	Toxicity of new components	5
			Soil toxicity	10
			Freshwater toxicity	10
			Sediment toxicity	5
		Complexity	Number of sub-components	5
		Bioavailability	Transport of soluble ingredients in soil	10
			Water solubility	5
			Solubility kinetics	5
			Abiotic degradation	10
			Biotic degradation	5
		Recyclability	Ease of extraction from the munition	10
			Ease of separation of components in the formulation	5
			Interest for the components	5
		Total		100
Cost	15	Units	Cost of ingredients	20
			Set-up cost	5
			Manufacturing costs	20
		Environmental cost of remediation		35
		Environmental cost of manufacturing	Manufacturing of ingredients	5
			Manufacturing of the formulation	5
			Casting operation	10
		Total		100
Technical feasibility	20	Raw material availability		25
		Ease of integration		25
		Requirement for a liner		20

(*Continued*)

Table 14.1. (*Continued*)

Criteria	Weight	Sub-criteria	Sub-sub-criteria	Sub-weight
		Availability of the production facilities		20
		Processing complexity		10
		Total		100
Performance	10	Supplementary charge		20
		Detonation pressure		20
		Detonation velocity		20
		Ageing characteristics		20
		Gurney velocity		20
		Total		100

planned only for the end of the RIGHTTRAC TDP, with rounds filled with the selected formulations. Instead, to compare candidates to the reference formulation and to discriminate between pre-selected candidates, in-house small-scale IM tests were designed for bullet impact (BI), fragment impact (FI), sympathetic detonation (SD), shaped charge jet (SCJ) and slow cook-off (SCO) [9, 10]. Early in the project, it was decided to not perform the fast cook-off because of the lack of in-house facility for the large pool of oil required for the burning. The variable confinement cook-off test (VCCT) was performed to provide a low cost, small-scale method for evaluating the response of energetic materials to thermal stimuli, as well as the large scale gap tests (LSGT), to measure the shock sensitivity of explosive formulations. A weight of 20% was allocated to BI, FI and SD. A lower weight (15%) was attributed to SCJ and SCO, because improving the level of reaction to SCJ and SCO was considered improbable. Indeed, Comp B already passes SCO. Conversely, no commercial explosive formulation was found to resist the shaped charge that was selected for the testing, based on a preliminary threat hazard assessment [23].

The IM tests performed were described in Brousseau *et al* [24]. Results and scores are reported in table 14.2. Results are expressed either in terms of the type of reaction obtained (NR: no reaction; Type 1, detonation; Type V, mild reaction) or in terms of gap cards (the lower the number of cards, the more IM the formulation). A score of X was allocated to formulations that did not pass the test, B when the test was passed with a reaction, and A when no reaction occurred.

The scoring results indicate that, on the IM side, CX-85 and GIM were better than Comp B, and that CX-85 was the best option. This corresponds to the properties observed in the field.

14.4 Environmental properties

The environmental and health impact of a given chemical depends on its toxicity to sensitive receptors, such as human, flora and fauna, and on its ability to reach those sensitive receptors. A complete risk assessment would require the evaluation of the

Table 14.2. IM results and scores.

Sub-criteria	Sub-weight	Results			Scores		
		COMP B	GIM	CX-85	COMP B	GIM	CX-85
BI[a]	20	Type I	Type V	Type V	X	A	A
FI[b]	20	No test	Type I	NR	NA	B	A
SCO[a]	15	Type V	Type V	Type V	A	A	A
SR[a]	20	Type III	Type III	NR	B	B	A
SCJ[a]	15	Type I	Type I	Type I	X	X	X
VCCT[a]	5	Type III	Type IV	Type V	B	C	A
LSGT[b]	5	216	188	162	X	C	B
Total	100	NA	NA	NA	29.85	66.35	83.3

NA: Not applicable; NR: No reaction.[a] [24];
[b] Unpublished results.

environmental and health impacts of all the chemicals in the formulation throughout all the energetic formulation life cycle [25], from the conception of individual ingredients to their end-of-life usage, e.g. use in service or demilitarization. Within the context of this project, this would have resulted in a tremendous amount of effort and required a substantial budget. Based on the fact that the availability of a toxicant contributes directly to the risk to the receptors, a more practical approach was adopted, which involved the evaluation of toxicity only for the chemicals demonstrating a potential to leach from the formulation and consequently pose a threat to the environment or the users' health.

Leaching was evaluated both in laboratory conditions and weathered in outdoor conditions. Results later demonstrated that the continuous dripping test represented the worst case scenario, and that outdoor conditions were much more representative of the actual fate of formulations in Canada. Following leaching, toxicity and transport studies were performed solely on mobile ingredients. A recycling study was also performed to evaluate end-of-life options.

The following sections briefly describe how each aspect was evaluated. More detailed technical information on the assessment of properties of greener formulation can be obtained in Brochu *et al* [25], while thorough technical information is provided in specific references associated with each test.

14.4.1 Human toxicity

The human toxicity was evaluated using the USEPA Health Advisories (HA) for a lifetime exposure, the drinking water equivalent (DWEL), the reference dose (RfD) and the carcinogenicity; definitions, results and scores are reported in table 14.3. Health Advisories are guidelines, not mandatory criteria to respect. The lifetime and DWEL refer to a concentration in drinking water that would cause no adverse non-cancer effects for a lifetime exposure, for a 20% and 100% exposure to the chemical, respectively. The RfD is the daily oral exposure acceptable during a lifetime. Those criteria were selected for the decision matrix because oral exposure reflects the main

Table 14.3. Components health advisories—results and scores.

Criteria	Weight	Health Advisories[a]	RDX	TNT	HMX	DEHA	COMP B	GIM	CX-85	Ref
							Scores			
					Results					
Human toxicity	10	Lifetime[b] (mg l^{-1})	0.002	0.002	0.4	0.6	X	C	B	[15]
		DWEL[c] (mg l^{-1})	0.1	0.02	2	20				
		RfD[d] (mg kg^{-1} d^{-1})	0.003	0.0005	0.05	0.4				
		Carcinogenicity[e]	Class C	Class C	Class D	Class C				
Total							0	3.3	6.6	

[a] Health Advisory (HA): An estimate of acceptable drinking water levels for a chemical substance based on health effects information;

[b] Lifetime HA: The concentration of a chemical in drinking water that is not expected to cause any adverse noncarcinogenic effects for a lifetime of exposure, incorporating a drinking water relative source contribution factor of contaminant-specific data or a default of 20% of total exposure from all sources.

[c] Drinking water equivalent level. A DWEL is a drinking water lifetime exposure level, assuming 100% exposure from that medium, at which adverse, noncarcinogenic health effects would not be expected to occur.

[d] Reference dose. An estimate (with uncertainty spanning perhaps an order of magnitude) of a daily oral exposure to the human population (including sensitive subgroups) that is likely to be without an appreciable risk of deleterious effects during a lifetime.

[e] Carcinogenicity: Class C is possible human carcinogen, Class D is not classifiable as to human carcinogenicity

civilian and military threat in the current context, as compared to skin exposure and inhalation, which are more much more significant for workers in production or manufacture facilities.

The results clearly indicate that the human toxicity increases in the following order: DEHA > HMX > RDX > TNT for non-cancer effects. Consequently, a score of B was allocated to CX-85, which contains HMX and DEHA. GIM was allocated a score of C because of TNT and HMX, and Comp B, a score of X, because of RDX and TNT.

14.4.2 Ecotoxicity

The ecotoxicity was assessed by evaluating the chronic and acute toxicity of contaminated soil, water and sediment toward Canadian relevant ecological receptors. As reported in table 14.4 and in Hawari *et al* [26], the toxicity of formulations to soil receptors was evaluated using rye growth and seeding inhibition, as well as earthworms lethality avoidance (the nominal concentration leading to 80% avoidance). The toxicity of amended soil elutriates (i.e. contaminated freshwater) were assessed on the bioluminescence of a bacteria (Microtox), on the growth of a freshwater green algae and on the survival of an aquatic plant. Amended artificial sediments were tested with mussels and crustacean amphipods, as well as by measuring the phagocytic activity. DEHA exhibits no acute toxicity effects to aquatic organisms and a low bioaccumulation potential [27].

The results, reported in table 14.4, indicate that the toxicity of the CX-85 formulation is extremely low, inferior to GIM's and Comp B's, which are similar.

14.4.3 Bioavailability

To estimate the potential exposure of sensitive receptors to chemicals leaching from the formulations, their transport was assessed using their water solubility, their soil/water distribution coefficient (K_d), their soil organic carbon/water coefficient (K_{oc}) as well as their octanol–water coefficient (K_{ow}). K_d represents the proportion of a chemical sorbed to soil matter to its concentration in water. The K_{ow} is defined as the ratio of a chemical's concentration in the octanol phase to its concentration in the aqueous phase of a two-phase octanol/water system. The K_{ow} is a measure of how hydrophilic a substance is; it gives an indication of how widely and quickly a chemical will move through different types of environmental matrix. Consequently, water-soluble chemicals with a low K_d or a low K_{ow} are more at risk of reaching sensitive receptors. Given their low K_d, the mobility of TNT, RDX and HMX in soils are governed by their water solubility [26]. The K_d of DEHA was not measured, but was evaluated to be in the range of 7700–7940 [27].

As seen in the results reported in table 14.5, HMX and DEHA are not considered a threat to sensitive receptors because of their low water solubility. Moreover, because of its high calculated K_d, DEHA is expected to strongly bind to the soil. Conversely, RDX and TNT are much more soluble. However, TNT tends to bind to the organic content of the soil (higher K_{ow}), and is seldom detected in groundwater

Table 14.4. Environmental impacts—ecotoxicity results and scores.

Criteria	Sub-criteria	Weight	Parameter	Results			Scores			Ref
				Comp B	GIM	CX-85	Comp B	GIM	CX-85	
Ecotoxicity	Soil toxicity	10	EC_{50}[a] Rye[b] (mg kg^{-1})	750	736	> 10 000	B	B	A	[26]
			Worms[c] lethality (%)	100%	100%	0%				
			Worms' avoidance (mg kg^{-1})	290	205	> 10 000				
	Freshwater toxicity	10	EC_{50} Microtox[d] (mg kg^{-1})	2.0	2.4	> 49.5	C	C	A	
			EC_{50} Algae[e] (mg kg^{-1})	0.96	1.07	> 90.9				
			Inhibition duckweed[f]	85%	97%	9%				
	Sediment toxicity	5	Mussels'[g] lethality (%)	50%	0%	??	B	B	A	
			Phagocytic efficiency	Decrease	30% Decrease	None				
			EC_{50} Hyalella[h] lethality (mg kg^{-1})	495	402	> 10 000				
			EC_{50} Hyalella[h] growth (mg kg^{-1})	514	255	> 10 000				
Complexity	Number of sub-components	5		3	3	6	A	A	B	
	Sub-total	30					18.2	18.2	28.3	

[a] EC_{50}: Half maximum effective concentration of a chemical inducing a response halfway between the baseline and maximum after a specified exposure time.
[b] Lolium perenne;
[c] Eisenia andrei;
[d] Vibrio fischeri;
[e] Pseudokirchneriella subcapitata;
[f] Duckweed Lemna minor,
[g] Elliptio complanata;
[h] Hyalella azteca.

Table 14.5. Environmental impacts—bioavailability results and score.

Criteria	Sub-criteria	Weight	Parameter	Results			Scores			Ref
				Comp B	GIM	CX-85	Comp B	GIM	CX-85	
Bioavailability	Transport of soluble ingredients in soil	10	K_d (l kg^{-1})	RDX: 0.3 (Sassafras Sand)[b]	TNT 0.19 (DRDC sand) 4.58 (organic clay)	HMX 0.07 (DRDC sand) 0.7 (Sassafras Sand)[2] 5.78 (organic clay)	X	B	C	[26], unless otherwise noted.
			Log K_{oc}[a]	RDX: 1.69–2.2	TNT: 3.2	HMX: 1.5–2.5 DEHA: 4.7				
			Log K_{ow}[b]	0.90 (RDX)	1.6 (TNT)	0.17 (HMX)				
	Water solubility (mg l^{-1})	5		55.56 (RDX)	131 (TNT) 4.80 (HMX)	4.80 (HMX) 0.48 (DEHA)	B	B	A	
	Leachability	5	Batch (% loss)	NA	2% (HMX, 80 d), 74% (TNT, 80 d)	16% (HMX, 160 d) 0.2% (DEHA, 160 d)	X	B	B	
			Dripping (% loss)	100% (TNT, 150 d)[c] 100% (RDX, 150 d)[c]	100% (TNT, 250 d) 59% (HMX, 350 d)	24% (HMX, 350 d) 59% (DEHA, 350 d)				
	Abiotic degradation	10	Photolysis of components ($t_{1/2}$)	0.8 d (RDX, water)[d]	0.21 d (TNT, water)	1.7 d (HMX, water)	B	B	C	
			Photolysis of formulations (% loss)	NA	19% (HMX) 29% (TNT) no mineralization(48 h)	13% (HMX) no mineralization (48 h)				
			Hydrolysis[d]	RDX: 112 ± 10 y (pH 7)[4] $t_{1/2}$: 10.6 min (pH 12)[b]		HMX 2400 ± 200 y (pH 7)[4] $t_{1/2}$: 2.5h (pH 12)[b]				
			Zero-valent iron	No mineralization (RDX)		No mineralization (HMX)				

(Continued)

Table 14.5. (*Continued*)

Criteria	Sub-criteria	Weight	Parameter	Results				Scores				Ref
				Comp B	GIM	CX-85		Comp B	GIM	CX-85		
	Biotic degradation	5	Ks		TNT: Ks = 0 (sand); no mineralization	HMX: Ks = 0 (sand)		C	B	C		
Post-detonation residues		5		NA	NA	NA		—	—	—		
	Sub-total	40						11.55	23.1	16.55		

a [28]
b [29]
c [30]
d [31].

or surface water in RTAs. The increasing order of mobility would thus be DEHA < HMX < TNT < RDX.

14.4.4 Leaching

The dissolution of each energetic formulation was studied in water batch and under constant dripping with a sandy soil typically found in Canadian RTAs. The dripping test represents the worst case scenario. Within 150 days, Comp B was completely dissolved (table 14.5), while CX-85 and GIM are not expected to dissolve completely, because of the protective layer of the polymeric binder (HTPB in CX-85 and ETPE in GIM). Given the total amount of dissolved material after a year, the potential for leaching is considered to increase in the following order: CX-85 < GIM < Comp B.

14.4.5 Degradation

The common degradation pathways (aerobic, anaerobic, photolysis, hydrolysis) of the ingredients which have leached from the formulations indicate that photolysis in water was found to be the fastest degradation process, with half-times less than two days (table 14.5) [26]. Some degradation was even observed for dry formulations exposed to sunlight. Hydrolysis in neutral water and biodegradation in sand was insignificant. Zero-valent iron led to ring cleavage, but not to mineralization. GIM was considered the best option with respect to degradation, because of the amount of energetic material degraded and the quickness of degradation of TNT. CX-85 and Comp B were considered to be on the same level. The biotic degradation of TNT, RDX and HMX was null in sand, but complete mineralization occurred for TNT in three weeks in a Webster clay loam (WCL) containing 28% clay, 39% silt, 2.39% total organic carbon, and a cation exchange capacity of 20 mequiv/100 g [26]. DEHA is readily degradable via hydrolysis and biotic processes [27].

14.4.6 Recycling

Demilitarization was included in this work as early as possible in the development process. A recyclability study was undertaken to evaluate the feasibility of recovering the energetic formulations from the munitions and the separation of the components from the formulations [32, 33]. The separation of the components from the formulation was done on small quantities using common laboratory equipment and solvents. Laboratory-scale tests were carried out to demonstrate that it was possible to recover some ingredients. The recovery of formulations from the munitions was straightforward for melt-cast candidates, which can be poured out of the munitions at temperatures higher than 80 °C, the melting temperature of TNT. However, cross-linked formulations had to be detached from projectile casings using high-pressure water jets, which resulted in a score of C for CX-85. The results and scores are reported in table 14.6.

Table 14.6. Environmental impacts—recyclability results and scores.

Criteria	Sub-criteria	Weight	Results			Scores		
			Comp B	GIM	CX-85	Comp B	GIM	CX-85
Recyclability	Ease of extraction from the munition	10	• Melt formulation and pour out of the munition • Easy cleaning of the equipment, tools and munitions	• Melt formulation and pour out of the munition • Harder cleaning of the equipment, tools and munitions	Water-jet	A	A	C
	Ease of separation of components in the formulation	5	• 100% of ingredients collected	• 100% of ingredients collected	• 84% of ingredients collected	A	A	A
	Interest for the components	5	• Recyclable and reusable • RDX is less expensive than HMX	• Recyclable and reusable	• HMX is costly to produce • Not recyclable, only reusable, because of crystalline form of HMX	A	A	B
	Sub-total	20				20	20	11.6

14.5 Costs

Three sub-criteria were considered to evaluate the cost: the unit cost, the environmental cost of remediation and the environmental cost of manufacturing. The results and scores are reported in table 14.7. The scores for the unit cost were allocated based on information obtained from GD. The elements that made a difference for this category are (1) the higher cost of HMX, ETPE and CX-85 binder; (2) the lack of CX-85 manufacturing facility in Canada; (3) the manufacturing process and related equipment, more complex for CX-85; and (4) the need of a liner for the GIM and CX-85 formulations.

The remediation of RDX-, HMX- or TNT- contaminated soil or water is a lengthy and cost prohibitive operation, whatever the remediation process employed. Greener munitions are meant to avoid resorting to remediation as much as possible. In other words, the remediation cost of HMX, RDX and TNT is similar, but the risk to resorting to remediation decreases in the order Comp B > GIM > CX-85. This explains the respective scores of X, C and B reported in table 14.7 for each formulation.

More detailed information cannot be included in this chapter because of confidentiality reasons. The cost of Comp B was hard to beat. CX-85 got a score of C, and GIM a B.

14.6 Technical feasibility

As seen in table 14.8, the technical feasibility was scored by evaluating the ingredients' availability, the ease of integration of all the components in the munitions design, the requirement for a liner, the availability of manufacturing facility and the processing complexity.

The technical feasibility category goes hand in hand with the cost category: the more complex the process, the more expensive it became. The ingredients' availability also played a significant role in the scoring process. Indeed, the commercially available ETPE did not meet the required specifications and had to be made in-house. Isophorone diisocyanate (for cross-linking) was hard to obtain and the binder had to be made in-house. The scores for the technical feasibility category followed the same trends as those of the cost category.

14.7 Performance

The need for a supplementary charge, the detonation pressure and velocity, the ageing characteristics and the Gurney velocity (ejection speed of fragments) were used to assess the performance category. The results and scores are reported in table 14.9. As expected, no significant difference was noted between the candidates, because an equivalent performance was a prerequisite for the preselection. GIM and CX-85 got a slightly lower score than Comp B because of the potential requirement for a supplementary charge.

Table 14.7. Cost—results and scores.

Criteria	Sub-criteria	Parameter	Weight	Results — Comp B	Results — GIM	Results — CX-85	Scores — Comp B	Scores — GIM	Scores — CX-85
Cost	Unit	Ingredients	20	RDX and TNT: low cost	HMX and ETPE: higher cost	HMX and binder: higher cost	A	B	C
		Set-up	5	Melt-cast: low cost	• Melt-cast, but more complex than Comp B • Need additional equipment	No CX-85 installation in Canada	A	B	X
		Manufacturing	20	Melt-cast: low cost	Same as Comp B	More costly than Comp B	A	A	B
Environmental cost of remediation			35				B	B	B
Environmental cost of manufacturing		Manufacturing of ingredients	5	• Discharge contaminated with TNT and RDX • Problem of pink water with TNT • Cost of green TNT prohibitive	• Discharge contaminated with TNT and HMX • Less mobile • Less toxic	• Discharge contaminated with HMX • Isocyanates extremely toxic	X	C	B
		Manufacturing of the formulations	5	Solvents, contaminated discharge, etc	Same as Comp B	Additional equipment for isocyanates handling	B	B	C
		Casting operations	10	Easy	Liner required	• Limited pot-life • Liner required	A	B	B
Total			100				81.4	74.5	54.45

Table 14.8. Technical feasibility—results and scores.

Criteria	Sub-criteria	Weight	Results			Scores		
			Comp B	GIM	CX-85	Comp B	GIM	CX-85
Technical feasibility	Raw material availability	25	• RDX and TNT readily available	• ETPE made in-house • Commercially available but very narrow specifications	• IDPI could be hard to obtain • Binder was made in-house	A	B	B
	Ease of integration	25	• Well-known process	• Well-known process	• Well-known process	A	A	A
	Requirement for a liner	20	• No liner	Liner required	• Liner required	A	B	B
	Availability of production facilities	20	• Available	• Same as Comp B	• No CX-85 installation in Canada—only a pilot plant	A	A	X
	Processing complexity	10	• Easy, well known	• More complex than Comp B due to ETPE	• More complex due to required cross-linking reaction	A	B	C
Total		100				100	81.3	58.0

Table 14.9. Performance—results and scores.

Criteria	Sub-criteria	Weight	Results				Scores		
			Comp B	GIM	CX-85		Comp B	GIM	CX-85
Performance	Supplementary charge	20	None required	Possibly required	Possibly required		A	B	B
	Detonation pressure	20	Well known	Same as Comp B	Within 10% Comp B's		A	A	B
	Detonation velocity	20	Well known	Same as Comp B	Same as Comp B		A	A	A
	Ageing	20	Good for decades	Excellent results from accelerated ageing	PBX are known to age well		A	A	A
	Gurney velocity	20	Easy, well known	More complex than Comp B due to ETPE	More complex due to required cross-linking reaction		A	A	A
Total		100	Well known	Same as Comp B	Same as Comp B		100	93.2	86.4

Table 14.10. Final scores for the explosive formulations.

Criteria	Weight	Comp B	GIM	CX-85
IM compatibility	25	7.46	16.59	**20.83**
Environment	30	16.43	**20.37**	19.41
Cost	15	**12.21**	11.18	8.17
Technical feasibility	20	**20.00**	16.26	11.60
Performance	10	**10.00**	9.32	8.64
Total	100	66.1	**73.72**	68.65

14.8 Final selection

Table 14.10 reports the final scores for each main criterion, as well as the total score. Overall, both candidates were better than Comp B with regard to environmental and health impacts, as well as insensitivity. GIM proved to be slightly better than Comp B with regards to the environmental and health properties, while conversely, CX-85 was a slightly better IM formulation. The cost, technical feasibility and performance of GIM were better than CX-85, but no better than Comp B. Based on these scores, GIM was selected for further tests in the TDP, and CX-85 was abandoned.

The decision could have been different if, for example, a PBX manufacturing facility had been available in Canada, which would have increased the technical feasibility. The CX-85 performance was also lower than the GIM's and the Comp B's. There was also uncertainty related to the DEHA leaching from the CX-85. DEHA is an oily liquid that is slightly water soluble, but weathering experiments [34] provided indications that DEHA had the potential to be transported with water. Further investigation would have been needed to obtain more accurate information and potentially allocate a better score to PBX. The recyclability of PBX was also an issue, because of the crystalline form of the recovered HMX that was different than the one required for reuse.

Additionally, the greener option (not specifically GIM or CX-85) was compared to the current munitions using a cost-efficiency analysis (CEA) on the whole life cycle of the munitions [30]. Several cost categories were considered for the CEA (e.g. range remediation, liability, manufacturing facility, health impacts, demilitarization, etc). However, obtaining reliable data for all categories proved to be an unreachable goal. Therefore, simulated data extracted from realistic baseline scenarios occurring in a hypothetical training facility were used to assess several cost categories. At the end, potential health hazards were not even considered because of the lack of information. Nevertheless, the results demonstrated that, on a 10-year basis, mean potential savings of several millions of dollars per artillery range could be reached just by using greener large caliber munitions. The status quo would thus be more expensive due to environmental hazards.

14.9 Conclusions

The Pugh decision matrix proved to be extremely useful to perform an impartial selection of the best energetic material formulation for a specific requirement when numerous parameters needed to be taken into consideration. This decision matrix is also a great tool to perform sensitivity analyses on selected parameters. Moreover, it can be adapted to each country's or projects' priorities by modifying either the scored criteria, sub-criteria or their respective weights. Whenever possible, data from reliable sources or specific to the project evaluated should be obtained.

Using this decision matrix, an informed decision was taken on the selection of the best explosive candidate for the RIGHTTRAC TDP. The Pugh decision matrix was also used to select the best propellant option by adapting a few sub-criteria. The use of this decision tool is recommended for munitions developers, stakeholders and buyers that wish to make an informed decision.

References

[1] Brinck T 2014 *Green Energetic Materials* 1st edn (New York: Wiley)
[2] USEPA (U.S. Environmental Protection Agency) Green Chemistry, https://epa.gov/green-chemistry (Access date: 14 May 2018)
[3] Jennings B, Kessinger M and Pack J 2007 *CPIAC* **33** 1–5
[4] Jenkins T F, Hewitt A D, Grant C L, Thiboutot S, Ampleman G, Walsh M E, Ranney T A, Ramsey C A, Palazzo A J and Pennington J C 2006 Identity and distribution of residues of energetic compounds at army live-fire training ranges *Chemosphere* **63** 1280–90
[5] Pennington J C *et al* 2006 Distribution and Fate of Energetics on DoD Test and Training Ranges: Final Report, ERDC TR-06-13 (Hanover, NH, USA: Cold Regions Research and Engineering Laboratory)
[6] Thiboutot S *et al* 2012 Environmental characterization of military training ranges for munitions-related contaminants *IJEMCP* **11** 17–57
[7] USEPA (Environmental Protection Agency) 1997 Region 1, Administrative Order, EPA Docket Nos. SDWA I-97-1019 and SDWA I-97-1030 https://semspub.epa.gov/work/01/448135.pdf
[8] Sabatini J 2014 Advances toward the development of 'green' pyrotechnics, chs 4 *Green Energetic Materials* Tore Brinck (New York: Wiley)
[9] NATO (North Atlantic Treaty Organization) 2010 Policy for introduction, assessment and testing of insensitive munitions (IM), STANAG 4439, Ed. 3
[10] U.S. DoD (Department of Defense) Test Method Standard 2011 2011, Hazard Assessment Tests for Non-Nuclear Munitions, Military Standard MIL-STD-2105D
[11] Walsh M R, Walsh M E, Poulin I, Taylor S and Douglas T A 2011 Energetic residues from the detonation of common US ordnance *IJEMCP* **10** 169–86
[12] SAIC (Science Applications International Corporation) 2006 Munitions baseline cost estimate, FOCIS Division Report No 06-5628-08-5842
[13] Brochu S *et al* 2014 Towards high performance, greener and low vulnerability munitions with the RIGHTTRAC technology demonstrator program *IJEMCP* **13** 7–36
[14] Pugh S 1991 *Total Design: Integrated Methods for Successful Product Engineering* (Addison Wesley)

[15] USEPA (U.S. Environmental Protection Agency) 2018 2018 Edition of the drinking water standards and health advisories tables EPA 822-F-18-001 https://epa.gov/sites/production/files/2018-03/documents/dwtable2018.pdf, Last accessed: 17 October 2018

[16] MSIAC (Munitions Safety Information Analysis Center) 2018 website, https://msiac.nato.int/, Last accessed: 17 October 2018

[17] ECA (European Chemicals Agency) 2018 European Union, http://echa.europa.eu, Last accessed: 19 October 2018

[18] GC (Government of Canada) 2018 List of Challenge substances, https://canada.ca/en/health-canada/services/chemical-substances.html, Last accessed: 14 May 2018

[19] ATSDR (Agency for Toxic Substances and Disease Registry) 2018 https://atsdr.cdc.gov/, Last accessed: 14 May 2018

[20] Ampleman G, Marois A and Désilets S 2003 *Energetic Copolyurethane Thermoplastic Elastomers*, *Can. Pat.* 2214729 European Patent Application, No 0020188.2-2115, US Pat. 6479614 B1

[21] Ampleman G, Brousseau P, Thiboutot S, Dubois C and Diaz E 2003 Insensitive Melt Cast Explosive Compositions Containing Energetic Thermoplastic Elastomers US Patent 6562159

[22] Ampleman G 2011 Development of new insensitive and greener explosives and gun propellants, NATO AVT-177/RSY-027 symposium, munition and propellant disposal and its impact on the environment Edinburgh, UK (17–20 October 2011)

[23] Lahaie N, Dietrich S, Fleury D, Gélinas R, Grigoriu M A, Pelletier P and Tran L B 2007 *Large Caliber Munitions Development and Integration*, GD-OTS Canada final report for contract W7701-045050/001/QCA-W7701-7-2411NG

[24] Brousseau P, Brochu S, Brassard M, Ampleman G, Thiboutot S, Côté F, Lussier L-S, Diaz E, Tanguay V and Poulin I 2010 RIGHTTRAC Technology demonstration program: preliminary IM tests *Insensitive Munitions & Energetic Materials Technical Symp. (Munich, Germany, 11–14 October 2010)*

[25] Brochu S *et al* 2013 Assessing the potential environmental and human health consequences of energetic materials: a phased approach, TTCP WPN TP-4 CP 4-42, Final report, TTCP TR-WPN-TP04-15-2014, DRDC Valcartier SL 2013-626

[26] Hawari J *et al* 2011 *Environmental Aspects of RIGHTRAC TDP - Green Munitions, Annual Report NRC #53361* (Biotechnology Research Institute of National Research Council)

[27] OECD SIDS (Organisation for Economic Co-Operation Screening Information Data Sets) 2018 Bis(2-ethylhexyl)adipate (DEHA), http://inchem.org/documents/sids/sids/103231.pdf, Last accessed: 29 October 2018

[28] HSDB (Hazardous Substances Data Bank) 2018 https://toxnet.nlm.nih.gov/cgi-bin/sis/htmlgen?HSDB, Last accessed: 18 October 2018

[29] Sunahara G I, Lotufo G, Kuperman R G and Hawari J 2009 *Ecotoxicology of Explosives* (Boca Raton, FL: CRC Press)

[30] Taylor S, Lever J H, Fadden J, Perron N and Packer B 2009 Simulated rainfall driven dissolution of TNT, tritonal, comp B and octol particles *Chemosphere* **75** 1074–81

[31] Perreault N, Monteil-Rivera N, Halasz A M and Dodard S 2017 Environmental fate and ecological impact of energetic chemicals—twenty-year study, summary report 1996-2016, National Research Council, NRC-EME-55873, DRDC-RDDC-2017-C256

[32] Poulin I, Brochu S and Diaz E 2011 New Canadian munitions: recycling tests for main charge explosive and gun propellant *42nd Int. Annual conf. of the Fraunhofer ICT (Karlsruhe, Federal Republic of Germany, 28 June–1 July 2011)*

[33] Poulin I, Brochu S, Tanguay V and Diaz E 2011 Demilitarisation of new insensitive and green munitions: recycling and open - detonation, NATO AVT – 177/RSY – 027 Symp., *Munition and Propellant Disposal and its Impact on the Environment, Edinburgh, UK (17–20 October 2011)*

[34] Côté S and Martel R 2011 Environmental behavior of green gun propellant and main charge explosive formulations, phase 3B, Preliminary Report. Institut national de la recherche scientifique - Centre Eau, Terre et Environnement, Research Report R-1231, DRDC CR 2011-572

[35] Sokri A 2015 Cost risk analysis of green and insensitive munitions, Defence Research and Development Canada, DRDC-RDDC 2015-R056